# Dynamic World

## Land-cover and Land-use change

**A. M. Mannion**

*Department of Geography*
*University of Reading*

A member of the Hodder Headline Group
LONDON
Co-published in the United States of America by
Oxford University Press Inc., New York

D1225533

First published in Great Britain in 2002 by
Arnold, a member of the Hodder Headline Group,
338 Euston Road, London NW1 3BH

http://www.arnoldpublishers.com

Co-published in the United States of America by
Oxford University Press Inc.,
198 Madison Avenue, New York, NY10016

*British Library Cataloguing in Publication Data*
A catalogue record for this book is available from the British Library

*Library of Congress Cataloging-in-Publication Data*
A catalog record for this book is available from the Library of Congress

ISBN 0 340 80678 8 (hb)
ISBN 0 340 80679 6 (pb)

1 2 3 4 5 6 7 8 9 10

Production Editor: Wendy Rooke
Production Controller: Martin Kerans
Cover Design: Terry Griffiths

Typeset in 10/11½pt Palatino by Phoenix Photosetting, Chatham, Kent
Printed and bound in Malta by Gutenberg Press

What do you think about this book? Or any other Arnold title?
Please send your comments to feedback.arnold@hodder.co.uk

Forward, forward let us range,
Let the great world spin forever down the ringing grooves of change.

From Locksley Hall

by
Alfred Lord Tennyson

1809–1883

# Contents

# Preface

Geography is all about people and place. There are specialisms within the subject that are concerned exclusively with people and those that are concerned exclusively with place, which is interpreted here in its widest sense to mean environment. This broad church is what makes geography so exciting and so appealing, but its polymathy is both its strength and its weakness. It can be argued that only in geography's core material (i.e. its unique focus on people–environment relationships) lies its true worth and distinction. The manifestation of these relationships as land-cover and land-use change in time and space, and their fundamental link with the global biogeochemical cycle of carbon is the subject matter of this book.

Following a scene-setting first chapter, Chapters 2 and 3 provide a temporal perspective on land-cover alteration. Palaeoenvironmental records provide base-line data on land-cover characteristics prior to significant human impact, and a record of how humans have manipulated the environment throughout prehistory and history. Reference is also made to various indices, such as the Ecological Footprint, which have been devised to describe human impact quantitatively or semi-quantitatively, and which can be used for comparative purposes. Chapters 4 to 9 examine specific human activities that have resulted in major changes on the Earth's surface. Agriculture is especially significant and is herein considered in relation to the developed and developing worlds. Chapters 6 and 7, on forestry and mineral extraction, reflect the impact of primary resource exploitation. Chapter 8, on the land-cover change associated with urbanization, is a first for a physical geographer, but nevertheless essential in any discourse on land-cover/land-use change; it proved a delight to research and was helped along by my personal experience of many of the examples quoted. Chapter 9 is a veritable pot-pourri of additional factors that have contributed to land-cover/land-use change, but that do not fit comfortably into the earlier chapters. Topics included are the impact of exotic plant invasions, problems of waste disposal, the direct and indirect effects of war and terrorism, as well as the impacts of sport and tourism. The final chapter presents not only a summary but also a prospect that focuses on the most likely stimuli that will affect land-cover/land-use change in the next few decades.

By the time I had finished writing this book I felt more convinced than ever of the wealth, practically and intellectually, of geography. Moreover, I was overwhelmed by the abundance of literature available from which to draw case studies, and it became increasingly evident that the peripheral aspects of geography, either physical or human aspects, are firmly linked to the core. The world is not a disjunct array of parts but an intimate and harmonious ensemble that is constant in its presence but ephemeral in its character. It all boils down to carbon in the end!

My intention is to share this experience with undergraduate students and their lecturers, whom I hope will be inspired to delve further into the intricacies of people–environment relationships than my 100,000 words allow.

*Antoinette M Mannion*
*Reading*
*December 2001*

# Acknowledgements

Many people have helped me to produce this book. I am grateful to Nicola Tuson, Christine Scurfield and Heather Howarth who typed the manuscript, and to Judith Fox for drawing the maps and diagrams. Many colleagues have been generous with their time, providing or suggesting relevant literature or websites; I wish to record my thanks to Sophie Bowlby, Erlet Cater, Geoffrey Griffiths, Sir Peter Hall, Sally Lloyd-Evans, Peter Merriman, Peter Pearson and Maria Shahgedanova. The biggest chore of all, the index, has been compiled by Mike Turnbull whose leisure time has been hijacked yet again.

Every effort has been made to trace copyright holders of material reproduced in this book. Any rights not acknowledged here will be acknowledged in subsequent printings if notice is given to the publisher.

# 1

# Introduction

## 1.1 Preamble

Land cover is a term used to describe the components characterizing the Earth's surface. For the most part these components comprise plants – as units of vegetation or ecosystems – soils, sediments, water and the built environment. The term land use differs from land cover in so far as it reflects the function of land units, notably the human use of land, which often has economic significance.

Both land cover and land use are dynamic. Change is occurring constantly on a temporal and spatial basis as a consequence of natural and/or cultural stimuli. Land cover and land use reflect natural factors such as climatic, soil, hydrological and geological conditions; these provide the environmental limitations within which life, including human life, operates. They also reflect human activity; land use especially is a direct manifestation of human endeavour, whilst land cover may result from natural processes alone or from human activity. The latter may exert a direct impact on land cover (e.g. through the removal of natural vegetation for agriculture) or an indirect impact (e.g. through the effect of pollution, which may spread far beyond its point of generation).

Apart from reflecting the state of the biosphere – the zone of land, water and air at the surface of the Earth in which life occurs – land cover and land use are also a manifestation of people–environment relationships. This is because they reflect the manipulation of biospheric resources for the necessities of life and for wealth generation. Such manipulation inevitably results in environmental change. Several quantitative measures have been devised recently to express this relationship, i.e. the degree of human impact. These include the Living Planet Index generated by the WWF (Worldwide Fund for Nature, 1998) and the Ecological Footprint (e.g. Rees, 1992); there is also the possibility of a Carbon Index (Mannion, 2000). These measures are discussed in Section 1.4 and in Chapter 3 (see also the comments below).

Equally important is the fact that land cover and land use enjoy a reciprocal relationship with the biogeochemical cycle of carbon. This is because most of the biosphere's components are organic, and so are comprised mainly of carbon, and because the existentialities of human existence and wealth generation mostly, though not exclusively, involve the manipulation of carbon. Moreover, the exploitation of carbon stores in the biosphere, especially in vegetation communities, and carbon stores, i.e.

fossil fuels, in the lithosphere (Earth's crust), themselves the product of past biospheres, is significant in terms of the regulation of climate. Climate and climatic change are, in turn, of prime importance in people–environment relationships.

The causes of land-cover and land-use change are many and varied. They may be natural or anthropogenic, and will operate at differing timescales. In general, natural change can either be very gradual or catastrophic, as illustrated by denudation processes and earthquakes respectively; naturally induced climatic variation and climatic change operate at scales from as little as a few years to as much as hundreds of millions of years. In contrast, changes generated through human activity are relatively rapid, especially in the context of geological time (i.e. the Earth's $5000 \times 10^6$-year history). Indeed, the catastrophism of earthquakes and volcanoes is often matched by human-induced changes; deforestation, urban construction and pollution events are cases in point. Other types of change, such as soil erosion and acidification, may be relatively insidious and almost imperceptible until thresholds are crossed and an impact becomes evident. The underpinning causes of human-induced change are due to either necessity, i.e. subsistence, or to wealth generation over and above necessity; in short, these stimuli can be described as need and greed. The manifestations of these stimuli can be classified under four headings: agriculture, mineral extraction, industrialization and urbanization.

Agriculture is the oldest of these stimuli with origins going back c. 10,000 years. Urbanization followed; subsequently mineral extraction (minerals other than stone) occurred and eventually, in the eighteenth century, the Industrial Revolution took place. Although urbanization had begun several millennia earlier, in association with early agriculture, it accelerated markedly with industrialization. Changes in land cover and land use thus reflect social/ cultural developments throughout prehistory and history; they are the key links between archaeology, history and geography. They can also be considered as the manifestation in time and space of the role of technology in people–environment relationships.

## 1.2 Terms and context

Definitions of land cover and land use are given in Section 1.1 above. As land cover encompasses a wide range of Earth-surface features, there are many ways in which different categories can be defined. In simple terms, for example, land cover can be subdivided into two categories: natural and anthropogenic. This subdivision is crude and is valuable only for large-scale studies at continental, hemispheric or global scales; nevertheless it reflects the degree of human modification of the Earth's surface. However, further subdivision of these categories is desirable for purposes such as conservation, planning and management at various scales. This is equally true of land use. Simple subdivisions are possible, e.g. agricultural land and non-agricultural land, but are only useful for large-scale studies where detail may be either unobtainable or even confusing. A classification scheme with numerous subdivisions is, however, preferable for regional, national and local studies.

With the advent of remote sensing, especially that using satellite imagery in the 1970s, the US Geological Survey (Anderson *et al.*, 1976) produced a land use and land cover classification system with a variety of subdivisions. Details of this scheme are given in Table 1.1. This is a multilevel scheme and was so constructed to accommodate many different types of remote sensing. Level I is the most crude, and Level II categories comprise coarse subdivisions (e.g. the Level I category of forest land can be subdivided into the Level II categories of deciduous, evergreen and mixed forests in order to afford increased detail). Levels I and II categories are particularly relevant to large-scale studies, e.g. global or regional. Additional refinements to Levels III and IV are possible. For example, croplands can be subdivided into Level III cereal crops and non-cereal crops; further subdivision into Level IV categories would involve the demarcation of

| Level I | Level II |
|---------|----------|
| 1 Urban or built-up land | 1.1 residential |
| | 1.2 commercial services |
| | 1.3 industrial |
| | 1.4 transportation, communications |
| | 1.5 industrial and commercial |
| | 1.6 mixed urban or built-up land |
| | 1.7 other urban or built-up land |
| 2 Agricultural land | 2.1 cropland and pasture |
| | 2.2 orchards, groves, vineyards, nurseries |
| | 2.3 confined feeding operations |
| | 2.4 other agricultural land |
| 3 Rangeland | 3.1 herbaceous rangeland |
| | 3.2 shrub and brush rangeland |
| | 3.3 mixed rangeland |
| 4 Forest land | 4.1 deciduous forest land |
| | 4.2 evergreen forest land |
| | 4.3 mixed forest land |
| 5 Water | 5.1 streams and canals |
| | 5.2 lakes |
| | 5.3 reservoirs |
| | 5.4 bays and estuaries |
| 6 Wetland | 6.1 forested wetlands |
| | 6.2 non-forested wetlands |
| 7 Barren land | 7.1 dry salt flats |
| | 7.2 beaches |
| | 7.3 sandy areas other than beaches |
| | 7.4 bare exposed rock |
| | 7.5 strip mines, quarries, and gravel pits |
| | 7.6 transitional areas |
| | 7.7 mixed barren land |
| 8 Tundra | 8.1 shrub and brush tundra |
| | 8.2 herbaceous tundra |
| | 8.3 bare ground |
| | 8.4 wet tundra |
| | 8.5 mixed tundra |
| 9 Perennial snow and ice | 9.1 perennial snowfields |
| | 9.2 glaciers |

TABLE 1.1 The United States Geological Survey (USGS) land-use/land-cover classification scheme (Anderson, *et al.*, 1976)

specific crops. This degree of detail is essential for medium- and small-scale studies at regional and local scales.

Many studies involving the monitoring of land cover and land use utilize this system of recording or related modified versions. On the one hand it can be argued that the system has a major inherent disadvantage in so far as the land-cover and land-use categories are inter-mixed. For example, Level III evergreen forest could represent plantation forestry or natural boreal forest. In this case, land cover may or may not reflect land use. On the other hand, because an ever-increasing area of the Earth's surface is being modified or changed by human activity, distinctions between, for example, natural land cover and human-modified land cover are becoming increasingly blurred. Moreover, because of accelerating human activity, land-use categories are more important

| | | | | |
|---|---|---|---|---|
| 1 | ARTIFICIAL SURFACES | | 2.4.4 | Agroforestry areas |
| 1.1 | URBAN FABRIC | | | |
| 1.1.1 | Continuous urban fabric (few areas of vegetation or bare soil) | | 3 | FORESTS AND SEMI-NATURAL AREAS |
| | | | 3.1 | FORESTS |
| 1.1.2 | Discontinuous urban fabric (vegetated areas and/or bare soil are significant) | | 3.1.1 | Broad-leaved forest |
| | | | 3.1.2 | Coniferous forest |
| 1.2 | INDUSTRIAL, COMMERCIAL AND TRANSPORT UNITS | | 3.1.3 | Mixed forest |
| | | | 3.2 | SHRUB AND/OR HERBACEOUS VEGETATION ASSOCIATIONS |
| 1.2.1 | Industrial or commercial units | | 3.2.1 | Natural grassland |
| 1.2.2 | Road and rail networks and associated land | | 3.2.2 | Moorland and heathland |
| 1.2.3 | Port areas | | 3.2.3 | Sclerophyllous vegetation |
| 1.2.4 | Airports | | 3.2.4 | Transitional woodland/shrub |
| 1.3 | MINE, DUMP AND CONSTRUCTION SITES | | 3.3 | OPEN SPACES WITH LITTLE OR NO VEGETATION |
| 1.3.1 | Mineral extraction sites | | 3.3.1 | Beaches, dunes and sand planes |
| 1.3.2 | Dump sites | | 3.3.2 | Bare rock |
| 1.3.3 | Construction sites | | 3.3.3 | Sparsely vegetated areas |
| 1.4 | ARTIFICIAL, NON-AGRICULTURAL VEGETATED AREAS | | 3.3.4 | Burnt areas |
| | | | 3.3.5 | Glaciers and permanent snowfields |
| 1.4.1 | Green urban areas | | | |
| 1.4.2 | Sport and leisure facilities | | 4 | WETLANDS |
| | | | 4.1 | INLAND WETLANDS |
| 2 | AGRICULTURAL AREAS | | 4.1.1 | Inland marshes |
| 2.1 | ARABLE LAND | | 4.1.2 | Peat bogs |
| 2.1.1 | Non-irrigated arable land | | 4.2 | MARITIME WETLANDS |
| 2.1.2 | Permanently irrigated land | | 4.2.1 | Salt marshes |
| 2.1.3 | Rice fields | | 4.2.2 | Salines |
| 2.2 | PERMANENT CROPS | | 4.2.3 | Intertidal flats |
| 2.2.1 | Vineyards | | | |
| 2.2.2 | Fruit trees and berry plantations | | 5 | WATER BODIES |
| 2.2.3 | Olive groves | | 5.1 | INLAND WATERS |
| 2.3 | PASTURES | | 5.1.1 | Water courses |
| 2.3.1 | Pastures | | 5.1.2 | Water bodies |
| 2.4 | HETEROGENEOUS AGRICULTURAL AREAS | | 5.2 | MARINE WATERS |
| 2.4.1 | Annual crops associated with permanent crops | | 5.2.1 | Coastal lagoons |
| 2.4.2 | Complex cultivation patterns | | 5.2.2 | Estuaries |
| 2.4.3 | Land principally occupied by agriculture, with significant areas of natural vegetation | | 5.2.3 | Sea and ocean |

**TABLE 1.2** The land-cover/land-use categories of the CORINE project (Wyatt *et al.*, 1993)

now in many areas than they were 30 years ago. Equally it could be argued that a scheme combining land cover and land use is more realistic and practical than two separate schemes.

Numerous other schemes to describe land cover and land use also exist. Another scheme similar to the USGS system is that devised for CORINE (Coordination of Information on the Environment). This programme was initiated by the Commission of the EU in the mid-1980s in order to collect information on the European environment. The CORINE scheme is given in Table 1.2, which shows that land-use categories are more numerous, and therefore facilitate the portrayal of more detail than the USGS scheme. In addition, many classification schemes are designed to be more specific than the USGS system (Table 1.1), which attempts to include all types of land cover and land use. One such example is that designed by the World Conservation Monitoring Centre for the WCFSD (World Commission on Forests and Sustainable Development, 1999: 4–5). The focus

of this publication is, obviously, forests, and the objective of these two maps (Figures 1.1 and 1.2) is to illustrate the difference between global forest extent 8000 years ago, i.e. the natural forest cover of the present interglacial (see Section 2.4), and that of the present, i.e. 1998. The maps are subdivided into six categories, of which four represent different types of forest (open, closed tropical and non-tropical forests) and two represent non-forest and water bodies. The past and present distribution of forests is reasonably well illustrated at this global scale, even with only four categories of forest. Elsewhere in the WCFSD report, the maps (prepared by the World Resources Institute, 1997) represent various regions (e.g. North America, Europe and Northern Asia, Southern Asia, Oceania, Africa and Latin America). All of these (see, for example, Figure 1.3) comprise two categories reflecting the degree of threat of destruction/modification with which forests are currently faced, and two categories representing forests for the degree of threat and modified, fragmented or planted forests. These

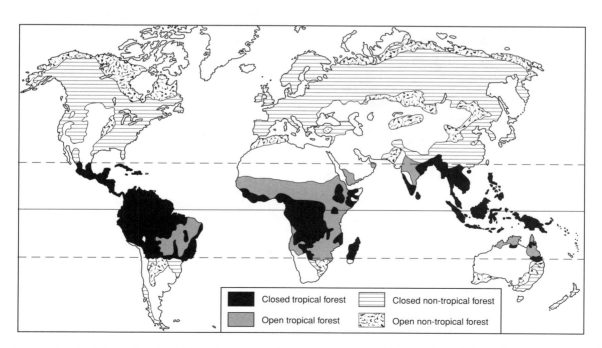

**FIGURE 1.1** Original global distribution of forests 8000 years ago. (Adapted from World Commission on Forests and Sustainable Development, 1999)

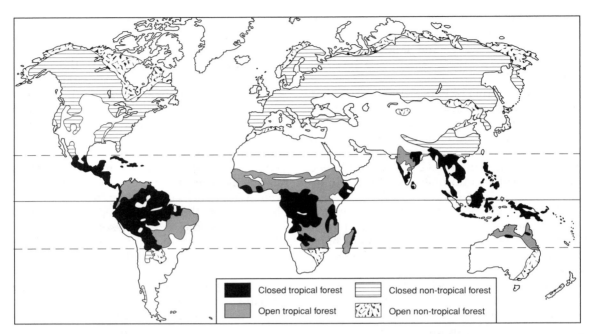

**Figure 1.2** Present global distribution of forests. (Adapted from World Commission on Forests and Sustainable Development, 1999)

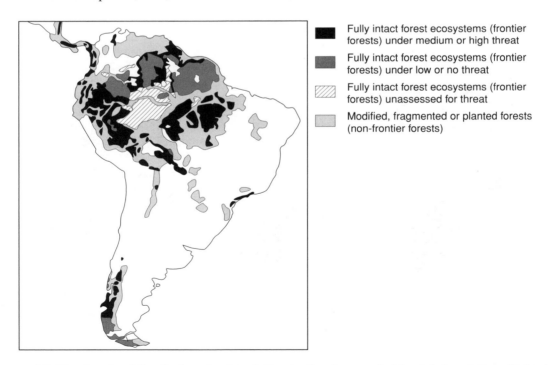

**Figure 1.3** The forests of South America in relation to the degree of risk of deforestation. (Adapted from World Commission on Forests and Sustainable Development, 1999)

**Urban characteristics**

| | |
|---|---|
| RESIDENTIAL | Residential accommodation<br>Institutional and residential<br>accommodation |
| TRANSPORT AND<br>UTILITIES | Highways and road<br>transport<br>Transport (other)<br>Utilities, e.g. gas/electricity<br>supply installations |
| INDUSTRY AND<br>COMMERCE | Industry<br>Offices<br>Retailing<br>Storage and warehouse |
| COMMUNITY<br>SERVICES | Community buildings<br>Leisure and recreational<br>buildings |
| VACANT | Land previously developed<br>Urban land not previously<br>developed<br>Despoiled land |

**TABLE 1.3** The classification of urban areas created for the collection of land-use change statistics (LUCS) by the UK's Ordnance Survey

maps provide a good visual impact of not only land cover but also include a measure of likely change in the near future. They also highlight the significance of forests in relation to non-forested areas on a global basis.

In contrast, the depiction of land use within a large metropolitan area such as London or New York can best be achieved using a wide range of categories and using a large scale. This is exemplified by the CORINE categories referred to above. Another example is given in Table 1.3. According to Perkins and Parry (1996), this scheme was devised by the UK's Ordnance Survey in 1985 and involves the collection of land-use change statistics. The detail afforded by the numerous categories is essential to appreciate the spatial variation in land use that occurs in a large city where land use can change street by street. This issue has also been discussed by Pauleit and Duhme (2000) who have developed a classification system with 24 separate units. Half of these units describe buildings, roads etc., whilst the other half describe 'green' areas, e.g. hedges, parks etc. This scheme is shown in Figure 1.4, which illustrates a small area of Munich, Germany.

Clearly, the depiction of land cover and land

**FIGURE 1.4** Transect showing pattern of urban cover types in Munich. (Adapted from Pauleit and Duhme, 2000)

use can be undertaken in numerous ways (see Parry and Perkins, 2000). One of the determining factors is the method of data collection or recording. For example, Level IV detail of the USGS is only possible with large-scale aerial photographs; at the other end of the spectrum, various types of satellite imagery, such as Landsat, are necessary for Level I classification. Of the examples given here, Figures 1.2 and 1.3 are derived from satellite imagery and Figure 1.4 is derived from aerial photography. The various possibilities are examined in Lillesand and Kiefer (2000). Other factors that contribute to depiction are purpose and scale; both are related to methods of data collection. As for the depiction of land cover in the past (e.g. Figure 1.1), there is less certainty than with the representation of present land cover/land use. This is because the latter can be observed in various ways from space and the reality, i.e. degree of accuracy, can be verified in the field. Reconstructions of past land cover are based on data collected in entirely different ways. There are two sources, one of which involves the analysis of plant remains from environmental archives such as peat deposits and lake sediments. The disadvantages of these types of data are numerous; for example, the spatial distribution of sites providing such data is limited, with an abundance in temperate areas and a dearth in tropical areas; and possibilities for testing the degree of accuracy of reconstruction by sampling modern-day ecological units is limited (see discussion in Prentice and Webb, 1998). A second source of information on past land cover, and past land use, is old maps and documents. Although these are reasonably reliable in terms of accuracy, inevitably coverage in both time and space is very limited.

Land cover and land use are dynamic; they change in space and time, sometimes rapidly and sometimes slowly, and not in unison globally. In terms of the recent past (for example, the last 20 years) enormous changes have occurred, mostly prompted by human factors. Records of land cover and land use are expressions of human impact on the Earth's surface in time and space, and thus they are records of environmental change in time and space. These themes are the primary focus of this book.

## 1.3 The dynamism of land-cover/land-use change

That land cover and land use are dynamic in time and space is indisputable. How, at what rate and why such change occurs is not always obvious or easy to determine. This is because change may be caused by many factors rather than a single factor, or because a single stimulus operates on several related factors in a feedback process that eventually becomes positive feedback. Land cover and land use are the manifestation of these processes or stimuli in the biosphere.

The dynamism of land cover and land use can be generated by natural or cultural factors, or a combination of the two. As discussed in Mannion (1997 and 1999a), the Earth's surface has been in a constant state of flux since it came into existence $c.$ $5000 \times 10^6$ years ago. Billions of individual imperceptible biogeochemical exchanges take place every second within and between the organic and inorganic components of the biosphere. At the heart of these exchanges is the element carbon and its biogeochemical cycle, as discussed in more detail in Section 1.5, but most naturally occurring elements have cycles or biogeochemical cycles. The latter involve the exchange of matter, i.e. carbon, nitrogen, iron etc., between the various components of the Earth's surface including the atmosphere and living organisms. This relationship is shown diagrammatically in Figure 1.5.

In turn, these exchanges influence life and climate, and are themselves conditioned by climate and life. This reciprocation is the essence of the Gaia hypothesis advocated by the independent British scientist James Lovelock in the 1970s (see Lovelock, 1995). The physics, chemistry and biology of the Earth are thus interrelated and are all equally important. They are the intrinsic factors that determine the nature of land cover and reflect the holistic nature of the Earth. Similarly, the physics, chemistry and biology of the Earth influence land use. Through climate, soils, geology and vegetation, the potential or capability of land is determined; these factors thus set the environmental limits for land use. Whilst human

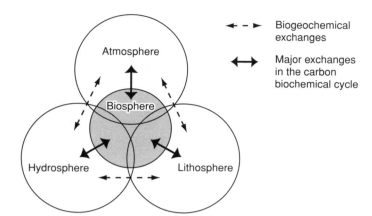

**FIGURE 1.5** The relationship between the various components of the Earth's surface and biogeochemical cycles

invention, i.e. technology, can temper such limits, limits nevertheless remain.

As well as these intrinsic components of Earth-surface processes, there are extrinsic factors. These are often referred to as astronomical factors and it is generally agreed that they influence climate. There are many such factors, e.g. alterations in the Sun's luminosity, including possible cyclic changes in sunspot occurrence, and the periodicity of the characteristics of the Earth's orbit around the Sun. Much has been written about all of these influences and their impact (see, for example, Bradley, 1999). Whilst a full appraisal of such possibilities is beyond the scope of this book, mention must be made of the influence of so-called Milankovich cycles on climatic change. They include changes in the elliptical path the Earth takes as it revolves around the Sun, i.e. orbital eccentricity with a periodicity of 96,000 years. Two other cycles also operate: the cycle of axial tilt with a periodicity of $c$. 42,000 years; and variations in the precessions of the equinoxes with a periodicity of 21,000 years. All three of these factors have had a significant impact on the biosphere, as illustrated in Figure 1.5. In the last $3 \times 10^6$ years, orbital eccentricity has played a major role in instigating major cold periods or ice advances of $c$. 100,000 years' duration; these are separated by warm periods known as interglacials and are relatively short-lived at $c$. 20,000 years. Temperature differences of $c$. 8°C to 10°C occurred globally as glacials ended and interglacials began. Inevitably, such climatic oscillations gave rise to land-cover changes; the land cover of cold/glacial periods globally was markedly different to that of interglacial periods. Moreover, the present interglacial – the Holocene – is quite different to its predecessors because of the substantial and intensifying impact of human activity. (These issues are discussed in detail in Chapter 2.)

## 1.4 People–environment relationships

Land cover and land use can also be described as representative of people–environment relationships. As discussed in Sections 1.1 and 1.2, land cover may be the manifestation of natural rather than anthropogenic factors, and it could be argued that where this obtains there is no reflection of people–environment relationships. This has some validity, but because of the holistic nature of the Earth, i.e. the interrelatedness/interdependence of the individual components which is itself dynamic, the extent and character of land cover resulting from purely natural factors is of considerable significance in terms of people–environment relationships. This view is legitimate on two counts: first, in a world that is increasingly being influenced by human activity, what remains of 'pristine nature' is itself a measure of people–environment relationships; second, the Earth's remaining natural environments are vital for the survival of humanity in so far as they provide essential ecosystem services such as carbon storage, biogeochemical regulation and hydrological functions etc. (The issue of biogeochemical regulation is discussed in Sections 1.1 and 1.5.)

The history of people–environment relationships encompasses a variety of approaches. The western intellectual approach, in many respects an attitude that epitomizes a nature *versus* people approach, does not appreciate holism. In effect, it espouses a clear distinction between people and environment or society and nature, and even advocates the domination of nature. Despite this, or perhaps because of it, the concept of environmental or geographical determinism rose to prominence in the late nineteenth century. It originated from the work of Charles Darwin (1809–82) and concerns the role of environment, especially climate, as the major control on human activity. Determinism as an expression of people–environment relationships, later fell into disrepute as did its companion philosophy, possibilism. This latter presented the view that humanity was capable of modifying environment, which thus could offer a range of possibilities, albeit within overall constraints, rather than a single opportunity for survival and development.

More recent approaches to describing people–environment relationships have involved the adoption of less polarized roles; for example, the systems approach advocates holism and recognizes the reciprocity that obtains between people and environment. The Gaia hypothesis is similar (see Section 1.3) as is the political economy approach, which considers that the social relationships of power are of primary importance in the use of environmental resources. These ideas have much in common with those of the Russian scientist Vladimir Ivanovich Vernadsky (1863–1945) who proposed the concept of the noosphere (Vernadsky, 1945; see also discussion in Frolova, 1998). This word derives from the Greek word *noos*, meaning mind; as a concept it embraces the idea of people and environment as mutually dependent components of the global environmental system. Vernadsky, along with the geologist and theologian Teilhard de Chardin, recognized the increasingly dominant role of society in the biosphere. Both believed, perhaps naively, that society could appropriate the Earth's resources for the greater good of both through science and technology, i.e. sustainable development, and thus foster a more symbiotic rather than combative relationship (see Section 1.6). Society, and its power relations, have much progress to make in this context.

In the last decade, attempts have been made to quantify people–environment relationships. These highlight difficulties of quantification but also reflect the variability of human impact. Such indices include the Living Planet Index recently proposed by the Worldwide Fund for Nature (1998) and the Ecological Footprint (see, e.g., Rees, 1992; Chambers *et al.*, 2000). Using data on production, consumption, waste production, biodiversity loss, water quality etc., these measures show which nations are living 'beyond their means', i.e. those exceeding their natural capital. In general, they show that environmental quality is changing rapidly and that the biosphere's resources are being depleted at an alarming rate. (These and other quantitative measures are discussed in Chapter 3.)

## 1.5 Land cover/land use and the carbon cycle

All types of land cover and land use are components of the global biogeochemical cycle of carbon and are thus linked with ecosystem services and climatic regulation. This has recently been examined by Schimel *et al.* (2001) in relation to global and regional patterns of carbon exchange by terrestrial ecosystems, which are considered to be the result of land-use changes over time. As Figure 1.6 illustrates, the major human influences on the carbon cycle at present involve the consumption of fossil fuel and the removal of the natural vegetation cover; both release considerable volumes of carbon (in various forms, such as carbon dioxide and methane) into the atmosphere. In relation to the overall cycle, the influences of

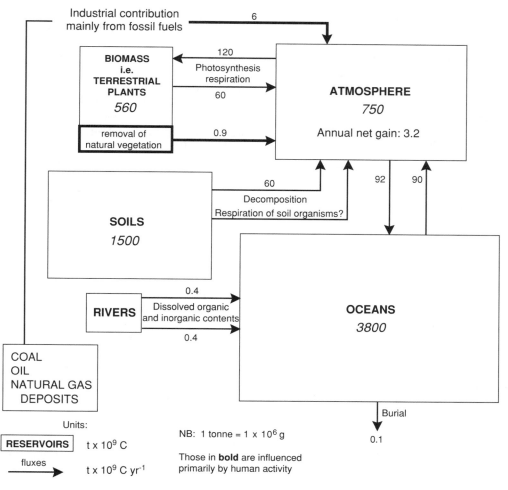

**FIGURE 1.6** The global biogeochemical cycle of carbon. (Based on data from Schimel, 1995; Schlesinger, 1997)

these two components is small in quantitative terms, yet both are highly significant as causes of environmental problems. Some of these problems are linked directly with carbon. For example, global climatic change is a response to increased carbon compounds in the atmosphere caused by a range of human activities, e.g. industry, cement production, ecosystem modification and destruction for agriculture. Further problems are linked indirectly with carbon; for example, acidification is caused by sulphurous and nitrous compounds, which are also generated by fossil-fuel use for a range of

purposes, including agriculture. Similarly, atmospheric lead pollution is a consequence of fossil-fuel use; lead has, in the past, been added to petrol to improve vehicle engine performance.

There are major differences in the temporal and spatial patterns of carbon appropriation. These have given rise to changes in land cover and land use, and are linked with stages of socio-economic development, often through the intermediary of technology. Indeed, the major advances in technology that have occurred have facilitated carbon appropriation

| |
|---|
| 1 The domestication of fire<br> *c.* $1.6 \times 10^6$ years ago |
| 2 The initiation of agriculture<br> *c.* $10 \times 10^3$ years ago |
| 3 The adoption of fossil fuels (Industrial Revolution)<br> *c.* 350 years ago |
| 4 A new threshold: biotechnology/genetic engineering currently occurring |

**TABLE 1.4** The major thresholds in carbon acquisition by society

and manipulation. These could be described as thresholds in human history and development, as well as thresholds in people–environment relationships. It is possible to recognize three major thresholds in the past and to suggest that society is beginning to cross a fourth (Mannion, 2000). These are given in Table 1.4.

The first is the harnessing of fire some $1.6 \times 10^6$ years ago by *Homo erectus*, an ancestor of modern humans (*H. sapiens sapiens*). As Pyne (1998) expresses so succinctly, 'The capture of fire by the genus *Homo* marks a divide in the natural history of the Earth. ... But from the time of *Homo erectus* one species acquired the capacity to start and stop fires, a niche filled by no other organism. A uniquely fire creature became bonded to a uniquely fire planet.' This unique ability to control fire gives humans what Pyne (1998) describes as a monopoly on fire. As a technology, fire can be used for numerous purposes but, however and wherever it is applied, it focuses on carbon manipulation. It requires a carbon source, and for much of prehistory and history that carbon source has been vegetation communities. Only some 300 years ago did attention turn to fossil fuels on a large scale (see below). Moreover, fire, as well as being reliant on a carbon source, is often used to manipulate carbon. In cooking food, for example, it renders carbohydrates, proteins etc. more palatable and digestible than those of uncooked food. Fire is also used to herd and trap animals, and so modifies vegetation cover. In agriculture, fire is often used to clear land

initially, a process that also produces nutrients by ashing biomass and thus assists in crop production. Today, fire remains a major agent for transforming land cover into land use, and for the management of numerous human-modified land-cover types.

The second major threshold in the land cover/land use–carbon relationship is the initiation of permanent agriculture. This began *c.* $10 \times 10^3$ years ago in the Near East and possibly in the Far East; other centres of innovation, though at later dates, include Central America, the Peruvian-Ecuadorian Andes and possibly sub-Saharan Africa. How and why agriculture developed are issues that remain enigmatic, as discussed in Mannion (1999b), but whatever the reasons, the objective was the appropriation of carbon in the form of food energy. From these beginnings myriad forms of agriculture have developed, all of which impinge to a greater or lesser extent on the carbon cycle. It is reasonable to describe agriculture as a carbon- and/or energy-processing technology, and one which remains a significant cause of land-cover alteration today. The environmental costs are high, not least of which is the fact that the resulting land is rarely capable of sequestering as much carbon as the ecosystem it replaced. Other problems also arise, e.g. loss of biodiversity, soil erosion, the impairment of water quality etc. Some of these problems are caused and/or intensified by the application of other forms of technology to enhance the carbon harvest, e.g. irrigation, agricultural chemicals and mechanization; these too may involve carbon as in the case of technologies driven by fossil fuels. (The role of agriculture in land-cover and land-use change is examined in Chapters 4 and 5.)

The third threshold in the land cover/land use–carbon relationship involves the adoption of fossil fuels *c.* 350 years ago, a development associated with the Industrial Revolution. Beginning with coal, fossil fuels underpinned the transformation of cottage industries into large-scale, mostly manufacturing, industries. This has to be considered as the most significant anthropogenic intervention in the carbon cycle so far. The use of coal, escalated through its role in wealth generation, was later usurped by natural gas and oil. All three are products of past ecosystems whose stored energy has been, and will certainly continue to be, used to transform the Earth's surface in many ways. This use was only possible through technology, e.g. the advent of steam power, coke production for iron smelting, the domestic and industrial gas industry based on the conversion of coal to gas, mechanization, the production of artificial fertilizers and many others. As in the case of agriculture, this manipulation of the carbon cycle has not been without cost, not least of which is global climatic change, which may eventually become a driving force in land-cover and land-use change. Environmental determinism is certainly not dead!

Finally, mention must be made of the fourth age of carbon. Whilst the thresholds described above involve carbon manipulation at the macroscale, recent developments in biotechnology and especially genetic engineering, involve carbon manipulation at the microscale, i.e. the molecular level. There are many applications of this new science in terms of agriculture and the environment, as well as medicine. It is currently contributing to land-use change in so far as genetically modified crops are already being grown in many countries. In agriculture alone, the possibility of producing drought-, disease- and pest-resistant crops and domesticated animals may be beneficial or detrimental; they may reduce the need for agricultural expansion into areas hitherto unsuitable or they may do quite the opposite and result in agriculture destroying yet more of the Earth's remaining and precious natural ecosystems. As discussed in Section 10.3.3., biotechnology – as in the case of all technologies – has potential disadvantages as well as advantages.

## 1.6 The causes of land-cover and land-use change

There are two major categories of causes of land-cover and land-use change: natural and anthropogenic (i.e. socio-economic). Natural change involves processes that are influenced by natural and not human factors. Until the present interglacial (last $c.$ $10 \times 10^3$ years) the impact of humanity on the Earth's surface had been limited. Even during the first half of this period (the Holocene) human impact, despite the advent of agriculture, was limited. Natural processes of environmental change thus dominated the biosphere. One of the most important of these is climatic change. It is perhaps ironic that the climatic changes characterizing the end of the last major glacial/cold stage may have been a major stimulus for the development of agriculture and its spread, thus making this interglacial unique. As discussed in Chapter 2, substantial changes in land cover occurred as the interglacial progressed. These changes reflect climatic change, and the synergistic response of organic and inorganic components of global ecosystems.

Against this backdrop of natural change, human influence on the biosphere has resulted in the transposition of wildscapes into landscapes, i.e. the transformation of natural land cover into modified land cover or land use. Both of the latter are also dynamic. Land use, for example, is not necessarily constant; socio-economic factors bring about land-use (and land-cover) change, and environmental processes, such as soil erosion and desertification, may be equally manipulative though these may themselves be underpinned by socio-economic factors.

There are many anthropogenic factors that cause land-cover and land-use change. Such factors are dynamic; what may be a stimulus at one point in time may cease to be important subsequently. An example of this is the role of agriculture in landscape change in the UK in the twentieth century. For most of the 1900s the emphasis has been on agricultural intensification in order to increase food production and self-sufficiency. However, since the 1980s, legislation through the European Union's (EU) Common Agricultural Policy (CAP) has attempted to reduce productivity through so-called set-aside policies. This has resulted in increased afforestation, the introduction of new crops, and alternative uses (e.g. recreation/ tourism for previously arable and pastoral land), as well as a downturn in the profitability of agribusinesses.

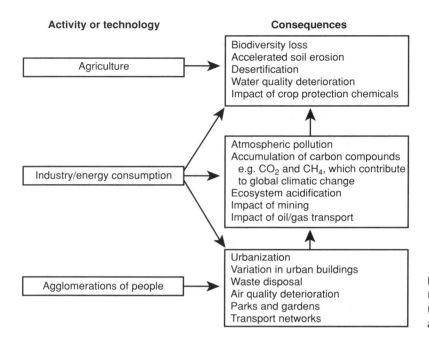

**Activity or technology**

Agriculture

Industry/energy consumption

Agglomerations of people

**Consequences**

Biodiversity loss
Accelerated soil erosion
Desertification
Water quality deterioration
Impact of crop protection chemicals

Atmospheric pollution
Accumulation of carbon compounds
  e.g. $CO_2$ and $CH_4$, which contribute
  to global climatic change
Ecosystem acidification
Impact of mining
Impact of oil/gas transport

Urbanization
Variation in urban buildings
Waste disposal
Air quality deterioration
Parks and gardens
Transport networks

**FIGURE 1.7** A summary of the anthropogenic factors that influence land cover and land use

The numerous anthropogenic factors that cause land-cover and land-use change can be grouped into three categories, as illustrated in Figure 1.7. All three are dynamic in time and space, and all are influenced by a wide range of factors that relate to subsistence, i.e. the existentialities of life, development and wealth generation. In each case the environment provides the broad limits for existence within which socio-economic factors, in combination with legislation and political policies, create the dynamics of land-cover and land-use change. In other words, they reflect the reciprocity between people and environment.

Sometimes this reciprocity is described as symbiosis, i.e. a mutually beneficial partnership as considered by Vernadsky in his description of the noosphere (see Section 1.4). However, this is far from the reality of the people–environment relationship. Whilst the world's population has increased rapidly to its present $5.9 \times 10^9$, environmental quality has deteriorated considerably, especially in the last 25 years (see Worldwide Fund for Nature, 1998). *H. sapiens sapiens* has become the most successful animal because of its intelligence and resulting ability to manipulate and control

the environment from which all resources derive. Other organisms and the environment itself have not fared as well and only now are they being recognized as vital components of human existence. In general, technology has facilitated human success at the expense of environmental quality. (The land cover and land use resulting from various forms of technology such as agriculture, forestry mineral extraction, industrialization and urbanization are examined in Chapters 4 to 9.)

## 1.7 Conclusion

Land cover and land use describe the characteristics of the Earth's surface. They reflect the operation of natural as well as anthropogenic processes, and are fluid in time and space. Anthropogenic changes in land cover and land use have, with the exception of catastrophic events such as earthquakes and volcanic eruptions, occurred much more rapidly than natural processes. Such changes have been generated through a range of socio-economic factors, e.g. population growth, wealth generation, political policies and basic subsistence needs. There are often indirect as well as direct changes, especially in land cover, that result from human activity; sometimes these take place over long periods and become manifest only after decades or centuries, e.g. ecosystem acidification.

Most land-cover and land-use change occurs as a consequence of technological innovation. Beginning with the use of stone, i.e. palaeolithic technology and the taming of fire, technology has become increasingly sophisticated. Agriculture metallurgy, industrialization and fossil fuel use are cases in point. All have played a pivotal role in people–environment relationships as, undoubtedly, will future technological developments. Moreover, most forms of technology are directed at appropriating and/or manipulating various forms of carbon. Consequently, land cover and land use are intimately linked to the carbon biogeochemical cycle. Such intervention has given rise to one of the most pressing environmental issues: global climatic change. Many other environmental problems also arise from this relationship, including the loss of biodiversity, which reflects humanity's success at the expense of other organisms. Land cover and land use are thus indices of a range of attributes that comprise people–environment relationship, and their adjustments in time and space.

## Further reading

**Ernst, W.G.** (ed) 2000: *Earth Systems. Process and Issues.* Cambridge University Press, Cambridge.

**Goudie, A.** 2000: *The Human Impact on the Natural Environment*, 5th edition. Blackwell, Oxford.

**Goudie, A. and Viles, H.** 1997: *The Earth Transformed*. Blackwell, Oxford.

**MacKenzie, F.T.** 1998: *Our Changing Planet*, 2nd edition. Prentice Hall, Upper Saddle River, New Jersey.

**Mannion, A.M.** 1997: *Global Environmental Change. A Natural and Cultural Environmental History*, 2nd edition. Longman, Harlow.

**Marsh, W.M. and Grossa, J.** 2002: *Environmental Geography: Science, Land Use and Earth Systems*, John Wiley and Sons, New York and Chichester.

# Land cover: some temporal perspectives

## 2.1 Introduction

Heracleitus, a philosopher of ancient Greece *c.* 500 BC is credited with saying 'All is flux, nothing is stationary.' This is particularly apt for describing land cover within a temporal framework. It is important to recognize that natural environmental change has sustained major alterations in land cover throughout geological time. Indeed, it is also important to note that such geological processes have generated resources from land-cover components, and oceanic biomass, that underpin many of society's economic activities today, e.g. fossil fuels, limestone, minerals and aggregates, to name but a few.

However, current patterns of land cover and land use are the products of natural and anthropogenic processes that have been operative over relatively short timescales when compared with the $5000 \times 10^6$ years of Earth history. Of particular importance is the last $3 \times 10^6$ years, i.e. the last $1.5 \times 10^6$ years of the Tertiary period and all of the Quaternary period. During this time period, the Earth experienced numerous climatic cycles each comprising a cold/glacial and warm/interglacial stage. The climatic changes, and especially temperature changes, associated with each shift from cold/glacial to warm/interglacial conditions had a global impact that was represented by alterations in land cover, as discussed in Section 2.2. The warm/interglacial stages of the last $3 \times 10^6$ years are characterized by minimal ice cover, high sea levels and maximum vegetation cover. For any given region, each warm/interglacial stage was similar in terms of vegetation succession in the wake of climatic amelioration, but different in the detail of the families and species involved. In temperate regions, for example, mixed deciduous forest predominated with species composition varying temporally, i.e. between interglacials, and spatially depending on regional and local conditions of topography, soils, etc. The overall sequence of change along with details of specific regions is examined in Section 2.3.

The present warm/interglacial stage, the Holocene, is similar in terms of natural change as the earlier warm/interglacial stages. It is, however, quite different in the context of anthropogenic change; no other warm/interglacial stage has been so influenced by human activity (see Section 2.4). This is despite the presence of humans, including modern humans (*Homo sapiens sapiens*) who probably evolved in Africa *c.* $250 \times 10^3$ years ago (see Section 2.5) in earlier interglacials. For whatever reasons, *Homo sapiens sapiens*, whilst continuing to rely on stone technology, gradually emerged as a major force in environmental systems (see Section 1.4). This began with the use of fire as well as scavenging and gathering, which progressed to hunting and gathering, and culminated with the inception of permanent agriculture. Thus it was that *Homo sapiens sapiens* became a significant controller of ecosystems.

The development of these early technologies represented major turning points in 'people–environment relationships' (Section 1.4) and in the role of humans in the carbon biogeochemical cycle (see Section 1.5). Later technological developments throughout prehistory and history, e.g. metallurgy, mining, pottery production, agricultural innovations, and eventual urbanization and industrialization, have continued to alter these relationships. In general, they have provided the wherewithal for human populations to survive and multiply, to expand their 'material culture' as development has proceeded and to create landscapes that bear witness to resource exploitation in time, and which are unique spatially. The land cover and land use characteristic of today's world is a culmination of the relationship between people and place in time and space.

## 2.2 The environmental changes of the last 3 × 10⁶ years

Until the mid-1800s, ideas about environmental change were limited, mostly relating to the supposed impact of the biblical flood, i.e. the diluvial theory. Recordings and observations of modern processes, especially in areas with active glaciers such as the Alps, were, however, beginning to generate new ideas about the environment as a dynamic entity. By the 1860s the glacial theory had been established and it was soon recognized that more than one major ice advance had occurred. In the 1870s, for example, it was being suggested that four glaciations had affected Britain and northern Europe. A century later, the extraction of ocean sediment cores offered yet more evidence for dynamism. Evidence from the marine sediment fossils such as foraminifera and their oxygen-isotope signatures, revolutionized ideas about environmental change yet again (see reviews in Lowe and Walker, 1997; Mannion, 1999a). In the last $1.8 \times 10^6$ years, for example, there is evidence for *c.* 32 climatic cycles each comprising a warm and cold stage. Similarly, in the preceding 1 to $2 \times 10^6$ years there were numerous warm/cold oscillations.

The last $3 \times 10^6$ years were, thus, far from static in terms of environmental change. With an average global temperature difference of 8–10°C between a warm and cold stage, not only was there repeated expansion and contraction of ice sheets but also significant rises and falls in sea level, and major changes in the architecture and composition of land cover. The major ecosystems of the world have assembled and disassembled numerous times in this most dynamic period of Earth history. Stability, consequently, becomes a relative term and depends on the timescales being considered. Each climatic cycle lasts *c.* $1.2 \times 10^5$ years, with the cold/glacial stage lasting *c.* $1 \times 10^5$ years and the warm stage lasting *c.* $2 \times 10^4$ years; warm/interglacial stages could thus be considered anomalous, and yet it is in this environment that humanity has thrived.

There has also been much debate as to the causes of these oscillations, the occurrence of which is confirmed by evidence from other archives of environmental history such as polar and alpine ice cores and loess sequences (see review in Mannion, 1999a). Numerous possibilities present themselves: tectonic uplift, variations in the Sun's luminosity, volcanic activity. Certainly, there is compelling evidence for the influence of the so-called Milankovich, or astronomical, factors that relate to the revolution of the Earth around the Sun (see Section 1.3 and Figure 1.3). Figure 2.1 gives details of these

**A  Orbital eccentricity**

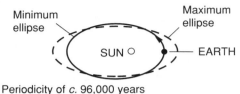

Periodicity of *c.* 96,000 years

The Earth's orbit around the sun varies and is elliptical rather than circular. When the Earth is furthest from the Sun, cooling occurs. The periodicity of orbital eccentricity is considered to be a major factor in the waxing and waning of ice ages.

**B  Axial tilt**

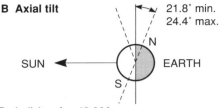

Periodicity of *c.* 42,000 years

The tilt of the axis around which the Earth rotates causes seasonality. It also determines the intensity of incident radiation. When the angle of tilt is at its minimum, 21.8°, incident radiation in the Northern Hemisphere is *c.*15% less than when the angle of tilt is at its maximum, 24.4°. Periods of minimum tilt therefore relate to cooling.

**C  Precession of the equinoxes**

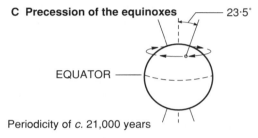

Periodicity of *c.* 21,000 years

This occurs due to the wobble of the Earth's axis. It controls the amount of solar radiation received at the Earth's surface by influencing the season in which the Northern Hemisphere is closest to the Sun. In particular, an ice age is likely to develop when the Northern Hemisphere is furthest from the Sun in summer.

**D  Variations in solar radiation resulting from A, B and C above**

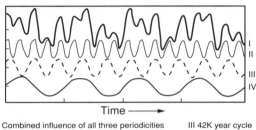

I Combined influence of all three periodicities

II 21K year cycle

III 42K year cycle

IV 96K year cycle

**FIGURE 2.1**  The astronomical forcing factors involved in the Milankovich theory (astronomical theory) of climatic change (Mannion, 1999a)

factors and shows that the major influence on the shift from cold/glacial to warm/interglacial conditions is orbital eccentricity.

This, therefore, is a major natural influence on land cover, though it is not the only influence at this crucial time of shift from cold to warm conditions. There is general agreement that orbital eccentricity does not of itself create a sufficient temperature difference. Most researchers agree that other elements must also be involved. Interestingly, one such candidate is the global carbon cycle of which land cover is an important component. Analysis of the composition of air trapped in bubbles in Antarctic ice cores (see Barnola *et al.*, 1987; Chapellaz *et al.*, 1990; Petit *et al.*, 1999) shows that the concentrations of both carbon and methane, the most common carbon compounds and heat-trapping gases present in the atmosphere, changed markedly as cold/glacial stages ended and warm/interglacial stages began and

vice versa, as shown in Figure 2.2. As a warm/interglacial opens, the concentration of carbon dioxide increases by *c.* 25 per cent and that of methane doubles. Similar trends occur in ice cores from Greenland, as Brook *et al.* (1996) have pointed out in relation to methane. These marked shifts indicate that Milankovich orbital eccentricity has a substantial impact on the global carbon cycle (the opposite is not possible). However, the question is whether such adjustments in the carbon cycle are forcing or reinforcing factors in climatic change, i.e. did they result from orbital forcing and the relatively weak but significant increase in temperature this promoted, but which was of sufficient strength to stimulate a reaction in the carbon cycle? The resulting release of carbon dioxide and methane would have enhanced the greenhouse effect, causing increased radiation to reach the Earth's surface. Thus the changes in the carbon cycle, with atmospheric enrichment

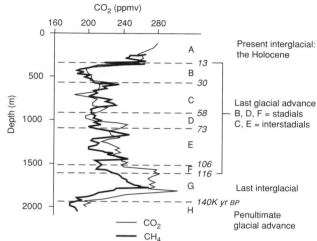

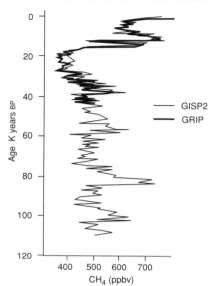

**FIGURE 2.2** The carbon dioxide and methane records from the Vostok ice core, Antarctica (based on Barnola *et al.*, 1987 and Chapellaz *et al.*, 1990) and the methane record from the Greenland GISP2 and GRIP ice cores. (Adapted from Brook *et al.*, 1996)

of carbon gases would have reinforced global warming and contributed to the initiation of an interglacial. This is one possible scenario, but notwithstanding the stimulus there is a clear link between the carbon cycle, climatic change and land cover under natural conditions. As Raynaud *et al.* (2000) have discussed, the expansion of wetlands in boreal regions was a likely source of the methane increase that occurred as the Holocene opened. It follows that modification of any one of these components of the biosphere by society will have repercussions; if land cover is altered substantially, then the carbon cycle will be altered too and so will the climate.

A further lesson provided by the environmental record of polar ice cores and ocean-sediment cores is the fact that climatic shifts can occur relatively rapidly. For example, Figure 2.2 shows that rapid increases in atmospheric carbon dioxide and methane concentrations occur as warm/interglacial conditions develop. These shifts parallel sharp changes in indicators of temperature such as oxygen-isotope and

deuterium concentrations (see Jouzel *et al.*, 1987; Lorius *et al.*, 1985; Von Grafenstein *et al.*, 1999). Similar patterns of temperature change are indicated by fossil coleoptera (beetles) in lake sediments in Britain, which were deposited in the transition phase from the last glacial stage to the Holocene. Figure 2.3

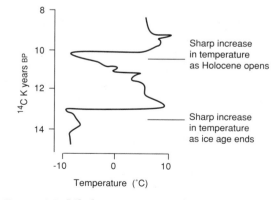

**FIGURE 2.3** Likely temperature changes during the late glacial period in Britain. (Adapted from Atkinson *et al.*, 1987)

illustrates temperature reconstructions from several sites in Britain. Steep increases in temperature are indicated at *c*. $15 \times 10^3$ years BP and $10 \times 10^3$ years BP. Such trends may involve rates of change of between 2.6°C per century and 1.7°C per century respectively (Atkinson *et al.*, 1987). Moreover, the trends in temperature change indicated at these sites show that climate was fluctuating considerably during this relatively short period (see Mayle *et al.*, 1999). Pollen and coleoptera are but two of the many indicators used to reconstruct this climatically complex period, a chronology for which has been derived not only from radiocarbon age estimation but also from varves and volcanic ash horizons (see Lowe *et al.*, 2001; Walker *et al.*, 2001 for details).

As a result, and because of sea level rises, plant and animal communities were changing rapidly as individual species responded to the changing environment. This was, therefore, a period of significant land-cover change. The lesson from this palaeoecological evidence is that climate can and has changed rapidly; global warming and cooling of sufficient magnitude to shift the Earth into its cold state or warm state can occur within a few centuries. Whilst current global warming, estimated to be *c*. 0.5°C in the 1900s (see Harvey, 2000; Intergovernmental Panel on Climate Change, 2001a for a review of the evidence), is considerably less than that which characterized the cold/glacial to warm/interglacial shift, its impact is still likely to be significant. Any intensification of this warming trend, which is likely if anthropogenic carbon dioxide and methane emissions are not curbed, will inevitably influence land cover and land use. Human-induced warming, which is unlikely to be globally uniform, is thus a new agent of land-cover and land-use change (see Section 10.3.1).

## 2.3 The interglacial cycle

As discussed in Section 2.2, the last $3 \times 10^6$ years have been characterized by alternating cold/glacial and warm/interglacial stages. Deposits relating to interglacials earlier than the present Holocene include lake and fluvial sediments that contain plant and animal remains. Many of these, especially in northwest Europe, have been investigated to reveal similarities between interglacials of the same age, but which are spatially disparate, and between interglacials of different ages. This was first recognized by Jessen and Milthers (1928) on the basis of their analysis of plant remains in Danish deposits. Subsequently, Iversen (1958) developed a model for interglacial ecosystem processes. A modified version of this is given in Table 2.1, which also includes schemes for other regions of Europe and for Florida. Each cycle comprises four stages, and whilst there is a gradual rather than sharp transformation between stages, this simplistic approach reflects changing land cover as an inherent characteristic of interglacial development.

The details given in Table 2.1 indicate that in areas subject to glaciation, e.g. the north temperate zone, initial plant colonization takes place as ice retreats leaving unconsolidated substrates that are often unstable and usually skeletal; they are mineral rich but lack cohesion and have little organic matter. Pioneer species such as shade-intolerant herbs, mosses and lichens colonized. These sparse, low-growing vegetation communities are characteristic of this cryocratic phase. Shrubs and pioneer tree species, such as birch and pine, invaded as soils began to develop in the protocratic stage. This is despite the fact that temperatures probably ameliorated considerably (see Section 2.2) as the cryocratic phase ended, and were sufficiently high to support a forest vegetation. That this did not occur quickly is likely to be due to incompatibility between temperatures and soil development; the former were conducive to forest cover but soils were immature with poor structure and poor water/nutrient-retaining capacities. As these properties improved, through the action of plants themselves whose organic matter was incorporated into the soil as they shed leaves and died, the vegetation characteristic of the mesocratic phase was able to invade. Brown earth mull soils supported a range of deciduous tree species (e.g. oak, elm, lime) whose success produced a dense canopy that discouraged many of the plant species characteristic of the cryocratic phase. The

| Phase of interglacial cycle | Faeroe and Shetland Islands | Western Ireland | Eastern Mediterranean | Florida |
|---|---|---|---|---|
| CRYOCRATIC | Sparse cover of Arctic-Alpine herbs Soils: Skeletal mineral | | *Artemisia*-goosefoot steppe | Open prairie vegetation Soils: Active sand dunes |
| PROTOCRATIC | Grassland with abundant tall herbs and ferns Soils: Unleached calcareous | Pine, birch, poplar, juniper woodland Soils: Unleached calcareous | Hollyoak-pistachio woodland Elm, lime, ash | Oak scrub with hickory Soils: Stable sandy |
| MESOCRATIC | Grassland with abundant tall herbs and ferns Soils: Fertile brown earths | Pine, oak and holly forest Soils: Brown earth | Elm, lime, ash Hornbeam and fir | Pine, oak woodland Soils: Stable sandy |
| OLIGOCRATIC | Acid heath and bog communities Soils: Podzols, peats | Fir, spruce, heaths, rhododendron Soils: Podzols, peats | Pine Holly, oak | Pine woodlands, bogs, swamps Soils: Leached soils and peats |
| TELOCRATIC | | | and pine | |

**TABLE 2.1** Characteristic plant assemblages of the interglacial cycle. (Adapted from Birks, 1986)

mesocratic phase is considered to represent the climatic optimum of an interglacial stage.

The final phase of the interglacial cycle is the telocratic phase, which is retrogressive. Soils have become podzolized, with peats in areas of high rainfall; these developments reflect nutrient depletion and natural ecosystem acidification. The dominant vegetation types are acidic woodland with conifers and ericaceous heaths. The telocratic phase is associated with the deteriorating climate that is a prelude to a cold/glacial period. However, Andersen (1966) has suggested that natural ecosystem acidification could have begun in the mesocratic phase independently of climatic regression. Consequently, he has revised Iversen's model to include an oligocratic phase, comprising the latter part of the mesocratic and early part of the telocratic phases (see Table 2.1).

These trends are evident from the many pollen diagrams and records of plant macrofossils from numerous interglacial sites relating to the last four or five interglacials in northwest Europe. They are also evident from Holocene records, which are discussed in Section 2.4, and which are unique in so far as they also contain a record of human activity in the last 10,000 years. An example of land cover during an interglacial period is given in Figure 2.4. Andersen and Borns (1994) have used pollen diagrams of Eemian interglacial age ($115 \times 10^3$ to $130 \times 10^3$ years BP), i.e. the last interglacial (known as the Ipswichian in the UK) in western Europe. During the mesocratic phase of this interglacial temperatures were probably warmer than today by *c*. 2°C; sea levels were, consequently, higher globally than they are today and sufficiently high to separate Fennoscandinavia as an island from Europe. In terms of land cover, forests extended further north then they do today. This example reflects not only similarities with other interglacial stages, such as the Holocene, but the fact that there were also significant differences.

As Birks (1986) has shown, the interglacial model of ecosystem development can be

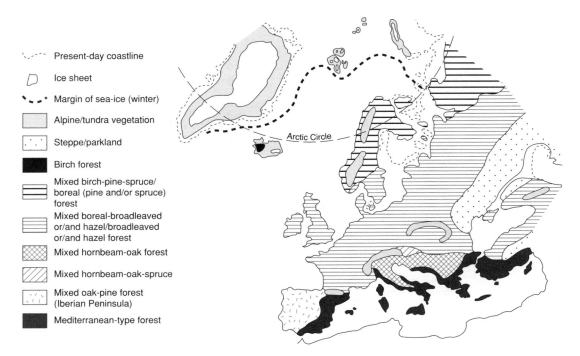

Present-day coastline

Ice sheet

Margin of sea-ice (winter)

Alpine/tundra vegetation

Steppe/parkland

Birch forest

Mixed birch-pine-spruce/ boreal (pine and/or spruce) forest

Mixed boreal-broadleaved or/and hazel/broadleaved or/and hazel forest

Mixed hornbeam-oak forest

Mixed hornbeam-oak-spruce

Mixed oak-pine forest (Iberian Peninsula)

Mediterranean-type forest

**FIGURE 2.4** Land-cover types during the Eemian interglacial. (Adapted from Andersen and Borns, 1994)

applied to areas outside northwest Europe, including areas that were never directly subject to glaciation and in which there is no true cryocratic phase. For example, in both the eastern Mediterranean and Floridan examples given in Table 2.1 open habitat species dominate the early phases with a vegetation succession of woodland reaching its climax in the mesocratic stage. Thereafter, regression occurred as acidification and climatic deterioration ensued. Recently, Okuda *et al.* (2001) have shown that the early stage of the last interglacial in the catchment of Lake Kopais, southeast Greece, was dominated by pioneer species such as birch and juniper; subsequently, deciduous oak forest developed with wild olive, and finally there was a reversion to pine and fir woodland. This pattern is similar to that proposed in the model of interglacial vegetation change. Although there are relatively few pollen diagrams from tropical regions, it is likely that the interglacial model can still be applied. For example, Hooghiemstra (1989) has

shown that in the High Plain of Bogotá, Colombia, variations in the tree line and in vegetation composition occurred during warm stages: grass paramo (a high-Andean vegetation type) was replaced with shrub paramo followed by forest. In Japan, the interglacial sequences of Lake Biwa reflect forest development as cool-temperate tree taxa are superseded by warm-temperate evergreen broadleaved trees or warm temperate evergreen coniferous tree species (Miyoshi *et al.*, 1999). Similar trends occurred in western Tasmania, Australia, where the last interglacial stage was characterized by cool-temperate rain forest (Colhoun, 2000), and in equatorial West Africa Dupont *et al.* (2000) have shown that savanna and dry forest were replaced by rain forest. However, the few pollen diagrams available from the Amazon region reflect the continuity of forest presence throughout the last glacial/Holocene period (Colinvaux and De Oliveira, 2001). Although there is no direct evidence of vegetation change during earlier interglacials, it seems likely that the vegetation sequence would not show such marked differences as it does elsewhere. The interglacial model is thus unlikely to be appropriate for this region.

The overriding control on the character and composition of pre-Holocene interglacial vegetation communities, and hence on land cover, is climate. Its impact is not always immediate in so far as the northwest European examples show that there may be a time lag between warming and the development of an equilibrium vegetation. This may not be the case in those areas not experiencing glaciation or periglaciation directly because soil processes are not disrupted so significantly. The changes discussed above once again reflect the dynamism of biospheric change and highlight the variable nature of interglacial land cover.

## 2.4 The Holocene

The Holocene is the current interglacial. It began c. $10 \times 10^3$ years ago as the last ice sheets finally retreated. It has much in common with earlier interglacial stages (see Section 2.3) in so far as it was characterized by an initial rapid amelioration of temperature, which promoted a reconfiguration of plant and animal communities against a backdrop of retreating ice and rising sea levels. Conversely, however, it is markedly different to earlier interglacials because it reflects considerable global change brought about by human activities. It is the period in which human communities have undergone major development at a rate faster than at any other time in the history of *Homo sapiens sapiens*. As discussed in Sections 1.4 and 1.5, the manipulation of resources facilitated through technological innovations during the Holocene has altered people–environment relationships profoundly. Similarly, technology has allowed humans to command a more significant role in the carbon cycle (Figure 1.6) than at any other time in the past. It is paradoxical that the more people are, apparently, divorced from the environment, e.g. the urban-industrial communities, the greater the impact they have

on the environment. This is because their existence and material well-being is heavily dependent on carbon consumption and manipulation, notably through agriculture and industry.

The early part of the Holocene, i.e. $10 \times 10^3$ to $5 \times 10^3$ years BP, was the mesocratic phase (Section 2.3) of the interglacial. Although agricultural innovations were taking place during this period (see Section 2.6) human activity was not exerting a major influence on land cover. Mesocratic landscapes developed as an expression of climatic climax vegetation communities. In a global context, Figure 1.1 depicts the extent and types of forest cover that were found on Earth *c.* $8 \times 10^3$ years ago. When compared with Figure 1.2, which shows the distribution of forest today, it is clear that all types of forest have diminished in extent. The temperate zone has been particularly severely affected and all types of forest have been subject to fragmentation. Sharma (1992, quoted in World Commission on Forests and Sustainable Development, 1999) has estimated that the original forested area of the Earth was $6.0 \times 10^9$ ha and is now only $3.6 \times 10^9$ ha, representing a massive 40 per cent loss. As discussed in Sections 2.6 and 2.7, there are many reasons for this loss, but the expansion of agriculture is a primary factor.

A map such as that shown in Figure 1.1 is compiled from pollen records extracted from a variety of sites (see comments in Section 1.2), each providing information about local vegetation change. A recent project, entitled BIOME 6000, has attempted to synthesize such published data from sites with good chronological control (usually through radiocarbon age estimation), to produce continental- or regional-scale reconstructions of land cover at $6 \times 10^3$ years BP, i.e. mid-Holocene (Prentice and Webb, 1998). The objective of this project is to provide a series of palaeovegetation maps for use in studies of climatic change wherein land cover is an important component of the atmosphere–lithosphere–biosphere relationship. Initially, the approach was used to derive a reconstruction of European vegetation cover (Prentice *et al.*, 1996). It relies on the construction of plant functional types (PFTs), categories

**A China**

   i   In eastern China forest zones were displaced northward, i.e. cool mixed forests extended northeast into what is now Taiga.
       Broad-leaved evergreen forests extended 300 km further north than their present limits.
       Temperate deciduous forest extended 500—600 km further north than their present limits.
   ii  In northwestern China, the extent of desert and steppe cover was reduced when compared with the present.
   iii On the Tibetan Plateau, forest extended to higher altitudes than today and the area of tundra was reduced as a result.

**B Former Soviet Union and Mongolia**

   i    West of the Urals there was an expansion of temperate deciduous forest northward and southward of its present range.
   ii   Cool mixed and cool conifer forests enjoyed a more northerly distribution than today.
   iii  Taiga was less extensive in European Russia but extended into the northwest where cold deciduous forest is now present.
   iv   The northern limit of the Taiga, i.e. the tree line, was higher latitudinally than at present, at the expense of tundra.
   v    Tundra was present in northeast Siberia.
   vi   Forest-steppe boundaries were similar to those of the present.
   vii  Dry conditions persisted in western Mongolia and north of the Aral Sea.

**C Africa**

   i   Madagascar, eastern, southern and central Africa were characterized by a biome distribution similar to that of the present with only minimal differences.
   ii  Major differences occurred in the region north of 15°N: steppe occupied sites at low altitude, which are now desert or temperate xerophytic (dry) wood/scrub, and warm mixed forest occupied the Saharan mountains.

**TABLE 2.2** The distribution of major vegetation units (biomes) at $6 \times 10^3$ years BP in China (Yu *et al.*, 1998), the Former Soviet Union and Mongolia (Tarasov *et al.*, 1998), and Africa and the Arabian peninsula (Jolly *et al.*, 1998)

of which are defined on the basis of plant features such as life form, leaf form and phenology, i.e. pattern of leafing, flowering, etc. The criteria defining PFTs are kept constant between regions to ensure consistency. Recently, results for regions outside Europe have been presented, e.g. for China (Yu *et al.*, 1998; see also Ren and Beug, 2002), the Former Soviet Union and Mongolia (Tarasov *et al.*, 1998), and for Africa and the Arabian peninsula (Jolly *et al.*, 1998). A brief summary of the results from these regions is given in Table 2.2, which shows that, in general, forests and woodlands occupied areas from which they are now absent and that this was the case both altitudi-

nally and latitudinally. All of these changes are ascribed by the various authors to differences in climate at $6 \times 10^3$ years BP when compared with the present. Once again, this emphasizes both the close link between climate and land-cover characteristics and the dynamic nature of this relationship.

The dynamism of natural vegetation change during the Holocene is well illustrated at the local scale. The results of pollen analysis from lake sediments and peats or mires provide an insight into such changes, as is illustrated in Figure 2.5. This diagram, from Hockham Mere, East Anglia, UK, shows that early Holocene vegetation communities underwent successional processes. In terms of forest development, open birch and pine woodland was superseded by climax forest comprising elm, oak, lime and alder by $c. 6 \times 10^3$ years BP. By 5 $\times 10^3$ years BP the impact of humans was becoming evident; in particular, the first agriculturalists were beginning to make their presence felt. (The role of humans in Holocene

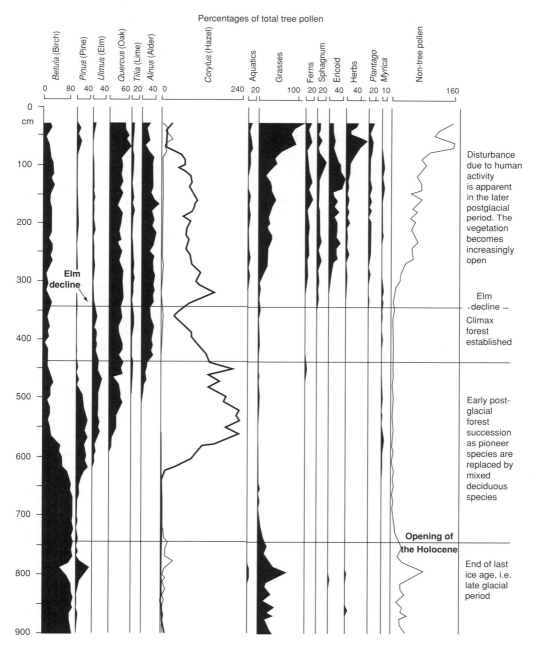

**FIGURE 2.5** Simplified pollen diagram from Hockham Mere, East Anglia, UK. (Adapted from Godwin, 1975)

environmental change is discussed in Sections 2.5 and 2.6, and in Chapter 3.) There are abundant publications from many parts of the world that reflect Holocene vegetation change (see Mannion, 1997 and 1999a for syntheses); all reinforce the dynamism of Holocene inter-glacial vegetation communities.

This dynamism is also reflected in studies that examine cold/glacial to warm/interglacial shifts in the carbon cycle. As explained in Section 1.5 (see also Figure 1.6), carbon is continually cycled between the atmosphere, biosphere and the oceans; it follows that as terrestrial biomes, especially forests, expanded following the end of a cold/glacial stage, carbon was transferred to the biosphere from another store or reservoir. Similarly, ice-core data (see Figure 2.2) show that there was a transfer of carbon, as carbon dioxide and methane, into the atmosphere (see comments in Section 2.2). Where did the carbon come from? Why did transfer occur? How much was trans-ferred? Why is it important to understand the components and processes involved? These are some of the questions that the palaeoenviron-mental record raises; they are particularly significant for understanding the dynamics of environmental change and hence for environ-mental management.

The oceans are the obvious source of the additional carbon in the atmosphere and bio-sphere, though how the carbon accumulated in the oceans during cold/glacial stages is enig-matic. Using two different approaches, Beerling (1999) and Francois *et al.* (1999) have estimated that carbon storage in the biosphere, i.e. in vegetation and soils, increased by as much as *c.* 668 Gt and 710 Gt respectively from the last glacial maximum to the mid-Holocene at $6 \times 10^3$ years BP. The transfer was effected mainly through increased photosynthesis as the atmos-phere became enriched in carbon dioxide and as the world's forest ecosystems expanded along with extensive peatland formation, espe-cially in northern Canada and Eurasia. Soil carbon storage also increased. The significant role of northern Eurasian ecosystems has been explored by Velichko *et al.* (1999), who have determined that 29.9 Gt and 131.4 Gt of carbon were stored at *c.* 18–20 $\times 10^3$ years BP (last

glacial maximum) and the mid-Holocene respectively. This represents 27 per cent and 120 per cent respectively of the present-day carbon storage. Since the mid Holocene, human activity has resulted in a diminution of the biosphere's carbon store as a whole, mainly through deforestation and the clearance of natural vegetation communities for agriculture (see Section 2.6).

## 2.5 Humans: a brief history

The story of human evolution is controversial and, like land cover and land use, is in a continuous state of flux. Traditionally, the story has been pieced together by the examination of the fossil remains of modern humans (*Homo sapiens sapiens*) and their ancestors; the emphasis has been on anatomical features in order to determine an evolutionary lineage. However, in the past 40 years, biomolecular evidence has become increasingly important, often raising more questions than it has answered. Biomolecular evidence, or molecular anthropology, involves analyses of blood proteins and genetic codes, i.e. the DNA (deoxyribonucleic acid) of primates, the nearest relatives of humans, and humans. There is now some agreement between traditional fossil evidence and molecular anthropology as to when the so-called hominine line (which even-tually gave rise to modern humans) diverged from other primates. The divergence between the African apes, i.e. gorillas and chimpanzees, the closest primate relatives of humans, occurred between $5 \times 10^6$ and $10 \times 10^6$ years ago.

These early hominines were quadrapedal (such as *Ramapithecus*, a hominoid of late Tertiary age, which disappeared from the fossil record *c.* $7 \times 10^6$ years ago, some $2 \times 10^6$ years before the first appearance of the Australopithecines in Africa). There is much debate as to whether these species were bipedal or capable of bipedalism when necessary. This is an important characteristic that requires a distinctive anatomy, and one that is considered to be more human than ape-like. It may also

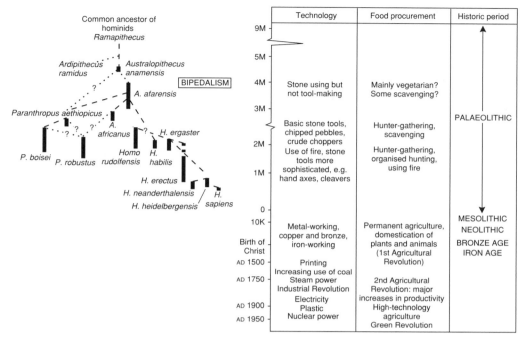

**FIGURE 2.6** Human evolution in relation to technology and food procurement strategies. (Adapted from Mannion, 1997)

imply an increased capacity to manipulate resources through the freeing of hands. Various fossils of Australopithecine species have been discovered; confined to Africa, they are the immediate precursors of *Homo* species, as illustrated in Figure 2.6.

By 2.5 × 10⁶ years BP the first hominine species had evolved. This is considered to be *Homo habilis*, so called because of the species' capacity to make tools. There is some debate as to whether all of the fossils, also found only in Africa, should be ascribed to *H. habilis* (Wood, 1992) or whether they in fact represent three different species, as shown in Figure 2.6. This aspect of fossil classification is another, often confusing and often changing, factor in accounts of human evolution. Nevertheless, these species of hominid had larger brains then their Australopithecine ancestors. It should also be noted that *H. habilis* demonstrated skills that set the species apart from other animals and that gave it a degree of control over resources.

*Homo erectus* was the successor of *H. habilis*; it

evolved c. 2 × 10⁶ years ago in Africa. It could use fire, make stone tools, and probably acquired food through active hunting and gathering rather that passive scavenging. This capability also reflects an increasing capacity to manipulate resources. The species has generated controversy in so far as *H. erectus* migrated out of Africa into Europe and Asia, and there is debate as to whether communication between individuals and groups occurred through language. It has been suggested that *H. erectus* did enjoy a primitive form of language (Cavalli-Sforza and Cavalli-Sforza, 1995), which probably contributed to the success of the species in food procurement and migration. This species migrated from its centre of origin in Africa's rift valley (in Kenya, Ethiopia and Tanzania) in three waves at c. 1700 × 10³, 1400 × 10³ and 800 × 10³ years BP. Why this should have occurred is a matter for debate but Bar-Yosef and Belfer-Cohen (2001) suggest that the absence of diseases transmitted to hominids from animals and insects may be the reason for the success of *H. erectus* in Europe and beyond.

As Figure 2.6 shows, the next stage in human evolution was the evolution of *H. sapiens*, *H. heidelbergensis* and *H. neanderthalensis*. Whilst the inter-species relationships are themselves interesting, the big questions surround the origins of *H. sapiens*, i.e. archaic modern humans. Did the species emerge from *H. erectus* populations in disparate parts of the world? This is the essence of the multiregion hypothesis. Alternatively, did the species evolve from *H. erectus* in Africa and then migrate into Europe and Asia, and eventually into Australasia and the Americas? The multiregion and single-region hypotheses, also known as the candelabra and Noah's ark models respectively, appear to be mutually exclusive. In general, the fossil evidence and biomolecuar evidence favour the single-region hypothesis (see discussion in Stringer and McKie, 1996 and in Pearson, 2001). Modern humans evolved between $400 \times 10^3$ and $200 \times 10^3$ years ago, but their attributes favoured their survival over their near relatives, the Neanderthals, who became extinct *c.* $35 \times 10^3$ to $45 \times 10^3$ years ago. The large brain size of *H. sapiens sapiens* facilitated language development, social organization, premeditated food procurement and migration. According to Sykes (1999; 2001), Cro-Magnons, anatomically modern humans, were present in Europe by *c.* $50 \times 10^3$ years ago. Their genetic linkages have been traced in modern Europeans to whose gene pools they contribute *c.* 10 per cent.

Certainly by *c.* $20 \times 10^3$ years ago, modern humans had developed considerable skills as hunter-gatherers and, by this time, had mastered artistic skills. Still stone using, modern humans precipitated a revolution of sorts $12 \times 10^3$ to $10 \times 10^3$ years ago when they began to select and domesticate plant and animal species (as discussed in Section 2.6). Sykes' (1999; 2001) analysis of European genetic data indicates that, by this time, the retreat of ice caps and glaciers had encouraged *H. sapiens* to emerge from its ice-age refugia and to spread once again into a Europe that was rapidly warming. Indeed, this population expansion has resulted in a contribution of *c.* 70 per cent to the modern European gene pool. The spread of agriculture via migration from the Near East contributed a further 20 per cent of the modern gene pool. Thereafter, technological developments unfolded rapidly in the context of the relatively short history of *H. sapiens sapiens* (see Figure 2.6).

## 2.6 The emergence of agriculture and metallurgy

As discussed in Sections 1.4 and 1.5, the emergence of agriculture marked a major turning point in people–environment relationships and in the role of humans in the global carbon cycle (see also Figure 1.6). The inception of agriculture represented a shift to a new technology that facilitated food procurement, i.e. the appropriation of edible carbon.

The first domestications of animals and plants occurred between $12 \times 10^3$ and $5 \times 10^3$ years BP, as shown in Table 2.3. Although domestication was a significant innovation, it developed in a context of already well-organized plant and animal manipulation. The preceding hunter-gatherer communities were skilled in the procurement of food, and domestication represents a transitory phase between hunting-gathering and permanent agriculture rather than a sharp boundary. Often referred to as the Neolithic Revolution, domestication and the inception of permanent agriculture represented the culmination of earlier people–environment relationships, an evolutionary process that gradually proceeded at various rates in different parts of the world. The repercussions of these developments were indeed momentous, not only in terms of people–environment relationships, but also in a socio-economic context and in terms of the power relations within and between human communities.

There is a considerable volume of information on where and when domestication occurred whilst why it occurred remains enigmatic (see discussion in Mannion, 1999b). The earliest sites of animal and plant domestication are in the Near East and Far East. In the former (see review in Smith, 1995) the dog was domes-

| Crop | Common name | Approx. date K years BP (uncalibrated radiocarbon years) |
|------|-------------|----------------------------------------------------------|
| **A   The Near east** | | |
| Avena sativa | Oats | 9.0 |
| Hordeum vulgare | Barley | 9.8 |
| Secale cereale | Rye | 9.0 See text |
| Triticum aestivum | Bread wheat | 7.8 |
| T. dicoccum | Emmer wheat | 9.5 See text |
| T. monococcum | Einkorn wheat | 9.5 See text |
| Lens esculenta | Lentil | 9.5 |
| Vicia faba | Broadbean | 8.5 |
| Olea europea | Olive | 7.0 |
| | | |
| **B   Africa** | | |
| Sorghum bicolor | Sorghum | 8.0 |
| Eleusine coracana | Finger millet | ? |
| Oryza glaberrima | African rice | ? |
| Vigna linguiculata | Cowpea | 3.4 |
| Dioscorea cayenensis | Yam | ? |
| Coffea arabica | Coffee | ? |
| | | |
| **C   Far East** | | |
| Oryza sativa | Rice | <11.5* |
| Glycine max | Soybean | 3.0 |
| Juglans regia | Walnut | ? |
| Castanea henryi | Chinese chestnut | ? |
| | | |
| **D   Southeast Asia and Pacific islands** | | |
| Panicum miliare | Slender millet | ? |
| Cajanus cajan | Pigeonpea | ? |
| Colocasia esculenta | Taro | 9.0 |
| Cocos nucifera | Coconut | 5.0 |
| Mangifera indica | Mango | 9.2 |
| | | |
| **E   The Americas** | | |
| Zea mays | Maize | 4.7 |
| Phaseolus lunatus | Lima bean | 5.0? |
| Manihot esculenta | Cassava | 4.5 |
| Ipomea batatus | Sweet potato | 4.5 |
| Solanum tuberosum | Potato | 5.0 |
| Capiscum annuum | Pepper | 8.5 |
| Cucurbita spp. | Various squashes | 10.7? |
| Gossypium spp. | Cotton | 5.5 |

Note: *Unconfirmed recent data (see Normile, 1997)

**TABLE 2.3** Some of the world's most important crop plants and their approximate dates of domestication

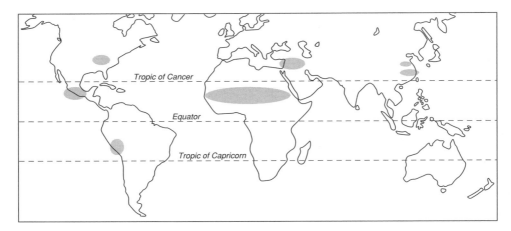

**FIGURE 2.7** The major centres of crop domestication. (Adapted from Smith, 1995)

ticated *c.* $12 \times 10^3$ years ago; this early domesticate was followed by pigs, sheep, goats and cattle between $9 \times 10^3$ and $8 \times 10^3$ years BP. As Table 2.3 shows, the Near East was also the region in which wheat, barley, oats, rye and a number of legumes were domesticated between $8.5 \times 10^3$ years and $9.8 \times 10^3$ years BP. However, recent research findings from Hillman *et al.* (2001) have pushed the dates for the earliest cereal domestication back by *c.* 1000–2000 years. Their results – based on cereal remains from Abu Hureyra, one of the earliest settlements in the Near East (see Section 8.2 for details) – indicate that domestication occurred *c.* $13 \times 10^3$ years ago, and that the species involved included rye (*Secale cereale*) as well as wheat. These animal and plant species remain the mainstays of agricultural systems today. In the Far East, there is preliminary evidence for rice domestication and cultivation in the middle Yangtze River Valley *c.* $11.5 \times 10^3$ years BP (Normile, 1997; Sato, 1997; Zhao, 1998). If the radiocarbon age estimates are confirmed, rice may be one of the first plants to be domesticated, and the agriculture of the Middle Yangtze River Valley may be one of the earliest agricultural systems initiated. This possibility is also advocated by Zhao and Piperno (2000), who have found further evidence for rice domestication in sediments of Diatonghuan Cave, 50 km southeast of the Middle Yangtze River.

Other centres of innovation include Mesoamerica, notably modern-day Mexico and the Andean region of what is today Peru (see Figure 2.7). In the former, maize was domesticated, notably *c.* $5 \times 10^3$ years BP, and thence spread into North and South America. In the Peruvian Andes, the potato was domesticated at about the same time; both of these crops have become staples throughout the world, having been introduced to Europe, and from there to Asia and Africa in the period of European colonization (see Section 2.7). Sub-Saharan Africa provided sorghum, the domestication of which occurred *c.* $4 \times 10^3$ years BP following the advent of a pastoral agricultural system focused on cattle (Harlan, 1992). Undoubtedly, domestications of other animals and plants occurred in these locations and elsewhere. As Jones and Brown (2000) have discussed, the modern and ancient genetic (DNA) characteristics of domesticated species is providing additional evidence for plant (and animal) domestication, including the possibility of a single domestication per species or multiple pathways. The latter is likely for barley, rice and emmer wheat, for example.

Why domestication and the initiation of agriculture began at all is, however, the most difficult question to address. The evidence for these events, mostly plant and animal remains preserved in archaeological sites, is inherently

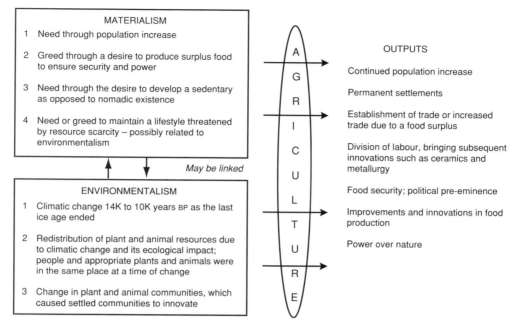

**FIGURE 2.8** Theories for the origins of agriculture. (Adapted from Mannion, 1999b)

unequivocal; the fossil remains bear witness to domestication through anatomical differences between domesticated and wild species. Age estimation is sometimes equivocal because of potential errors associated with radiocarbon (see discussion in Fritz, 1995, in relation to maize domestication) but despite ongoing research, new discoveries and increasingly sophisticated methods of palaeoenvironmental reconstruction, there is little that can be stated definitively about why people abandoned what Harlan (1992) describes as the Golden Age to encumber themselves with hard work and considerable responsibility. Hunter-gatherers generally acquire sufficient food with less effort than agriculturalists.

Several models can be proposed to explain why the switch occurred. Those involving materialism and environmentalism are illustrated in Figure 2.8. The first possibility, materialism, involves stimuli that emanate from human communities. They include so-called need and greed (Mannion, 1999b) requirements. The stimuli of necessity include population increase and development requirements to encourage a sedentary rather than nomadic lifestyle; it is easier to share and benefit from resources and expertise in nuclear rather than dispersed human groups. Conversely, the greed aspect of materialism could have been just as important as the need aspect. The production of a surplus, especially a surplus that could be anticipated and planned, confers a degree of power and certainly provides commodities with which to barter and thus to acquire other resources. The alternative driving force may have been environmental change. As discussed in Section 2.3, the transition phase between the last cold/glacial stage and the present warm/interglacial stage (the Holocene) was characterized by rapidly changing temperatures. It is interesting that this period corresponds chronologically to the initial domestications in the Near East, and possibly the Far East (see above). This could be coincidental or causal; if the latter obtains, then the role of environmental change in the origins of agriculture is an example of environmental/climatic determinism. It must also be considered that both materialism and environmentalism played significant roles in the origins of agriculture.

Whatever the reason or reasons, the repercussions were profound. Natural land cover was modified or altered entirely to make way for agriculture. Land cover was transformed into land use as agricultural practices spread from the centre of innovation (see Figure 2.7). The more carbon that was appropriated as food energy, the more substantial was the impact per unit area and the greater the modification or transformation of natural land cover. This process has continued throughout prehistory and history; intensive and extensive agricultural systems have developed within the constraints of the physical environment. The stimuli underpinning this process are the same socio-economic factors that are referred to above in relation to the origins of agriculture. (Further discussion can be found in Chapters 4 and 5.)

The initiation of agriculture, notably in the Near East, was accompanied or followed closely by a number of other developments that altered people–environment relationships. For example, it was probably paralleled by permanent settlement, as hunter-gatherer communities became increasingly sedenterized, and pottery production also began. Both reflect increasingly sophisticated lifestyles and an enhancement of material culture. Almost certainly, division of labour characterized these settled communities as a proportion of the population could be freed from food production. Other innovations followed, possibly as a result of this organization. One such innovation was the development of metallurgy. Beginning with copper, later bronze and eventually iron, humans learnt how to tap another range of resources. When and how metalworking technologies emerged is not clear and it is likely that there were numerous centres of innovation.

However, the earliest metal objects have been found in the Near East; these were hammered copper objects dated at $9.5 \times 10^3$ years BP (De Laet, 1994). Evidence for actual metal smelting from Çatal Hüyük in central Anatolia is later at $8 \times 10^3$ years BP. From this centre, copper use spread and, later, bronze production began.

Bronze is an alloy of copper and arsenic, or copper and tin, although it is not known how the combinations were discovered to produce durable and preferable alternatives to copper. It may have been accidental – many of the copper ores on Anatolia, for example, contain arsenic. Many further centres of innovation emerged in Europe, e.g. the Balkans, Italy, Spain and southern France (De Laet, 1994). From these, and probably others, bronze production spread throughout Europe; from the Anatolian centre, metallurgy spread into Asia where independent centres also developed.

By the time bronze working was beginning in Britain c. $4.5 \times 10^3$ years BP, iron working had already begun in the Near East. This was particularly important because iron deposits are widespread and iron was, thus, a more readily available raw material than copper. By c. $2.5 \times 10^3$ years BP iron was in widespread use in Europe where many independent centres of production had developed, as also occurred in India and the Far East. Africa may have been influenced by European centres, especially the cultures of the Mediterranean coast, with independent centres elsewhere.

Not only do these developments reflect increasing sophistication in technology, as iron smelting requires more efficient furnaces than copper and requires the addition of carbon to reduce brittleness, they also reflect a broadening of the resource base. The production of iron implements in particular was an important development because they were available to all. In terms of people–environment relationships, metal implements in general, and iron implements in particular, provided a means of modifying the environment with greater efficiency than earlier stone implements. However, the latter were not abandoned but were complemented by metal implements. Iron ploughs, for example, were more efficacious than antlers or bone for tilling the soil, and iron axes were more effective than stone axes for tree felling. Thus the advent of metallurgy assisted in the transformation of land cover into land use, once again altering people–environment relationships. Certainly, there is abundant evidence for deforestation during the Bronze and Iron Ages of Europe (see review in Mannion, 1997).

During these periods, civilizations such as the Egyptians, Ancient Greeks, Phoenicians and Assyrians rose and fell; their development must have been underpinned by their command of resources and their degree of carbon appropriation as food.

## 2.7 Technological innovations of the last 2000 years and their impact

All manner of innovations have occurred in the last 2000 years, an arbitrary starting point for an assessment of technological impact. The Romans were amongst the most significant technologists/engineers and their impact was widespread as their empire expanded. Any innovations in agriculture influenced land use and land cover; examples include the spread of irrigation systems and the draining of wetland, e.g. parts of the Fens in East Anglia, UK, and the Po Valley in Italy. The Romans also introduced crops into non-indigenous areas; for example, the vine was grown in England for wine production and other species, such as the walnut, sweet chestnut, various species of cherry and herbs such as sage and dill (Campbell-Culver, 2001), were introduced. They also had specific ways of subdividing lands, e.g. centuriation (the creation of regular square fields separated by roads and paths).

The expansion of Europe after 1500, facilitated by military capacity and associated technology, as well as improvements in sailing ships and navigation procedures, was another major development in people–environment relationships. For the first time the Old and New Worlds were linked and, over a period of c. 400 years, flora, fauna, crops and agricultural systems were exchanged between the two. European countries benefited the most. The original objective of securing precious metals from their colonies was achieved but perhaps the bigger prize was the redistribution of people and biomass. Unlike most developing countries today, European nations were able to export their people and import biomass-based goods. The Spanish, British and French, for example, annexed the Americas; in so doing wheat, cattle, sheep etc. were introduced to new lands and unfamiliar agricultural systems were superimposed on native populations along with new European-dominated land ownership arrangements. The shipment of agricultural products from the New World to the Old World accelerated in the 1880s with the advent of refrigerated transport. This also facilitated Britain's importation of grain and wheat from Australia and New Zealand.

In the opposite direction, the New World provided maize, potato, tobacco and a number of other crops, such as peppers and gourds, which were adopted in European agricultural systems. Of these, maize and potato have become the mainstays of agricultural systems worldwide. In addition, nitrates from Chile were imported into Europe, beginning in the 1830s, and thus provided much-needed nutrients that had hitherto been provided through crop rotations, animal manure, seaweed etc. The import of nitrates increased agricultural productivity prior to the advent of the Haber-Bosch process for producing artificial nitrate fertilizers in the early 1900s.

These exchanges not only altered land cover and land use in the Americas and Europe, as will be discussed in Chapters 4 and 5, but they fuelled Europe's development, including the Industrial Revolution, at a time of rapid population increase. The Industrial Revolution had been preceded by various agricultural innovations in the 1700s, e.g. new crop rotations and new sources of nutrients such as marl. These are all components of the inherently technological nature of agriculture, which was being increasingly influenced by other technological developments. These include the beginning of mechanization involving the use of threshing machines and harvesters (the first combine harvester was produced in the USA in 1838). This, in combination with the Haber-Bosch production of artificial nitrate fertilizers, represents a significant alteration of agricultural practices in so far as fossil fuels were beginning to make their mark.

Agriculture was becoming industrialized, a process that has since intensified with the introduction of artificially produced crop-protection chemicals. Coupled with the need, worldwide, to increase food production in the face of rapidly growing populations, these technological innovations changed both the face of agriculture and the face of the Earth. Such technology, and expanding irrigation systems, as well as need, encouraged the spread and intensification of agriculture. As discussed in Section 3.4, the period 1700 to 1900 witnessed massive expansion of croplands and pasturelands at the expense of natural ecosystems. Technology, population increase and colonialism conspired to effect land-cover and land-use change on a large scale (see reviews in Mannion, 1995; 1997), the outcome being the increasing involvement of humans in the global carbon cycle (Figure 1.6) and thus in patterns of carbon exchange between the atmosphere, the oceans and the biosphere (see Schimel *et al.*, 2001).

Today, another major development is taking place. Biotechnology, and its subdiscipline of genetic engineering, is about to create another 'agricultural revolution'. This time it is likely to be a true revolution as it is taking place quickly; within a decade of the first genetically modified crops being produced, substantial hectarages are being grown in the USA, Australia and China. This is another landmark stage in carbon manipulation (see Section 1.5) and its likely ramifications for land-cover and land-use change are examined in Section 10.3.3.

ations of the biosphere that have been precipitated by human activity. The latter rivals climate as an agent of land-cover and land-use change. In a temporal context, the emergence of agriculture was a major threshold in environmental and human history. It altered people–environment relationships profoundly and represented an increase in the degree of involvement of humans in the carbon cycle.

Evidence from palaeoenvironmental and archaeological studies in a wide range of contexts indicates that environmental change can proceed rapidly as a result of either climatic change or human impact. The greater the involvement of humans in the carbon cycle, the more likely they are to generate climatic change. The sensitivity of the global climatic system to human interference in the global carbon cycle is intimately related to land-cover and land-use characteristics. In turn, land-cover and land-use characteristics are linked with technological innovations, of which agriculture is one.

The driving forces generating change, other than the natural processes of climatic change, include a range of socio-economic factors that are similar to those that may have stimulated the onset of agriculture initially, e.g. population increase, resource scarcity, the need to generate products for trade. These are just as relevant now as they were *c*. $10 \times 10^3$ years ago.

## 2.8 Conclusion

The last $3 \times 10^6$ years have witnessed a remarkable continuum of environmental change; land, ocean and atmosphere are linked in a dynamic relationship that is characterized more by shift than stasis. The details of this relationship are revealed in palaeoenvironmental records from archives that include polar ice, temperate peats, tropical sedimentary sequences and oceanic sediments. The present interglacial, the Holocene, is especially dynamic in terms of land-cover and land-use change. This is because it shares the dynamism characteristic of the natural environmental changes of a warm/interglacial stage and encompasses the extensive global-scale alter-

## Further reading

**Fagan, B.M.** 1998: *People of the Earth. An Introduction to World Prehistory*, 9th edition. Longman, New York.

**Lowe, J.J. and Walker, M.J.C.** 1997: *Quaternary Environments*, 2nd edition. Longman, Harlow.

**Mannion, A.M.** 1999: *Natural Environmental Change*, 2nd edition. Routledge, London.

**Williams, M., Dunkerley, D., De Dekker, P., Kershaw, P. and Chappell, J.** 1998: *Quaternary Environments*, 2nd edition. Arnold, London.

**Wilson, R.C.L., Drury, S.A. and Chapman, J.L.** 2000: *The Great Ice Age. Climatic Change and Life*. Routledge and the Open University, London and New York.

# Land use: some temporal perspectives

## 3.1 Introduction

Land use, by definition, involves human manipulation of, and impact on, land cover (see Sections 1.1 and 1.2). People use land and land cover for a wide range of purposes which can be described under two categories: existentialities, or the necessities of life, and wealth generation. Both involve the exploitation of resources that may be organic or inorganic. Organic resources comprise plants and animals, the production of which is the primary objective of agriculture, whilst inorganic resources include minerals and water. These are components of the biosphere and lithosphere. Links between the two and the atmosphere are effected through biogeochemical cycling in which land cover and land use play a significant role. Moreover, fossil fuels, the products of ancient biospheres, which are now sequestered in the lithosphere, are increasingly being used for both existentialities and wealth generation.

In a temporal context, it is difficult to determine when the earliest land use came into being. The hunter-gatherer precursors of the first farmers were adept at resource manipulation,

but whether they created pastures within the areas roamed by target species is debatable. It seems more likely that these skilled people manipulated land cover rather than transformed it. However, as discussed in Section 2.6, the advent of agriculture $c. 10 \times 10^3$ years BP, and its subsequent spread, began the subjugation of nature and the initiation of control systems, a process that was, and still is, manifest as land use. For at least the second half of the Holocene, i.e. since $5 \times 10^3$ years ago, there is abundant evidence for land-cover alteration and land use. Much of this evidence derives from palaeoenvironmental investigations, using techniques such as pollen analysis of peat deposits, lake sediments and archaeological sites. Examples from a range of sites are given in Section 3.2, the geographical distribution of which attests to the significance of early agriculture as a widespread cause of Holocene land-cover alteration. As in the case of the inception of agriculture itself, the palaeoenvironmental record does not reveal motives; what the stimuli were for early agriculture and its expansion remain enigmatic, although both 'need and greed' (see Section 2.6) were almost certainly involved.

The historic period prior to 1700 also witnessed considerable transformation of land cover into land use. Europe in particular experienced substantial changes in its vegetation cover, especially the replacement of forests with agricultural land. No other continent has experienced such a major degree of land-cover alteration or destroyed so much of its forest resource. Much of this, though not all, occurred before 1700, by which time industrialization was beginning and colonies were being annexed (see Section 2.7). The resulting movement of people into the Americas, Australasia, Asia and Africa set in train a further pattern of land transformation with a huge expansion both of croplands and arable land. This also resulted in adjustments in the global carbon cycle. (These trends are examined in Sections 3.3 and 3.4.)

Land-cover alteration and land-use change is continuing apace today. In general, there are two underlying trends: in the developed world there is little need for the expansion of agricultural systems and in some regions, e.g. Europe, agriculture is contracting; conversely, in much of the developing world, agricultural systems are intensifying and expanding. The underpinning causes of these trends are socio-economic, and include rates of population growth, the need to provide biomass products for export, and resettlement programmes. (These issues are discussed in Chapters 4 and 5.)

In the last decade, as environmental issues have become important in mainstream politics and as awareness has been heightened through the media, several attempts have been made to categorize and quantify types, rates and costs of land-cover/land-use alteration. For example, several indices have been devised to express the degree of alteration or degree of human impact. Such indices include the Ecological Footprint, which has widespread application. This and other indices, including a possible Carbon Index are the subject of Sections 3.5 and 3.6.

## 3.2 Land use in the Holocene ($8 \times 10^3$ to $2 \times 10^3$ years BP)

The period at the end of the last ice advance was particularly dynamic in terms of climatic change (see Sections 2.2 and 2.3). Changes in average annual temperature regimes generated shifts in ecosystem characteristics. This may have caused crises in human communities in sensitive areas, and prompted, along with other pressures, animal and plant domestication, and the advent of agriculture. As this extended beyond centres of innovation, modification and clearance of natural vegetation occurred. Evidence for this is widespread, especially in the temperate and Mediterranean zones, where palaeoenvironmental work has traditionally been concentrated. Moreover, such studies are now being pursued in the tropics and subtropics, and are revealing the spatially disparate impact of human communities in the early part of this interglacial period.

Figure 3.1 provides a generalized scheme for the temporal spread of agriculture in the early Holocene. From the major centre of origin in the fertile crescent of the Near East, agriculture, involving a crop complex of wheat/barley/legumes and animals such as cattle, sheep, goats and pigs, spread north and then northwest to reach the periphery of western Europe by $c. 5 \times 10^3$ years ago. Evidence from human genetic characteristics, notably the variation that exists across Europe (Cavalli-Sforza and Cavalli-Sforza, 1995), indicates that agriculture probably spread as migration occurred from the fertile crescent region into peripheral areas, i.e. the process of demic diffusion. However, recent work by Sykes (1999; 2001; see also Section 2.5) confirms that demic diffusion occurred but that it accounts for only $c. 10$ per cent of genetic variability in modern Europeans rather than the $c. 80$ per cent proposed by Cavalli-Sforza et al. (1994). Ideas, crops and animals were probably also disseminated through trade contacts. These developments represented the beginning of the rise to dominance of human communities within ecosystems, a characteristic unprecedented in earlier warm/interglacial stages. In its early stages, the northwestward spread of agriculture appears to have been selective in terms of favoured soils; these were mainly alluvial soils in southeast Europe and loessic soils in northwest Europe (loess is a fine-grained yellow sediment of aeolian origin). The former are advantageous because they retain water and thus help to counteract the effect of the long dry season that characterizes the Mediterranean climatic zone. Beyond central Europe, notably the Hungarian Plain, loessic soils were initially favoured. They may have been selected because they are nutrient rich and easily worked. The different environments encountered by early agriculturalists required adaptations to local environments and in some parts of temperate Europe it is likely that some form of shifting cultivation (swidden) was common, which allowed human communities to exploit a range of environments with different biotic resources. These factors conspired to produce varied effects, spatially and temporally, on early mid-Holocene vegetation cover. Any appraisal of land-cover change in the early mid-Holocene can only be undertaken on a site-by-site basis. This in itself introduces bias because the evidence derives from non-randomly distributed sites, i.e. archaeological sites and/or

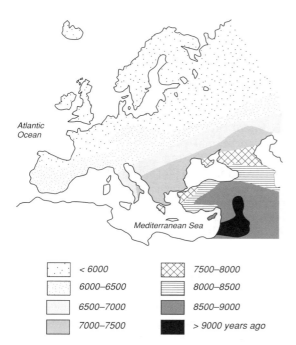

Atlantic Ocean

Mediterranean Sea

| | | | |
|---|---|---|---|
| ⋅⋅⋅ | < 6000 | ⨯⨯ | 7500–8000 |
| | 6000–6500 | ≡ | 8000–8500 |
| | 6500–7000 | | 8500–9000 |
| | 7000–7500 | ■ | > 9000 years ago |

**FIGURE 3.1** A schematic representation of the spread of agriculture from the Near East into Europe. (Adapted from Cavalli-Sforza and Cavalli-Sforza, 1995)

archives such as lake sediments and peats. This, however, is the only evidence available, so the reconstruction is of necessity piecemeal.

Reference has already been made (in Section 2.6) to the discovery of early domesticated cereals at Abu Hureyra, Syria. The evidence of preserved plant remains derives from habitation layers and thus reflects events within the settlement. Environmental change in the region has, however, been addressed by Willcox (1996) who has reconstructed the vegetation history of the middle Euphrates on the basis of botanical remains at three early Neolithic sites near Aleppo, some 130 km west of Abu Hureyra and also in Syria. Age estimates based on radiocarbon indicate that the materials investigated belong to the ninth millennium BP and the absence of pottery indicates that the three sites were established in the Pre-Pottery Neolithic. The plant remains show that, between $9.8 \times 10^3$ and $9.2 \times 10^3$ years BP, the local steppe vegetation contained wild cereals and pulses as well as a wide variety of other plant resources, e.g. wild almond and pistachio, which were exploited by the local population. However, by $c.\ 9.7 \times 10^3$ years BP, the botanical remains from one site, Halula, include domesticated emmer wheat, naked wheat and barley. Although wild cereals were also present, these data probably reflect a shift from land cover to land use. Willcox also notes that these results are paralleled by those from other Pre-Pottery Neolithic discoveries in the central Turkey/Middle East region.

Another study, based on pollen analysis of a variety of sites in the mountainous region of northwest Iberia, has shown that after $5 \times 10^3$ years BP tree pollen began to decline, representing the opening up of the alder/oak/elm/hazel forest (Santos et al., 2000). A summary of the recorded characteristics is given in Table 3.1, which shows that the impact of human activity, i.e. burning and forest clearance, is evident between $4 \times 10^3$ and $5 \times 10^3$ years ago. From $c.\ 2.5 \times 10^3$ years ago, deforestation is continuous until the present day; by this time cereals such as rye were being cultivated. Similar dates for disturbance due to clearance for agriculture have been recorded from a valley mire at Espinosa de Cerrato, in the

| Years BP $\times 10^3$ | Laguna Lucenza (Courel Sierra) 1420m a-s-l | | Fraga (Queixa Sierra) 1360m a-s-l |
|---|---|---|---|
| 2.2–0 | All forest species reduced c. 2.2 $\times 10^3$ years (Roman times).Increase in charcoal. Open habitat species, e.g. heaths, grasses dominate. Cultivated plants include walnut and rye. | 2.0–0 | All forest species decline whilst herbs/shrubs increase. High incidence of charcoal reflects increasing fire. |
| 4.0–2.2 | Tree pollen declines; grasses expand. Evidence for increasing frequency of fire. | 4.0–2.2 | Decrease in tree pollen reflecting a reduction in forest extent. Evidence for local fire and local erosion, probably due to human impact. |
| 7.5–4.0 | High concentrations of tree pollen, notably alder, elm, oak etc. By c. 4.5 $\times 10^3$ years ago grasses, docks expand indicating a reduction in the forest cover due to anthropogenic influence. | 7.0–4.0 | Initial high concentrations of tree pollen, representing optimum Holocene forest cover; forest decline began c. 6 $\times 10^3$ years ago. This coincides with an increase in charcoal and the pollen of open habitats, e.g. grasses, heaths, including *Calluna*. |
| 9.3–7.5 | Increase in tree pollen reflecting forest development. | <8.0–7.0 | Initially open habitat pollen spectra are replaced by tree species, notably birch, pine, oak. |

**TABLE 3.1** Land cover change in northwest Iberia, based on pollen analytical investigations at Laguna Lucenza and Fraga. (Adapted from Santos *et al.*, 2000)

Spanish Northern Mesetas (Mugica *et al.*, 2001). Here, cereal pollen grains have been found in horizons dated to *c.* 4.5 $\times 10^3$ and 2.7 $\times 10^3$ years ago as clearance of oak woodland and later pine woodland occurred.

A similar date for the onset of land-cover disruption has been reported by Ramrath *et al.* (2000) for the catchment of the Lago di Mezzano in central Italy. From 3.7 $\times 10^3$ (cal) years BP to present there is evidence for human activity, notably two settlement periods, in the catchment. The earliest of these, reflected in an increased incidence of charcoal and a 50 per cent decline in tree pollen percentages and concentrations, is associated with a Middle Bronze Age settlement. Further disturbance is indicated at 3.3 $\times 10^3$ (cal) years BP, which is associated with a later phase of Bronze Age settlement. Within another thousand years, at 2.5 $\times 10^3$ (cal) years BP, the construction of a

Roman village brought another influx of sediment caused by erosion and deforestation. Two additional phases of later, major disturbance are associated with renewed settlement construction *c.* AD 900 and in the last 200 years.

By 6.5 $\times 10^3$ years ago, agriculture had reached much of western Europe and had become established in Britain by 5 $\times 10^3$ years BP. Pollen diagrams from many parts of not only Britain but other areas of western Europe indicate that land-cover change was characterized by an opening up of the formerly dense climax forest. This opening is characterized by a decline in elm followed by forest clearance, and often evidence for cereal cultivation and the weeds of cultivation. Various hypotheses have been advanced for explaining the elm decline (see review in Mannion, 1997 and Parker *et al.*, 2002); these include the possibility of climatic change such as an increased incidence of spring frosts, competition from other species or clearance by early agriculturalists who might have sought out the fertile soils preferred by elm. Currently, the most favoured explanation for this enigmatic land-cover change involves the incidence and rapid spread of a pathogen such as Dutch elm disease. The fact that there is a high degree of synchroneity from region to region, and that there was no impairment of other frost-sensitive tree populations such as lime, lend support to this possibility, which was first suggested by Troels-Smith in 1960. However, the elm decline may still have been associated with agriculture. Peglar's (1993) detailed work on the laminated sediments of Diss Mere in East Anglia has shown that the elm decline occurred in only six years (this precision is possible because each distinct layer of sediment – a varve – was deposited annually), a rapidity similar to that which characterized the recent outbreak of Dutch elm disease. Nevertheless, the Diss Mere elm decline coincides with evidence for early agriculture. Consequently, Peglar has suggested that woodlands were disturbed initially by early agriculturalists, whose migration probably helped to spread the pathogen. Whatever the cause, or combination of causes, the elm decline represents a significant alteration of land cover in the mid-Holocene.

Many palaeoenvironmental studies in western Europe show that alteration or removal of the forest cover was often short-lived as well as small scale in extent. In many areas, these so-called 'landnam' phases were perhaps as of little as 50 years' duration and no longer than 200 years, after which forest regeneration often occurred; some areas experienced more than one landnam phase. For example, Wimble et al. (2000) have recorded several small-scale but significant forest clearance phases during the Bronze Age, beginning c. $3.5 \times 10^3$ years BP in south Cumbria with a major clearance phase occurring c. 2000 years ago, i.e. the Romano-British period and later in the historic period (see Section 3.2). In the chalk region of southeast England, land-cover alteration occurred earlier. According to the analysis of molluscan remains (by Preece and Bridgland, 2000) of sediments at Holywell Coombe, near Folkestone in Kent,

opening of the forest canopy began c. $6.5 \times 10^3$ years BP, i.e. in association with Mesolithic and Neolithic activity, with major clearance in the early Bronze Age c. $3.5 \times 10^3$ years BP. At Caburn in East Sussex, a pollen chronosequence presented by Waller and Hamilton (2000) indicates that humans were influencing the vegetation cover as early as $6.35 \times 10^3$ (cal) years BP. This and subsequent changes are summarized in Table 3.2, which shows that several phases of forest disturbance/deforestation due to human impact occurred.

Beyond Europe there is also abundant evidence for land-cover change due to human activity. In lowland Amazonia, for example, a region that has remained uninvestigated until recently, Bush et al. (2000) have presented evidence for human activity c. $5.5 \times 10^3$ and $3.35 \times 10^3$ years BP. Using pollen and diatom remains, sediment chemistry and charcoal analysis in conjunction with radiocarbon age estimation from two lakes, Bush et al. (2000) have suggested that the earlier episode was caused by burning, possibly to encourage the growth of or to facilitate the planting of various palms, e.g. Mauritia and Mauritiella, which were used for a range of purposes including food. The second episode, also associated with a prior increase in burning frequency, was characterized by the presence of pollen of Zea mays; this is clear evidence for agriculture involving maize cultivation. These episodes reflect land use in the mid-Holocene by indigenous people in the lowland Amazon though at no time did complete destruction of the forest cover occur. Moreover, there is evidence for maize cultivation much earlier, at $6 \times 10^3$ years BP, in the region (see Bush et al., 1989), reflecting a shift from land cover to land use. Clement and Horn (2001) have also presented pollen-analytical evidence for maize cultivation and forest clearance from Laguna Zoncho in southern Costa Rica. Their data show that maize cultivation has been significant for the last three millennia and that fire was used to clear forest for agriculture. A decline in agricultural intensity is recorded about 500 years ago, possibly because European colonization resulted in a major population decline caused by disease and labour appropriation.

| Age × 10³ (cal) years BP | Actual stratigraphic changes | Likely reconstruction |
|---|---|---|
| 4.15 | Increased incidence of tree pollen, e.g. lime, yew, hazel and declining grasses. | Woodland regeneration and reduction in open habitats. |
| 4.50 | Reduced incidence of lime and yew pollen. Abundant grass pollen. | Clearance of secondary woodland; cereal cultivation. |
| 5.40 | Increases in pollen of hazel, yew, lime and elm. | Secondary forest succession; deciduous woodland is re-established. |
| 5.70 | Declines in pollen of elm, ash, lime, oak. | Increased pollen of open-habitat species such as grasses. Primary elm decline (see text for explanation). |
| 6.10 | High frequencies of lime and ivy pollen with subsequent increase in oak and decline in grasses. | Possibly woodland management involving selective lopping; possible use of leaves for animal fodder. |
| 6.35 | Increasing incidence of open habitat species such as grasses and Asteraceae. Declining hazel and variable lime pollen concentration. | Some limited woodland management/manipulation. |

TABLE 3.2  A synopsis of mid-Holocene vegetation (land-cover) change at Caburn, East Sussex, UK. (Adapted from Waller and Hamilton, 2000)

Further afield, in the region of China north of the Yangtse River, there is pollen-analytical evidence from 22 sites from mid- to late-Holocene forest history, as Ren (2000) has recently reviewed. The pollen diagrams, in conjunction with radiocarbon age estimation, show that the earliest forest disturbance occurred $c.$ $5 \times 10^3 - 6 \times 10^3$ years ago (5500–6500 calendar years) in the middle and lower reaches of the Yellow River. As Ren points out, this region has been recognized as the heart of ancient Chinese civilization, though the radiocarbon age estimates suggest an earlier origin than hitherto supposed on the basis of documentary evidence. From this centre, human impact radiated into China; in the north, northeast and northwest, forest destruction occurred progressively later; in the far north of northeast China and west China little forest demise has occurred. Ren has also addressed the question of whether the reduction in forest cover could have been climatically driven. Apart from the metachronous relationship referred to above, which is unlikely to be due to climatic change, Ren demonstrates that the forest decline was not selective, but involved all types of tree pollen. This too is likely to be caused by human rather than climatic impact. If the latter occurred, individual species rather than entire communities would have responded.

## 3.3 Land use in the historic period: pre-1700

As agriculture intensified and continued to spread, the natural land cover of the Holocene interglacial experienced further modification.

Whilst it is arbitrary when prehistory ended and history began, a useful starting point is the admittedly Eurocentric focus of the Greek and Roman eras. These began *c.* $4.5 \times 10^3$ and $2.5 \times 10^3$ years BP respectively. Both empires, in common with their Egyptian and Sumerian predecessors, achieved considerable political and economic domination within and beyond their Mediterranean origins. This would only have been possible with adequate and possibly even surplus, food supplies; agriculture must, thus, have been a prime activity within the home lands and those under their control. Similarly, the use of metals, especially iron, was of vital importance for a variety of reasons. Together, agriculture and metallurgy led the onslaught on natural vegetation, and especially forest cover: clearance was essential for agriculture whilst wood, to produce charcoal, was essential for a range of activities including metal smelting.

One region that has experienced a long history of land-cover/land-use change is Greece and its islands. Generalizations are difficult to make for a region characterized by such a diverse array of landscapes, but the island of Crete provides a valuable case study to illustrate the nature and direction of land-use change since the end of the last ice age. The conclusions from three major palaeoecological studies are summarized in Table 3.3. This shows that woodland had declined in extent by *c.* $5 \times 10^3$ years BP and that cultivation of the olive subsequently became significant, with a peak in the Bronze Age *c.* $2.5 \times 10^3$ years BP. Following a probable decline in olive cultivation, further periods of deforestation characterized the early historic period as cereal cultivation and grazing increased.

The last $5 \times 10^3$ years of land-cover change in the catchment of the Lac d'Annecy in the French Alps has been reported by Noël *et al.* (2001), who show that a major deforestation episode accompanied the Roman invasion *c.* 1700 years ago. Similarly, in the Lower Rhone Valley, southern France, Andrieu-Ponel *et al.* (2000), using pollen and coleopteran (beetle) analyses with radiocarbon age determination, have demonstrated that the basal sediments of La Calade, formerly an alluvial marsh, correspond to the Greco-Roman period. Even by this time, considerable forest removal had occurred in the surrounding area to provide marsh grazing. The results of this study are summarized in Figure 3.2. Throughout the historic period, the region was characterized by agricultural landscapes, which replaced previous Holocene forests of pine, oak, fir and beech, as is indicated by palaeoenvironmental research from elsewhere in the region (references are quoted in Andrieu-Ponel *et al.*, 2000). Interestingly, the pastures associated with phases A and B most probably contributed to wool production whilst the evidence for the cultivation of teasel (a plant with hooked bracts) also relates to fibre production as teasels were used for combing wool. This production of wool under the auspices of monastic institutions was not confined to the Lower Rhone. It was prevalent in the first millennium AD elsewhere in northwest Europe; there are abundant written records that attest to the significance of this activity in the UK, for example, from a variety of sources (see comments in Mannion, 1997). In many upland regions of the UK the expansion of sheep farming exacerbated deforestation.

Between the twelfth and seventeenth centuries, the temperate forest of European Russia underwent considerable alteration, as Serebryanny (2002) has reviewed. He suggests that the initial mid-Holocene opening of the forest canopy began *c.* $5 \times 10^3$ years ago, about the same time as it occurred in northwest Europe (see Section 3.2) with subsequent periods, on a site-to-site basis, of regeneration and clearance. However, the long-fallow system of agriculture, which developed in the Middle Ages (*c.* the tenth to twelfth centuries), involved deforestation to accommodate strip cultivation on a shifting basis. A later Tartar invasion caused the abandonment of some agricultural land, and forest regrowth ensued until the fourteenth and fifteenth centuries, when this land was once again brought into cultivation. Thus, an arable land cover replaced scrub that had earlier been arable land or fallow grassland. A major change occurred after the defeat of the Kazan Tartars in 1552; forest clearance resumed and continued apace into

**SITE: Asi Gonia (Atherden and Hall, 1999)**

| PERIOD | CHARACTERISTICS (pollen assemblages) |
|---|---|
| Late Turkish–present 1898– | Decrease in heath species, some recovery of oak until recent years when it declined. |
| Late Turkish 1770–1898 | Heath species increase, then decrease Some recovery of oak and reduced grasses/sedges. |
| Turkish 1650–1770 | Heath species remain dominant but grasses, sedges and oak increase. |
| Late Venetian–Early Turkish 1500–1650 | Increase in heath and maquis taxa. |
| Early Venetian 1210–1500 | Tree and shrub pollens decrease as heath/grass/sedge taxa increase: open habitats expand at the expense of woodland. |
| Second Byzantine– Late Second Byzantine 961–1210 | Major decrease and then increase in tree pollen as woodland declines and then recovers. Cereal cultivation. |
| Early Byzantine and Saracen c. 380–961 | Oak woodland predominant. |

**SITE: Tersana (Moody et al., 1996)**

| PERIOD | |
|---|---|
| Late Bronze Age c. 1450–1150 BC | No information available. |
| Middle Bronze Age c. 1900–1450 BC | Decline in olive cultivation; oak maquis recovers, probably due to reduction in human impact. |
| Early Bronze Age c. 2900–1900 BC | Widespread and probable peak of olive cultivation. |
| Neolithic c. 6100–2900 BC | Reduction in oak woodland, possibly for increasing cultivation of olive. |

**SITE: Ayia Galini (Bottema, 1980)**

| PERIOD (note: dates are BP) | |
|---|---|
| 4650 BP | Tree pollen declines, though oak remains the dominant tree taxa; increasing open habitats. |
| 7300 BP | Oak predominates with little pine; ivy and vines are significant. |
| 8000 BP | Increasing open habitats as oak declines. |
| 9500 BP | Pine is a significant component of the vegetation: oak/pine woodland. |
| 10,090 BP | Pine woodland dominates with low oak. |

**TABLE 3.3** Environmental/land-cover changes in Crete, Greece, in the last $10 \times 10^3$ years BP. (Adapted from Atherden and Hall, 1999; Moody *et al.*, 1996; Bottema, 1980)

the seventeenth century, by which time the southern forest limits had been reached. Serebryanny (quoting French, 1963; 1983) states that between 1696 and 1796 the forest had been reduced by $47 \times 10^3 \text{km}^2$, leaving *c.* 53 per cent of the land forested as compared with 60.6 per cent in the 1600s. Apart from the need for land

for agriculture, additional economic pressures on Russia's forests included potash production as well as the provision of fuel and building materials.

Recent work by Taylor *et al.* (2000) has also used pollen analysis – along with data on lake levels and river discharge, and archaeological

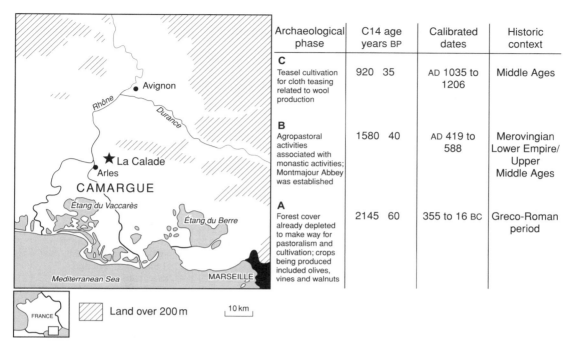

| Archaeological phase | C14 age years BP | Calibrated dates | Historic context |
|---|---|---|---|
| **C** Teasel cultivation for cloth teasing related to wool production | 920   35 | AD 1035 to 1206 | Middle Ages |
| **B** Agropastoral activities associated with monastic activities; Montmajour Abbey was established | 1580   40 | AD 419 to 588 | Merovingian Lower Empire/ Upper Middle Ages |
| **A** Forest cover already depleted to make way for pastoralism and cultivation; crops being produced included olives, vines and walnuts | 2145   60 | 355 to 16 BC | Greco-Roman period |

**FIGURE 3.2**  The last $2 \times 10^3$ years of land-cover change at La Calade, Lower Rhone Valley, southern France. (Adapted from Andrieu-Ponel *et al.*, 2000)

evidence – to reconstruct environmental change in western Uganda for the last *c.* 3000 years BP. Their results show that substantial natural and cultural changes occurred during the early historic period. For example, the introduction of iron and, possibly, new agricultural crops, took place between *c.* 500 BC and AD 350–540. Increased aridity in combination with these innovations caused forest disturbances and degraded soils. What occurred in the ensuing *c.* 500 years is obscure but by AD 1078 aridity was decreasing and the iron industry was either re-established or intensified as new people moved into western Uganda. This resulted in the development of relatively large nucleated settlements and further forest disturbance in middle-montane altitudes. Between AD 1326 and the AD 1700s aridity increased again but agricultural/economic activities continued to cause forest demise until the end of this period, when nucleated settlements declined and were replaced by homesteads.

In their review of environmental change in Switzerland, based on some 23 pollen diagrams from sites in Switzerland, Van der Knaap *et al.* (2000) have pointed out that Swiss forests have experienced several phases of deforestation. The most important phases began in medieval times (the fifth to fifteenth centuries) up to the late nineteenth century. Fir and beech were consistently reduced at all sites whilst alder, birch and hazel were affected in different ways at different times; in the catchments of most of the sites investigated, the pollen diagrams indicate that the tree cover was replaced with grassland cover. Van der Knaap *et al.* suggest that human impact has been the major factor in land-cover change in Switzerland. Pollen analysis of Cerná Hora Bog in the Krkonose Mountains of the Czech Republic has shown that the onset of significant deforestation also occurred in the Middle Ages (Speranza *et al.*, 2000). The results are summarized in Figure 3.3. Bog growth began *c.* $2.1 \times 10^3$ years ago and,

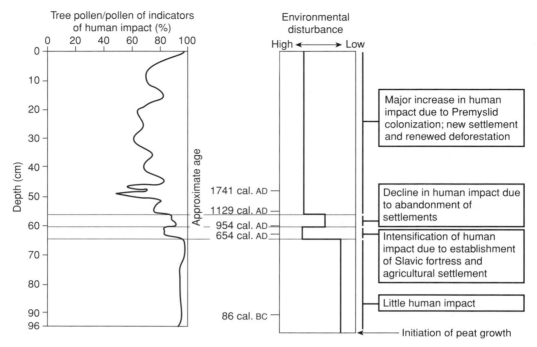

**FIGURE 3.3** A summary of environmental/land-cover change in the vicinity of Cerna Hora Bog, Czech Republic. (Adapted from Speranza *et al.*, 2000)

for a further 700 years, there appears to have been little human influence in its vicinity. At 1.4 × 10³ years BP deforestation occurred. Later abandonment of settlements in the vicinity of the bog allowed some re-establishment of forest, but by *c*. 900 BP colonization of the area by the Premyslids led to the establishment of new settlements and the intensification of agriculture, which resulted in renewed deforestation.

The historic period was also one of substantial vegetation change in south Cumbria, UK (see Section 3.2 for reference to details of earlier environmental change), as Wimble *et al.* (2000) have shown. The late Iron Age-Roman period, approximately 2.5 × 10³ years BP, witnessed the first phase of substantial woodland clearance, with some recovery after the departure of the Romans. Subsequent clearance occurred *c*. 1.1 × 10³ years BP, which may be associated with Norse colonization, a process recorded elsewhere in the region. After some recovery,

further clearance characterized the region by late Medieval times *c*. AD 1424–1652; woodland clearance is reflected in declines in tree pollen and increases in the pollen of open habitat species. Wimble *et al.* (2000) suggest that this change in land cover resulted from the intensification of land use associated with arable agriculture and especially sheep rearing for wool, which intensified with the establishment of Furness Abbey in AD 1127. In the sixteenth century, the Abbey was dissolved but the apportionment of land to yeoman farmers then led to a shift to a mixed agricultural economy with increased emphasis on arable agriculture.

The examples referred to above, and in Section 3.2, reflect varied and often substantial changes in land cover and land use in the pre-1700 period. The case studies also highlight the value of pollen analysis as a powerful tool for environmental reconstruction in a temporal context but at the same time they reflect its inadequacies for spatial reconstruction.

## 3.4 Land use in the historic period: post-AD 1700

By AD 1700 Europe was intensifying its expansion. In the preceding two centuries exploration had been increased and new lands had been discovered in the east and west and in the Southern Hemisphere. The 'Flat Earth' concept was no longer accepted and major strides had been made in navigation techniques, although the elusive and crucial key to longitude was not discovered until the 1730s (Sobel, 1995). This latter improved navigation substantially and played a major role in facilitating the movements of goods and people between Europe and its colonies from the 1800s onward. This was not entirely a one-way movement. Whilst Europe exported its agricultural practices along with its search for wealth through the acquisition of precious metals, it also imported components of New World agricultural systems that altered those of the Old World. Of the exchanges that occurred, the introduction of wheat, barley, cattle, sheep and goats to the New World was of major significance; equally significant was the introduction of maize, potato and tobacco to the Old World. Agriculture globally was transformed as a result of these, and other, exchanges and as Europeans flooded into Europe's colonies in search of new lives and the prospect of improved fortunes.

The period since 1700 has witnessed intense environmental and land-cover change as a consequence of the development referred to above. Most importantly, it has been a period of significant land-cover alteration as natural ecosystems have been converted to agricultural systems, i.e. land use. Various sources of information on these changes, notably inventories of cropland/agricultural land, have provided the basis for several studies focusing on quantitative assessments of the extent of land-cover change. Such studies include those of Williams (1990) and Richards (1990) who have derived statistics on forest change and land-cover change respectively. More recently, Ramankutty and Foley (1999) have reconstructed changes in global croplands for the 1700 to 1992 period on

the basis of a comparison between a land-cover classification of the Earth's natural vegetation undisturbed by human activity, i.e. potential vegetation cover (this was derived from remotely sensed satellite imagery), and cropland inventory data from a wide variety of sources, e.g. the Food and Agriculture Organisation and numerous sources of national data (see Ramankutty and Foley, 1999 for details), which provide data on a temporal basis. This approach has facilitated the generation of data on cropland extent for 1700 and at 50-year intervals by region. A summary of changing cropland extent is given in Table 3.4. This shows that the global extent of cropland increased relatively gradually between 1700 and 1850 with the most significant increases occurring in the Old World, notably in Europe, the Former Soviet Union and China. From the early 1800s the extent of croplands began to increase markedly. As Table 3.4 shows, major increases occurred in North America, Asia and the Pacific developed countries (Australia and New Zealand).

The take-off points for cropland expansion, and hence the reduction of natural vegetation cover, for the regions referred to in Table 3.4 are given in Figure 3.4. This shows that cropland expansion occurred earliest in Europe and China with considerable increases from 1700 (and probably prior to 1700); tropical Africa and the Former Soviet Union experienced notable increases in cropland between 1700 and 1850 but on a reduced scale compared with the former. Almost everywhere in the Old World substantial increases occurred from 1850 as they did in parts of the Americas, notably the USA, particularly east of the Mississippi, and Canada. For the USA west of the Mississippi, and Central America/Mexico, expansion accelerated c. 1900 reflecting the slightly later influx of migrants/population growth. Other parts of South America, the Middle East and tropical Africa experienced take-off points between 1900 and 1950, whilst in Australasia cropland expansion began c. 1870 and accelerated c. 1940. Globally, croplands expanded from c. 4.05 $\times$ $10^6$ km$^2$ in 1700 to 17.92 $\times 10^6$ km$^2$ in 1990, an increase of c. 342 per cent (see Table 3.4). As Ramankutty and Foley (1999) have demon-

| | 1700 | 1750 | 1800 | 1850 $10^6 km^2$ | 1900 | 1950 | 1990 |
|---|---|---|---|---|---|---|---|
| Canada | 0.0 | 0.03 | 0.05 | 0.07 | 0.19 | 0.48 | 0.54 |
| USA (east of Mississippi) | 0.02 | 0.18 | 0.33 | 0.48 | 0.98 | 0.98 | 0.98 |
| USA (west of Mississippi) | 0.00 | 0.01 | 0.02 | 0.03 | 0.72 | 1.27 | 1.34 |
| Mexico and Central America | 0.04 | 0.06 | 0.09 | 0.11 | 0.19 | 0.35 | 0.45 |
| S. America north | 0.02 | 0.03 | 0.04 | 0.05 | 0.09 | 0.34 | 0.93 |
| S. America Uruguay and Chile | 0.02 | 0.03 | 0.04 | 0.05 | 0.09 | 0.44 | 0.44 |
| Tropical Africa | 0.74 | 0.82 | 0.89 | 0.92 | 0.97 | 1.16 | 1.52 |
| N. Africa and Middle East | 0.15 | 0.17 | 0.19 | 0.20 | 0.27 | 0.48 | 0.65 |
| Europe | 0.73 | 0.99 | 1.26 | 1.52 | 1.68 | 1.77 | 1.65 |
| Former Soviet Union | 0.53 | 0.86 | 1.16 | 1.51 | 2.08 | 2.72 | 3.54 |
| China | 0.66 | 0.97 | 1.30 | 1.63 | 1.93 | 2.33 | 2.09 |
| S. Asia | 0.96 | 1.07 | 1.19 | 1.40 | 1.80 | 2.16 | 2.41 |
| S.E. Asia | 0.11 | 0.13 | 0.14 | 0.17 | 0.31 | 0.58 | 1.04 |
| Pacific Developed Countries | 0.06 | 0.06 | 0.07 | 0.07 | 0.13 | 0.23 | 0.50 |
| WORLD | 4.05 | 5.41 | 6.78 | 8.21 | 11.44 | 15.22 | 17.92 |

**TABLE 3.4** The changing extent of cropland since 1700 by region. (Adapted from Ramankutty and Foley, 1999)

strated, this was achieved at the expense of forests/woodlands, as well as grassland/savannas and steppes, the extents of which have decreased by $8.80 \times 10^6 km^2$ (16.7 per cent of the extent in 1700) and $5.61 \times 10^6 km^2$ (17.4 per cent of the extent in 1700) respectively. Clearly, this represents a considerable alteration of the Earth's land cover and its conversion to land use. Moreover, it must also be remembered that these data relate only to cropland expansion; the creation and expansion of pastureland also occurred during this period to modify the Earth's natural ecosystems further.

At continental, regional and local scales, the impact of European annexation and migration was considerable. Williams (1989), for example, provides a wealth of data on the demise and alteration of North American forests as the frontier was driven ever westwards. Such developments inevitably altered the fluxes

within the carbon cycle (see Section 1.5), the magnitude of which has been examined by Houghton and Hackler (2000) and Houghton *et al.* (2000) in terms of changing forest cover in the USA since 1700. Land-cover/land-use change between 1700 and 1900 altered the fluxes of carbon from the biosphere to the atmosphere considerably; rates of flux increased from $10 TgC yr^{-1}$ in 1700 to *c.* $400 TgC yr^{-1}$ in 1880, and then decreased to nil by 1950. This reflects the huge alteration in land cover that occurred during this period, during which 32.6 Pg of carbon was released into the atmosphere as the area of forests and woodlands declined by $160 \times 10^6 ha$, the major source of the carbon released.

However, land-cover changes were often more subtle than straightforward deforestation as Dull (1999) has shown for meadows in the Sierra Nevada, California. His analysis of

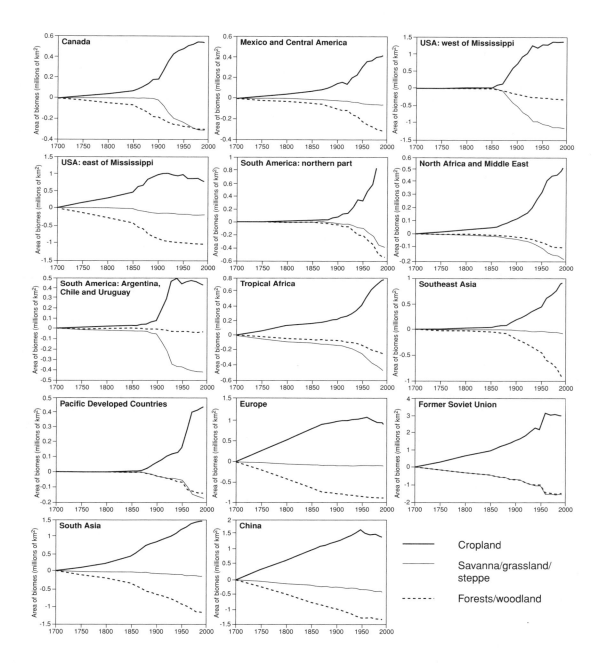

**Figure 3.4** The timing and extent of cropland expansion and its replacement of natural vegetation communities. Reproduced by permission of American Geographical Union from Ramancutty, N. and Foley, A. 1999: Estimating historical changes in global land cover: croplands from 1700 to 1992. In *Global Biogeochemical Cycles* 13(4), 997–1027. © 1999 American Geographical Union.

pollen and spores in soils from Monache Meadows on the Kern Plateau shows that grassland composition altered significantly with the introduction of livestock species, notably cattle and sheep, in the post-1850 period. This was the time of the California gold rush; it encouraged rapid population growth through migration and this led to an increase in the demand for animal products. As a result new pastures were sought beyond the valleys and foothills. Dull's study shows that in pre-1850 times *Riccia* (a liverwort), *Poaceae* (species of grass) and *Asteraceae* (composites) dominated the upper meadow communities but, after 1850, *Riccia* became less significant as *Cyperaceae* (sedges), *Artermisia* (sage brush) and *Rosaceae* (species of the rose family) increased. In the lower meadow, *salix* (willow) decreased markedly with the change in grazing regime as *Cyperaceae* increased.

The scale of land-cover alteration was just as varied in Australia, as authors such as Heathcote (1988; 1994), Jeans (1987) and, more recently, Young (1996) have recounted. The introduction of European livestock and crops altered much of the land cover in the periphery of this continent, especially in the southwest and southeast. The introduction of sugar cane, cotton, camels and mangos transformed it further, as have accidental introductions of non-agricultural alien species. According to Heathcote (1994) the period between *c.* 1830 and *c.* 1920 was the 'golden age' of ranching; by 1888 there were $80 \times 10^6$ sheep, $8 \times 10^6$ cattle and $1 \times 10^6$ horses. All hard-hooved, these animals were the economic basis of agricultural enterprises in the semi-arid region of Australia south of the Tropic of Capricorn. By this time arable agriculture had also expanded; this involved $2.8 \times 10^6$ ha, mostly in coastal areas and the plains of Victoria and New South Wales. Much of this development was stimulated by the rapid growth in markets at home and in Europe; the former were growing because of accelerating immigration, whilst the latter increasingly demanded meat and wool.

Large-scale and small-scale changes occurred in Australia's land cover as a result of this agricultural development. Table 3.4 and Figure 3.4 document the magnitude and timing of overall

change in croplands, forests etc. Young (1996) reports that the *c.* $43 \times 10^6$ ha of natural forests that remains in Australia represents *c.* 62 per cent of the forest cover in 1788 when Europeans first colonized. Some 38 per cent, or $26 \times 10^6$ ha, has disappeared beneath the plough since then. Such changes have been dramatic in a relatively short space of time. Other changes have been more subtle but just as real, as land cover has been altered to accommodate human activity. This is reflected in the work of Witt *et al.* (2000), who have examined the pollen content of sheep faeces deposited over a 50-year period in a shearing shed in southwest Queensland. The pollen represents the diet of the sheep and hence the composition of the pastures grazed. The results show that there was a major increase in grass from the mid-1950s as pasture herbs declined; this reflects declining biodiversity. Some alteration probably began prior to European settlement because it is well established that Aboriginal people used fire as a means of manipulating vegetation communities for various purposes, as Bowman (2000) has discussed in relation to Australia's rain forests.

Fire remains a management technique in Australia's rangelands and in some cases Aboriginal firing regimes have been abandoned in favour of methods 'better' suited to pastoral agriculture. This is the case in the Cape York Peninsula of northeast Australia, for which area Crowley and Garnett (2000) have recounted fire management practices between 1623 and 1996. They show that pre-European burning took place throughout the dry season, i.e. May to October, but from *c.* 1913, pastoralists of European extraction limited burning to a shorter period between May and early August. The former regime may have maintained a grassland cover whilst the latter regime maintained a good supply of forage and controlled cattle movements. Overall, Aboriginal fire regimes appear to have contributed to the maintenance of habitat diversity. Similarly, Hill *et al.* (2000) have shown that in the Wet Tropics World Heritage Area of the northeast coast of Queensland, Aboriginal burning regimes contributed to rain forest survival prior to the introduction of sugar cane farming, and have also played a role in rain forest recovery since 1945.

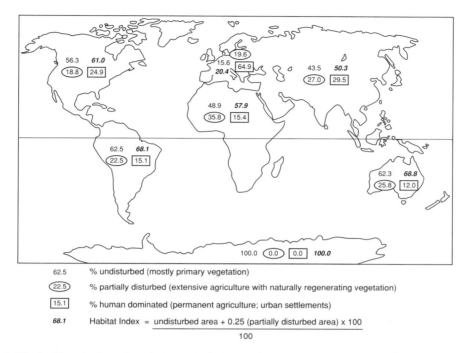

**FIGURE 3.5** The habitat index of environmental change by continent. (Adapted from Hannah *et al.*, 1994)

## 3.5 Indices of environmental change 1: the Habitat Index and the Living Planet Index

As the above sections attest, land-cover and land-use change can occur rapidly. Both national and cultural factors conspire to maintain this dynamism. This renders the quantification of land-cover and land-use change, which are indicators of environmental change, difficult. Nevertheless the increasingly significant role of environmental change in political agendas at all scales, and the need for base-line data for management policies, has led to the formulation of a range of indices. These include the Habitat Index of Hannah *et al.* (1994) and the Living Planet Index devised by the Worldwide Fund for Nature (1998). Land-cover/land-use characteristics play a significant, but not exclusive, role in the derivation of these indices, which are discussed below. Two further indices – the

Ecological Footprint and a Carbon Index – are considered in Section 3.6.

The Habitat Index is a direct measure of land cover/land use. It reflects the degree of disturbance of natural land cover, a characteristic that Hannah *et al.* (1994) have recorded from a range of sources, including atlases, vegetation maps and remotely sensed data. The minimum area mapped was $40 \times 10^3$ ha so the results provide a generalized rather than detailed portrayal of human impact. Figure 3.5 gives a summary by continent of the degree of disturbance of natural land cover. It shows that Europe has the lowest Habitat Index, reflecting its greater extent of disturbed land cover in comparison with other continents; Australia and South America have the highest Habitat Indices as they have the highest percentages of undisturbed land cover. Overall, this index indicates that *c.* 52 per cent of the Earth's surface remains undisturbed, though this falls to *c.* 27 per cent if bare rock, ice and barren areas are excluded. The corresponding Habitat Indices are 5.60 and

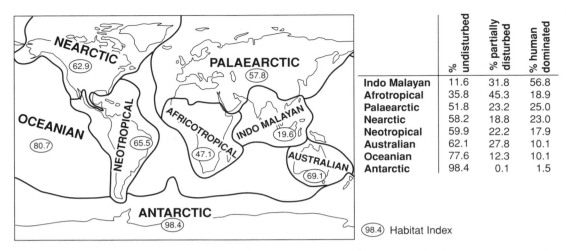

| | % undisturbed | % partially disturbed | % human dominated |
|---|---|---|---|
| **Indo Malayan** | 11.6 | 31.8 | 56.8 |
| **Afrotropical** | 35.8 | 45.3 | 18.9 |
| **Palaearctic** | 51.8 | 23.2 | 25.0 |
| **Nearctic** | 58.2 | 18.8 | 23.0 |
| **Neotropical** | 59.9 | 22.2 | 17.9 |
| **Australian** | 62.1 | 27.8 | 10.1 |
| **Oceanian** | 77.6 | 12.3 | 10.1 |
| **Antarctic** | 98.4 | 0.1 | 1.5 |

(98.4) Habitat Index

**FIGURE 3.6** The Habitat Index (devised by Hannah *et al.*, 1994) for the Earth's biogeographic realms, as defined by Udvardy, 1975

36.2 respectively, indicating the considerable impact of humans globally.

Hannah *at al.* (1994; 1995) have also pointed out that the Habitat Index can be applied in a specifically biogeographical context. For example, they have derived Habitat Indices, using the same method as that given in Figure 3.5, for the Earth's biogeographic realms. These are continental-scale units with a high degree of structural uniformity in relation to the floral and faunal elements. The Habitat Indices for these realms are given in Figure 3.6, which shows that the Indo-Malayan realm is the most disturbed; in a more detailed analysis, Hannah

*et al.* (1994) have shown that 10 of the 20 Indo-Malayan biogeographic provinces that comprise the Indo-Malayan biogeographic realm have Habitat Indices below 10. These include Java, Bengali rain forest, Southern Chinese rain forest, Philippines and Ceylon rain forest. The least-disturbed realms are those of Australia, Oceania and Antarctica, though there is considerable variation within all the realms except the latter. Although the Habitat Index does not involve a measure of extinction, the two are, inevitably, related. The overall high degree of disturbance, whether considered on a continental basis (Figure 3.5) or in relation to

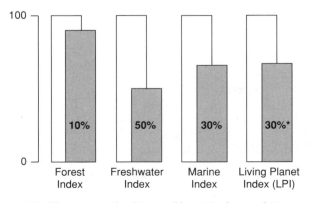

* The 30% decline in the LPI since the base-line year of 1970 indicates that *c.* 30% of the Earth's natural wealth (or natural capital) has been lost

**FIGURE 3.7** Changes in the Living Planet Index and its components between 1970 and 1995. (Adapted from Worldwide Fund for Nature, 1998)

biogeographic realms (Figure 3.6), is an indication that human-induced extinction rates are high. This, in turn, has implications for ecosystem services, such as carbon cycling and climatic regulation (see the reference to the role of forest demise in the USA in global carbon cycling since 1700 in Section 3.4).

The Worldwide Fund for Nature (1998; 1999) has recently formulated an index – the Living Planet Index (LPI) – to express declining environmental quality. This reflects increasing concern with human impact on the Earth's surface and its emerging role as a political issue. However, the LPI is more complex than the Habitat Index discussed above; it comprises measures of forest loss, of changes in freshwater organism populations and changes in marine organism populations. The LPI's baseline date is 1970, so it provides a measure of environmental change for the last 25 years based on forest, freshwater and aquatic ecosystem alteration. Whilst changes in the biota of the latter two habitats were undoubt-

edly significant between 1970 and 1995, reflecting increasing human pressure on aquatic resources, they are not direct indicators of land-cover/land-use change. Rather, changing aquatic biotas are a consequence of direct use of water resources (including waste disposal) and are the indirect result of catchment/continental land-cover/land-use change. Thus the LPI is a composite index in so far as it is a combined measure of land-use/land-cover change and its impact.

Figure 3.7 illustrates changes in the LPI and its components between 1970 and 1995. It shows that the index fell by 30 per cent during this period. This is an alarming rate of natural capital loss and, of course, there had already been a substantial loss by 1970 (see Table 3.4 and Figure 3.4). Of the components of the LPI, the greatest decline occurred in freshwater ecosystems, with activities such as pollution, siltation, overfishing and shoreline development constituting the main causes; similar activities were responsible for the decline in the

| A   REGIONAL VALUES | Consumption units Based on I = world average per person per year in 1995 |
|---|---|
| Africa | 0.55 |
| Middle East/Central Asia | 1.11 |
| Asia/Pacific | 0.83 |
| Latin America/Caribbean | 0.95 |
| North America | 2.70 |
| Western Europe | 1.72 |
| Central and eastern Europe | 1.29 |
| **B   EXAMPLES OF NATIONAL VALUES** | per capita |
| USA | 2.74 |
| Japan | 2.35 |
| UK | 1.43 |
| China | 0.85 |
| Laos | 0.63 |
| Kenya | 0.59 |
| India | 0.47 |
| Niger | 0.34 |
| Rwanda | 0.28 |

**TABLE 3.5** Variations in consumption pressure as calculated by the Worldwide Fund for Nature (1998)

Marine Index (Figure 3.7), which declined by *c*. 30 per cent. Interestingly, the Forest Index of the LPI registered a decline of *c*. 10 per cent, although this appears to be a low value in relation to the aquatic ecosystems. The Worldwide Fund for Nature records an actual loss of 3.68 $\times$ $10^6$ km² between 1970 and 1995, bringing global forest cover to less than *c*. 50 per cent of its original extent. Activities such as logging, mining, fuelwood consumption and various forms of agriculture, ranging from ranching to shifting agriculture, underpinned much of this decline (see Chapters 4 and 5).

Whilst global population growth is one factor in consumption pressure causing this decline in the Earth's natural resources, so too are the high standards of living that characterize much of the developed world and that are increasing in parts of the developing world. As Table 3.5 shows, consumption pressure is highest in Europe and North America, and lowest in African countries. The measure is based on grain, marine fish, wood consumption, freshwater use, carbon emissions and cement production. All of these relate either directly or indirectly to land-cover/land-use change, and thus consumption pressure, and its temporal and spatial variations, must be considered as a major stimulus to Earth-surface alteration.

Both the Habitat Index and the LPI are sensible means of expressing the intensity of environmental change. Land-cover and land-use change are important components of both indices but dominate the Habitat Index in comparison with the LPI, which includes other, albeit related, measures of biodiversity loss. Two further measures are examined in Section 3.6.

## 3.6 Indices of environmental change 2: the Ecological Footprint and a potential Carbon Index

The Ecological Footprint (Rees, 1992; Wackernagel *et al.*, 1997; 1999; Wackernagel and Rees, 1996) shares features with the LPI (see Section 3.5) in so far as both focus on resource consumption, notably forest/wood-land resources. However, the Ecological Footprint is calculated on the basis of a broader spectrum of consumption requirements, e.g. the consumption of agricultural products, than the LPI, and takes direct account of waste production (see Chambers *et al.*, 2000). Essentially, it calculates the area of land required to support a given population in a given area through the provision of resources and capacity for waste disposal. Thus the Ecological Footprint is a measure of sustainability. As such, it has gained considerable credibility as a management technique and can be applied at various scales and in relation to various economic activities. All of the indices referred to in Section 3.5, as well as the Ecological Footprint, involve resource manipulation and rely on various ways of expressing this quantitatively. These resources are almost all carbon based, i.e. forest products, agricultural produce, fish, and fossil fuels, and many waste products are carbon based. Consequently, a further index is here suggested based on carbon consumption and production. Such an index also has merit for environmental management.

According to Rees (1992), an Ecological Footprint is an estimate of the ability of an area of land – including all its assets such as plants, animals and soils – to support a given population. In many cases, especially in relation to areas of high population concentration such as cities, the area of supporting ecosystem will exceed considerably the area of occupation. In addition, the area of supporting ecosystem will be required to absorb waste materials as well as to provide food, fibre, mineral resources etc. Where the area of supporting ecosystem cannot fulfil these roles, the system cannot be sustained, as in the case of cities, for example, unless there is recourse to resources and waste disposal further afield. On a regional or national basis, the total population may be sustainable if the Ecological Footprint does not exceed the regional/national biocapacity, i.e. its ability to provide the necessary resources and absorb waste. The regions or nations that exceed their biocapacity generate an ecological deficit that can be expressed in terms of hectares per capita.

| | Ecological Footprint (ha per capita) | Biocapacity (ha per capita) | Ecological deficit (ha per capita) |
|---|---|---|---|
| **Europe** | | | |
| Belgium | 5.0 | 1.2 | −3.8 |
| Germany | 5.3 | 1.9 | −3.4 |
| Portugal | 3.8 | 2.9 | −0.9 |
| Norway | 6.2 | 6.3 | 0.1 |
| Russian Federation | 6.0 | 3.7 | −2.3 |
| UK | 5.2 | 1.7 | −3.5 |
| **Asia** | | | |
| Bangladesh | 0.5 | 0.3 | −0.2 |
| China | 1.2 | 0.8 | −0.4 |
| Hong Kong | 5.1 | 0.0 | −5.1 |
| India | 0.8 | 0.5 | −0.3 |
| Indonesia | 1.4 | 2.6 | 1.2 |
| Japan | 4.3 | 0.9 | −3.4 |
| Malaysia | 3.3 | 3.7 | 0.4 |
| Philippines | −1.5 | −0.9 | −0.6 |
| Singapore | 6.9 | 0.1 | −6.8 |
| **Americas** | | | |
| Canada | 7.7 | 9.7 | 1.9 |
| USA | 10.3 | 6.7 | −3.6 |
| Brazil | 3.1 | 6.7 | 3.6 |
| Chile | 2.5 | 3.2 | 0.7 |
| Mexico | 2.6 | 1.4 | −1.2 |
| Peru | 1.6 | 7.7 | 6.1 |
| Venezuela | 3.8 | 2.7 | −1.1 |
| **Australasia** | | | |
| Australia | 9.0 | 14.0 | 5.0 |
| New Zealand | 7.6 | 20.4 | 12.8 |
| **World Average** | **2.8** | **2.1** | **−0.7** |

**TABLE 3.6** Examples of Ecological Footprint, biocapacity and ecological deficit. (Based on Wackernagel *et al.*, 1997)

Table 3.6 gives examples of Ecological Footprints, biocapacity and ecological deficits in relation to average global values. According to Wackernagel *et al.* (1997), there is 1.7 ha of biologically productive land available per person on a global basis. They have also shown that 42 nations, out of 52, have Ecological Footprints that exceed 1.7 ha per capita. The implication is that current patterns of consumption coupled with waste production are unsustainable globally. The data in Table 3.6 also reflect considerable variations between nations. For example, the developed nations of the UK and USA have similar ecological deficits. Although the former's Ecological Footprint is only half that of the USA, its biocapacity is only 20 per cent of that of the USA. Ethiopia, one of the world's poorest nations, has a low Ecological Footprint, but it still has an ecological deficit because it has low biocapacity. Moreover, ecological deficits can be reduced if biocapacity is 'imported', i.e. if goods and waste disposal are appropriated elsewhere. In Europe, for example, ecological deficits must be partly due to the lack of forests to absorb carbon dioxide. Forests elsewhere must, therefore, perform this function. A corol-

| | $CO_2$ emissions mt $\times$ $10^3$ | Agricultural production[1] mt $\times$ $10^3$ | Wood production m³ $\times$ $10^3$ | Paper production t $\times$ $10^3$ | Total carbon index t $\times$ $10^3$ | Population $\times$ $10^3$ | *Per capita* carbon index for 1995 |
|---|---|---|---|---|---|---|---|
| Ethiopia | 3525 | 9808 | 46,038 | 8 | 59,379 | 55,053 | 1.07 |
| China | 3,192,484 | 585,303 | 326,060 | 25,468 | 4,129,315 | 1,221,464 | 3,38 |
| New Zealand | 27,440 | 1,187 | 19,352 | 868 | 48,847 | 3,575 | 13.66 |
| Singapore | 63,669 | NIL | 145 | 93 | 63,907 | 2,848 | 22.44 |
| UK | 542,140 | 29,446 | 10,154 | 5,777 | 587,517 | 58,258 | 10.08 |
| USA | 5,468,564 | 347,417 | 606,489 | 85,261 | 6,508,028 | 263,250 | 24.72 |
| World | 22,714,561 | 2,649,165 | 3,804,994 | 272,082 | 29,440,802 | 5,716,426 | 5,15 |
| Year compiled | 1995 | *Average for 1994–96* | *Average for 1993–95* | *Average for 1993–95* | | 1995 | |

*Note:* [1] cereals + roots/tubers + pulses (data From World Resources Institute, 1996; 1998)

**TABLE 3.7** Examples of Carbon Indices

lary of this is that increased afforestation is necessary to reduce European ecological deficits. Technology also has the capacity to increase or decrease the Ecological Footprint. For example, mechanization and electrification have played a major role in the adoption of fossil fuels, emissions from which intensify the Ecological Footprint, yet emission controls on power stations and vehicles can reduce this impact considerably. Similarly, biotechnology has the potential for reducing the Ecological Footprint through increasing agricultural efficiency and reducing the environmental impact of agriculture.

However, it must also be noted that calculation of the Ecological Footprint can be problematic. For times past, for example, data may be inadequate, inappropriate or incompatible within and between nations or regions. Calculations of current Ecological Footprints can also be criticized; as van den Bergh and Verbruggen (1999) point out, trade is not considered adequately. Whilst it is not a perfect tool for assessing sustainability, the Ecological Footprint remains a valuable indicator of the degree of human impact. It also provides a quantitative measure of the relationship between people and environment, and can be used in a practical context to improve the efficiency of resource use.

As well as a means of expressing the environmental impact of a nation or city, the Ecological Footprint can be used to assess the impact of a specific activity. This is illustrated by Haraldsson *et al.* (2001), who have compared the Ecological Footprint of so-called eco-living with traditional living in two towns, Toarp and Oxie, in southern Sweden. There was only *c.* 10 per cent difference in the Ecological Footprint of the building materials between the two, despite the Toarp house being an eco-house, but overall the Ecological Footprint was found to be 2.8 ha per person in Toarp and 3.7 ha in Oxie. The largest component of the Ecological Footprint in both towns was domestic food and energy consumption at 75 per cent. Another example of the application of the Ecological Footprint has been provided by Bunting (2001), who has used it to assess the impact of aquaculture, while Barrett (2001) has determined the Ecological Footprint of Guernsey in the Channel Islands. The latter has also used it in the context of reducing the impact of passenger transport.

In view of the factors involved in the calculation of the Ecological Footprint, and the indices examined in Section 3.5, it should be possible to derive another index based on the common denominator in these indices, notably carbon (see Section 1.5). This occupies a pivotal role in people–environment–technology relationships through its appropriation as food and fuel energy; food security, development, wealth generation and political influence all derive from carbon appropriation (see Section 1.5). A Carbon Index could, thus, be a useful tool for environmental management.

At a basic level, data on greenhouse gas emissions (notably carbon dioxide) are available on a national per capita basis and can be used as surrogate measures for fossil-fuel consumption (e.g. World Resources Institute, 1998). Similarly, data on agricultural production – notably cereals, root and tuber crops, and pulses – are available. These data reflect the impact of agriculture on the global carbon cycle. Wood and paper production are also significant, especially the major products of roundwood, sawnwood and paper. These three elements comprise the most important ways that humans manipulate carbon and so influence the carbon biogeochemical cycle. The compilation of a Carbon Index requires that each of these factors should be combined to give an overall value; this can be divided by the population to give a per capita value per nation. A comparison of national per capita indices with the average world index gives a measure of each nation's role in carbon manipulation and appropriation. The examples given in Table 3.7 show that the variation in the Carbon Index is similar to that of the Ecological Footprint (Table 3.6), but that the range of values is greater. As Table 3.7 shows, the average world Carbon Index is 5.15. At one extreme is Ethiopia with a Carbon Index of only c. 20 per cent of the world average, whilst the USA has a Carbon Index of almost 500 per cent of the world average. Although only a few examples are given in Table 3.7, the data show that the poorest nations have the least impact on the global carbon cycle, whilst the richest nations have the greatest impact.

This is a relatively simple approach to the compilation of a Carbon Index; it uses information only on the main activities of humans and ignores minor activities, e.g. the production of agricultural commodities other than cereals, roots, tubers and pulses. It could be more precise if all types of agricultural production were considered. The degree of precision does,

however, relate to data availability and compatibility. Similarly, data on greenhouse gas emissions could be substituted by data on coal, oil and natural gas consumption. In some respects simplicity is a key characteristic, especially as internationally compatible data must be used. This problem is similar to that inherent in the calculation of the Ecological Footprint (see above). The index could be calculated to be more or less sophisticated but, for comparative purposes, data must be compatible in relation to spatial and temporal scenarios. In terms of the role of these indices for environmental management, they focus directly on carbon management and thus address problems that relate to human intervention in the global biogeochemical cycle of carbon.

## 3.7 Conclusion

The present interglacial, which began c. $12 \times 10^3$ years BP, is unique in relation to the many preceding interglacials because of the influence of human activity. Whilst pre-agricultural hunter-gatherer societies were adept at manipulating biotic resources, the transition to permanent agriculture, beginning c. $10 \times 10^3$ years BP, was a landmark development. The emergence of agriculture represented an increasing degree of control by humans over their environment; it set in train modification and alteration of land cover on a scale hitherto unprecedented. Such alteration and modification has continued until the present. Agriculture remains the most important agent of environmental/land-cover change today; agricultural land use is also the most extensive type of land use on the Earth's surface today.

Evidence for past land-cover/land-use change can be derived from a variety of archives. For the early part of this interglacial, pollen analysis of peats and lake sediments is a major source of information, whilst written records and maps provide detailed information on change in the last 300 years. The land transformation that has occurred in the past has also contributed substantially to anthropogenic alteration of the global carbon cycle; it has caused a release of carbon, as carbon dioxide, from the biomass of land cover into the atmosphere where it is contributing to global warming.

Loss of natural ecosystems, the problem of global warming, and the inclusion of such issues into political agendas has resulted in recent innovations to quantify the degree of land-use/environmental change. To this end, and for purposes of environmental management, several indices have been proposed. These include the Habitat Index, the Living Planet Index and the Ecological Footprint. All have advantages and disadvantages. Another index, the Carbon Index, is also proposed; it involves basic data on fuel and food energy production and wood products; it has drawbacks but its merits lie in its simplicity. Its focus on carbon-based consumption highlights the significance of the carbon cycle as a vital link between people and environment.

# Further reading

**Chambers, N., Simmons, C. and Wackernagel, M.** 2000: *Sharing Nature's Interest. Ecological Footprints as an Indicator of Sustainability.* Earthscan, London.

**Evans, J.G.** 1999: *Land and Archaeology: Histories of Human Environment in the British Isles.* Tempus, Stroud.

**Head, L.** 2000: *Cultural Landscapes and Environmental Change.* Arnold, London.

**Mannion, A.M.** 1997: *Global Environmental Change. A Natural and Cultural Environmental History*, 2nd edition. Addison Wesley Longman Ltd, Harlow.

**Turner II, B.L., Clarke, W.C., Kates, R.W., Richards, J.F., Mathews, J.T. and Meyer, W.B.** (eds.) 1990: *The Earth as Transformed by Human Action.* Cambridge University Press, Cambridge.

# Land use related to agriculture: the developed world

## 4.1 Introduction

Agriculture has always been, and remains, the most important cause of land-cover alteration and the most important type of land use globally. Historically (see Section 2.6 and Chapter 3), the middle latitudes have experienced the greatest extent of land-cover transformation into agricultural land use. This is particularly true of the latitudes north of the Tropic of Cancer; Europe has been profoundly influenced by the centre of agricultural innovation in the Near East (Figure 2.7). The Near East centre, for example, produced wheat, barley, rye, oats and several leguminous crops, and these rapidly became the mainstays of arable agricultural systems in temperate Europe and western Asia, whilst cattle, sheep, goats and pigs came to dominate pastoral agriculture. The domestication of rice, buckwheat and soya provided arable crops for eastern Asia (see Chapter 5). Nomadic herding has also been a component of high latitudes in the developed world of the Northern Hemisphere. Examples include the herding of reindeer in northernmost Europe and nomadic pastoralism based on the saiga antelope in Central Asia.

Today's agricultural systems remain dominated by the crop and animal complexes referred to above. However, interactions between the New and Old Worlds in the 1600s and 1700s, as a result of the so-called expansion of Europe, meant that New World crops were introduced into Europe and Asia (see Section 3.4). Maize and potato in particular were incorporated into European and Asian agricultural systems. Other introductions, such as soybean into Europe from China and tobacco into Europe and Asia from tropical America, also contributed to altering the character of land use.

In the Southern Hemisphere the extent of land in the developed nations is much more limited in than it is in the Northern Hemisphere (see Figure 4.1). There is no equivalent of the nomadic pastoralism of northern Europe, and the agricultural systems of the middle latitudes are largely the product of European influence. As Table 3.4 shows, cropland expanded considerably between 1850 and 1920 in Australia and New Zealand. In this region the major crops grown are similar to those of the Northern Hemisphere and the dominant domesticated animals are cattle and sheep.

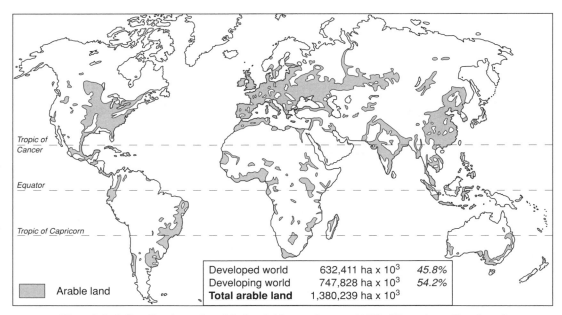

Developed world        632,411 ha x 10³    45.8%
Developing world       747,828 ha x 10³    54.2%
**Total arable land**  1,380,239 ha x 10³

Arable land

**FIGURE 4.1** The global distribution of arable land (data relate to 1998). (Data from Food and Agriculture Organisation, 2001a)

There is unlikely to be much expansion of agricultural land in the short or medium term in the high or temperate latitudes, especially when compared with low latitudes (see Chapter 5 and Section 10.3.3). In much of Europe, for example, set-aside policies are in operation to reduce agricultural production and, in many countries, most suitable land is already in production. This does not mean that the patterns of land cover and land use in these regions will remain unchanged. The environmental influences on agriculture alter, as do the socio-economic factors that stimulate agriculture. The outcome is a dynamic situation even where agriculture is well established. Currently, the most important stimuli to change are socio-economic and/or political, though land degradation and conservation priorities also play a role; in the next few decades, environmental factors will become increasingly significant especially as global climatic change occurs. (The latter is discussed in Chapter 10.) The operation of these factors is examined below and their impact in terms of land-cover

and land-use change in the recent past is illustrated with reference to case studies.

## 4.2 Types of agriculture and their distribution

Figure 4.1 illustrates the distribution of the world's arable land. It shows that the greatest extent of arable land lies in the middle latitudes of the northern hemisphere; it extends across Europe, from Ireland through central Europe and into Central Asia. North America, especially the USA (and central China) also have a substantial proportion of the world's arable land. These are the regions in which much of the world's staple crops – notably wheat, maize and potatoes – are produced. Further detail on agricultural systems in the developed world is given in Figure 4.2. Apart from illustrating the distribution of arable systems, Figure 4.2 also shows that ranching/herding is predominant in the continental interiors of Asia and North

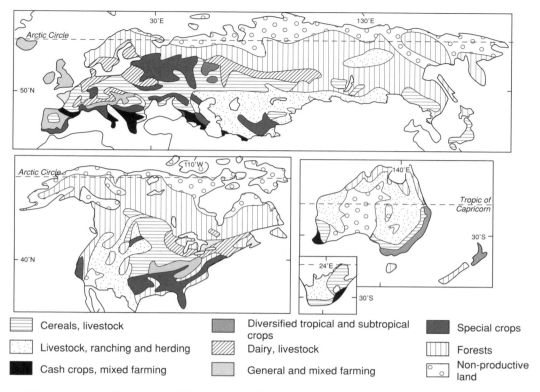

| | Cereals, livestock | | Diversified tropical and subtropical crops | | Special crops |
| | Livestock, ranching and herding | | Dairy, livestock | | Forests |
| | Cash crops, mixed farming | | General and mixed farming | | Non-productive land |

**FIGURE 4.2** Agricultural systems of the developed world. (Adapted from the *Oxford Hammond Atlas of the World*, 1993)

America, and that cotton production predominates at similar latitudes but at low altitudes in North America and Central Asia.

Many factors influence the specific character of land used for agriculture at any given time; these are summarized in Figure 4.3. Of these, the environment sets the physical limits in which agricultural systems have developed; the overriding control is climate, as it dictates water availability and length of growing season. Other factors include soil depth and nutrient status which, in turn, are a function of underlying geology. Equally important influences are socio-economic and political factors; these include rates of population change and aspirations for wealth generation, as well as the production of goods for export. Innovations, such as the adoption of new rotations, and scientific and technological advances have also

exacted significant changes in agricultural systems and thus on land-use characteristics. Some changes brought about by science and technology may be the result of innovations affecting downstream, i.e. on-farm, components, e.g. crop/animal breeding programmes, the application of pesticides and on-farm mechanization. Other changes will result from upstream, i.e. off-farm, developments such as the invention of refrigeration, and improvements in transportation and food processing. 'Beyond the farm' influences are also many and varied; they include government or politically-based polices such as the promotion of agricultural intensification or set-aside, both of which have altered agriculture in Europe profoundly since *c*. 1940. Other such factors include changing markets, subsidies for either specific crop production or for so-called land improve-

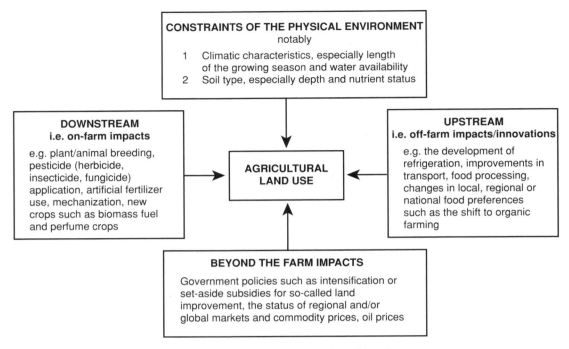

**FIGURE 4.3** A representation of the factors that affect agricultural land use

ments, e.g. the drainage of wetlands and hedge removal.

The operation of all these factors (see Figure 4.3), the intensity of which will have varied spatially and temporally, has given rise to the agricultural systems, and hence the agricultural land-use patterns that characterize the developed world today (Figure 4.2). Indeed, it is these factors that have made a major contribution to the transformation of developed regions from wildscapes into landscapes, or from natural landscapes into human-dominated landscapes. This is reflected in the Habitat Index (as discussed in Section 3.5 and illustrated in Figure 3.5), which shows that Europe is the most altered of all the developed regions. Whilst Figure 4.2 provides a visual image of the distribution of the agricultural systems of the developed world, Tables 4.1 and 4.2 provide a snapshot of the area and the output of arable systems engaged in the production of wheat, maize, rice and potatoes, and livestock population respectively for the mid- to late 1990s. The

data reflect slight downward trends in all of the crop and livestock types, and marginal increases in the area of arable and pasture land. This is in contrast to trends in the developing world (see Chapter 5, Tables 5.1 and 5.2).

In the developed world in general, the area of arable land is equivalent to c. 82 per cent of the entire area of Australia or c. 60 per cent of the area of Europe. The amount of land given over to permanent pasture is equally substantial, being equivalent to c. 115 per cent of the area of Europe or c. 160 per cent of the area of Australia. These statistics highlight the role of agriculture as a major agent of land transformation in the developed world, despite the often common perception, especially in Europe, that industry, plus its associated activities, and urbanization predominate. Moreover, the statistics given in Tables 4.1 and 4.2 show that the agricultural systems of the developed world account for c. 36 and c. 46 per cent of the world's pasture and arable land respectively.

| | 1995 | 1998 | 2000 | World (1998) |
|---|---|---|---|---|
| **Arable land ha × 10³** | 630,050 | 632,411 | N/A | N/A |
| **CROP TYPES** | | | | |
| **Area harvested 10 ha × 10³** | | | | |
| **Wheat** | 117,780 | 119,538 | 115,285 | 225,883 |
| **Maize** | 44,154 | 47,756 | 46,827 | 138,926 |
| **Rice** | 4,360 | 4,130 | 4051 | 151,934 |
| **Potatoes** | 11,078 | 10,536 | 10,494 | 18,784 |
| Data for comparison | | | | |
| Area of Europe | 1,049,800 | ha × 10³ | | |
| Area of Australia | 768,700 | ha × 10³ | | |
| Total land area of the world | 7,279,100 | ha × 10³ | (excluding Antarctica) | |
| Total arable land area of the world | 1,380,239 | ha × 10³ | | |

**TABLE 4.1**  Statistics on agricultural land use in developed countries (Food and Agriculture Organisation, 2001a)

| | 1995 | 1998 | 2000 | World total 1998 |
|---|---|---|---|---|
| **Permanent pasture ha × 10³** | 1,213,253 | 1,219,466 | N/A | N/A |
| **LIVESTOCK** | | | | |
| **Cattle** | 366,838,236 | 339,128,375 | 326,826,815 | 1,334,370,820 |
| **Sheep** | 439,815,180 | 394,876,176 | 387,098,653 | 1,056,644,070 |
| **Goats** | 32,060,482 | 30,513,465 | 29,120,543 | 695,967,753 |
| **Pigs** | 302,703,653 | 290,417,004 | 289,742,929 | 875,179,322 |

Data for comparison
Area of Europe = 1,049,800 ha × 10³
Area of Australia = 768,700 ha × 10³
Total land area of the world = 7,279,100ha × 10³ (excluding Antarctica)
Total pasture land area of the world = 3,429,531ha × 10³

**TABLE 4.2**  Statistics on livestock populations in developed countries (Food and Agriculture Organisation, 2001a)

## 4.3 Nomadic pastoral agriculture and its alteration of land cover

Extensive pastoral agriculture is a significant characteristic of the developed world, not only as a means of food production but also as a distinct way of life often associated with nomadism. It makes use of land that cannot be cultivated, but that can capitalize on animals' ability to convert various types of vegetation into protein. It predominates in high latitudes, notably the boreal and tundra zones of Europe, Asia and North America, where it involves reindeer herding. The most well-known example is that of northern Scandinavia where

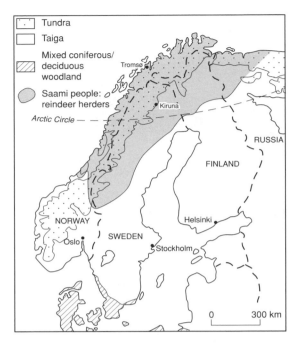

**FIGURE 4.4** Area of reindeer herding in northern Scandinavia in relation to the major ecosystem types

the reindeer herders are known as Saami (see Figure 4.4); some 50,000 of these people occupy an area centred on the Arctic Circle that is often referred to as Lapland. The animal herded is *Rangifer tarandus*. Many other herding groups occupy northern Asia as far east as the Bering Strait. In Eurasia there is a long history of herding, whereas in Greenland, Canada and Alaska it was introduced in the 1950s as a supplementary activity to the eskimo hunting tradition. Herding is not, however, confined to high latitudes in the developed world. It is also significant in the lower latitudes of the arid and semi-arid region of Central Asia, notably in Kazakhstan, Russia and Mongolia where the species herded is the saiga antelope (*Saiga tartarica*).

The history and economy of the Saami have been well documented (e.g. Beach, 1988; Paine, 1994). It is likely that reindeer herding emerged from reindeer hunting, probably in the mid-Holocene as food-procurement strategies became increasingly planned and controlled.

Local variations in topography and vegetation determine patterns of reindeer herding, which utilizes forest, tundra and coastal zones of the Arctic environment (as shown in Figure 4.4). Sometimes herds utilize the primary productivity of all three environments. One of the most important sources of reindeer forage is lichen, the distribution of which in terms of quality and quantity is of primary consideration for determining carrying capacity. Consequently, the impact of reindeer herding on land cover is manifest in lichen communities. Reindeer herders determine patterns of migration on the basis of lichen quantity and quality, and if overgrazing occurs lichen pastures may take many years to recover. As a result, land cover will vary seasonally, annually and interannually. Apart from pasture characteristics, social and economic factors also influence herding strategies. Such factors include access to markets, and slaughterhouses, the sharing of pastures between neighbouring herds and protection against predators such as wolves. Moreover, what was once a subsistence type of agriculture has become increasingly orientated towards a market economy. This, plus the fact that nomadism is declining as the families of herders now occupy permanent settlements, means that the *genre de vie*, or the cultural tradition, of the Saami, has changed rapidly in the past 40 years. However, the herders themselves have espoused technology to improve productivity, marketing etc. Their use of snowmobiles, light planes, helicopters, radio and satellite communications, as well as information technology has introduced a fossil-fuel subsidy into this agricultural system, in contrast to most extensive pastoral systems elsewhere (see below and Section 5.2), and it has led to the intensive use of lichen pastures.

The reciprocal relationship between lichen pastures and reindeer density is exemplified by the investigation of Kumpula *et al.* (2000) into Finnish reindeer management areas. Their results show that reindeer densities were highest in areas with between 20 per cent and 30 per cent lichen cover. More importantly, the study showed that lichen pastures must produce at least $1000\,kg/ha^{-1}$ of biomass (dry matter) for sustainable use; this would require a winter reindeer density of less than five to seven reindeer per km². Kumpula *et al.* (2000) also show that after forest fires lichen ranges require *c.* seven years free of grazing in order to recover to the productivity level of 1000 kg of dry matter per hectare. This contrasts with 18 years, which is the average time necessary for the recovery of full production of 7000 kg of dry matter per hectare. Thus land-cover change is effected through firing, recovery and then grazing over a relatively short timescale.

Reindeer herding utilizing lichen pastures is also widespread beyond Scandinavia, throughout Eurasia. Many indigenous groups are involved and many experienced considerable changes in their lifestyles following the discovery and exploitation of oil and gas fields in the 1960s and 1970s, as discussed below. Further adjustments have characterized the post-Soviet era (since *c.* 1990) with economic recession. All of these factors have influenced the management of reindeer herding with consequences for land cover/land use. Vilchek (2002) has recently reviewed the role of the development of the hydrocarbon industry in causing degradation of Siberia's lichen pastures. He states that these pastures comprise 350 × 10⁶ ha, which support 1.7 × 10⁶ domesticated reindeer and more than 1 × 10⁶ wild reindeer. Much of this remote region is undisturbed but in the hydrocarbon-producing centres, especially Yamal (see Figure 4.5), reindeer herding is a primary activity. The Yamal area is occupied by some 35 per cent of the total native population of the region (i.e. 154,000); it contains *c.* 54 per cent of all tundra pastures and some 47 per cent of all reindeer. The animal herds follow south-to-north routes, grazing the forest-tundra zone in the winter and migrating to the central pastures of Yamal in spring; they then take advantage of the short summer growing season to graze coastal pastures. These extensive migratory paths and pastures occupy hydrocarbon fields, and are crossed by pipelines and railways. Vilchek has estimated that *c.* 5 × 10⁶ ha are no longer used for reindeer herding because of oil and gas extraction and prospecting, and that pollution is a major cause of ecosystem damage.

This has compounded an already existing problem of pasture degradation caused by overgrazing. Even in the 1800s the Yamal region was used by indigenous people known as the Nenets for large-scale reindeer herding, in contrast to small domestic herds elsewhere in Siberia. There was then a huge increase of reindeer from 200,000 in 1922 to 508,000 in 1996 (data given in Vilchek, 2002) in parallel with Soviet policy, as state ownership replaced private ownership of herds. At the same time, pastures were assigned to farms, diminishing the capacity for free-ranging herd movements to accommodate variations in climate, pasture quality and quantity, and reindeer numbers.

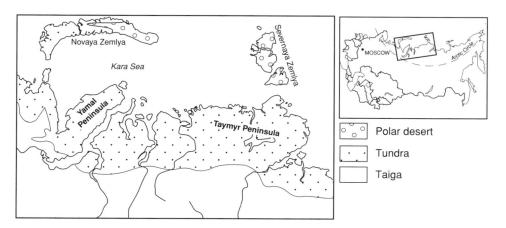

**FIGURE 4.5** Area of reindeer herding in the Yamal and Taymyr Peninsulas, Siberia. (Adapted from Vilchek, 2002)

Between the 1930s and 1950s carrying capacity had declined by almost 30 per cent because of this, and because of the use of land for oil and gas production, which curtailed the use of fallow periods to allow pastures to recover. Today, pollution through sulphur dioxide emissions continues to impair pastures from which sensitive lichens are disappearing.

The changing political regime of Central Asia in the last century has also influenced Kazakhstan's land use, which is associated with the herding of the saiga antelope (*Saiga tatarica Scapital*). This species feeds mainly on grasses and is utilized for its meat, horns and hide. It occurs in the desert, semi-desert and steppe ecosystems of Central Asia, where three separate populations of saiga exist (as shown in Figure 4.6). This separation has been accentuated through land-use change as Bekenov *et al.* (1998) observe. Factors such as the spread of arable farming, the construction of irrigation systems, mining and settlement have all contributed to determining modern distributions and movements, whilst excessive hunting

and poaching as well as periods of deep snow cover (*dzhuts*) have frequently caused major population changes in the past 200 years. Of particular note is the impact of the so-called 'Virgin Lands' campaign, which began in the 1950s under Khrushchev's plan to develop uncultivated land in the eastern republics of the USSR. The pastures on some of the saiga's migration routes were replaced with arable fields and the animals began to graze the edges of the cropfields, a situation exacerbated in the 1974–77 period by high saiga populations and drought. In the pastures themselves, Bekenov *et al.* (1998) note that parts of the saiga's winter ranges (see Figure 4.6) are subject to pasture degradation and thus land-cover change. This is because of high numbers of both saiga and domestic animals, e.g. sheep, cattle, horses and camels.

Whilst little attention is paid to the vast regions associated with the high latitudes of Eurasia, or the dry continental interior of Asia, the above examples testify to the dynamism of environmental change, which is manifest as

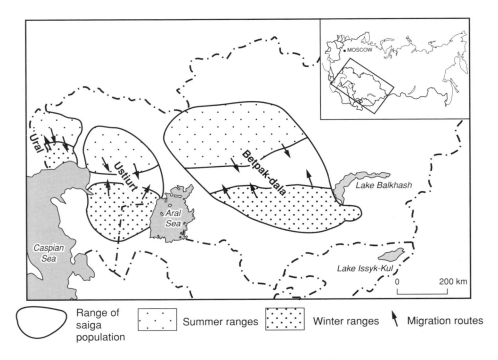

**FIGURE 4.6** The saiga pastures of Kazakhstan. (Adapted from Bekenov *et al.*, 1998)

land cover and/or land use. Moreover, it is apparent that changes in political regime, especially in the region of the Former Soviet Union, have been major stimuli to change in both the agroecosystem components and indigenous peoples.

## 4.4 Sedentary pastoral agriculture and its alteration of land cover

Table 4.2 gives details of livestock populations in the late 1990s in the developed world, where sheep and cattle dominate on the 1,219,466 × $10^3$ ha of permanent pasture. Moreover, this area of pasture is approximately 36 per cent of the world total, and by definition, pastureland comprises mainly grasses with herbs, the composition of which varies according to location and management factors such as grazing intensity, herbicide and fertilizer applications. Many areas of pasture are not natural climax vegetation but the product of human activity throughout prehistory and history, as discussed in Sections 3.2, 3.3 and 3.4. The downlands of southeast England, for example, have their origins in the establishment of agriculture at the beginning of the Neolithic period c. $5 \times 10^3$ years ago. Numerous socio-economic factors have also played a significant role in determining the intensity and extent of pasture in space and time. As discussed in Section 3.3, a major factor determining the extent of pasture in western Europe in the twelfth to the fourteenth centuries was the burgeoning demand for wool to satisfy cloth-producing industries. In recent years subsidies for upland sheep farming via the Common Agricultural Policy (CAP) of the European Union (EU) have caused grazing intensities to increase and species composition to change as a result.

In the developed world cattle and sheep farming are undertaken on two bases: extensive grassland farming or ranching, and intensive grassland farming. The former is characteristic of the semi-arid rangeland of central and southwest USA whilst in the temperate zone animal farming is intensive. In most animal production systems in the developed world, either exten-sive or intensive, there is a fossil-fuel subsidy (see Section 1.5) through the use of herbicides, animal health products and, sometimes, artificial fertilizers. Consequently, these activities contribute to the environmental problems associated with fossil-fuel consumption, including global warming, which is itself likely to bring about land-use change in the future (see Chapter 9). Environmental and socio-economic factors continue to influence land-cover and land-use in relation to pastoral agricultural systems as the examples below illustrate.

The extent of rangeland to facilitate cattle ranching, along with variations in grazing management and competition from possible alternative land-use types has varied considerably in the recent past in response to a variety of pressures. Such dynamism is illustrated by the work of Smethurst (1999) who has examined land-use change in El Dorado County in the Central Sierra Nevada of California during the period 1957 to 1997. These changes are summarized in Table 4.3, which shows that the most significant land-use change is the decline in ranchland and the increase in residential land use, which includes vast shopping malls as well as houses. Smethurst points out that the livestock industry began in the early 1900s to supply local mining communities; by 1930 more than 95 per cent of the land in El Dorado County was given over to livestock production, including all of the national forest land, which was used for summer grazing, along with at least 80 per cent of farmland. Between 1950 and 1960 a number of developments began to curtail livestock ranching. Fire-suppression policies, for example, allowed trees to increase, shading out the ground flora used for browsing and hindering cattle movements, which were also curtailed by the abrasive seal used on roads. Moreover, the construction of a dam to create Edison Lake provided a water supply for Georgetown but in so doing flooded meadows previously used for cattle grazing. Although some ranchers bought land in Oregon to where cattle were transported for half of the year, the transport costs eventually became prohibitive. These changes were reflected in the revenues derived from livestock productions using figures adjusted to a constant; Smethurst states

| Land use | 1954 | 1996 | % Change |
|---|---|---|---|
| | Acreage | Acreage | |
| Public timber[1] | 315,000 | 308,000 | −2.2 |
| Ranch land | 168,236 | 77,979 | −53.6 |
| Farm land[2] | 25,314 | 21,912 | −13.4 |
| Private timber[3] | 359,000 | 238,000 | −33.7 |
| Residential | 69,000 | 217,332 | +315 |

Data are from various sources quoted in Smethurst (1999)

Notes:
[1] Public or national forest
[2] Arable production
[3] Woodland/forest on private land

**TABLE 4.3** Land-use change in El Dorado County, Central Sierra Nevada, California, 1954–96. (Adapted from Smethurst, 1999)

that revenues declined from US$3.37 × 10⁶ in 1957 to US$1.44 × 10⁶ in 1994.

Other factors that have contributed to these land-use changes include an increased demand for residential property, as the population of northern California has grown and wealth has increased. Coupled with declining prices for cattle, such alternative land-use, fuelled by demand, is increasingly attractive. Moreover, the cost of raising cattle has risen more than the prices fetched by cattle. For example, grazing fees for the use of federal land have increased and environmental legislation means that the time available for grazing must be reduced; there is also increased uncertainty about grazing permit renewals, even for those families that have held such permits for a century or more. In contrast, legislation has favoured forests in the area and, as Table 3.4 shows, the decline in public forests especially has been much more limited when compared with the decline in ranchland. This particular case illustrates well the operation of, and interaction between, numerous socio-economic and legislative factors, and their impact on land-use change in the Central Sierra Nevada.

Major land-cover changes have also occurred in the prairie-forest ecotone of the Southern Great Plains. Both pastoral and arable agriculture have contributed to the creation of a mosaic of land-cover types within this transi-tional zone. According to Pogue and Schnell (2001), there has been an overall loss of natural prairie, a reduction in the size of remaining patches of prairie and enhanced isolation of the fragments. The land-cover patterns in this region are the result of a c. 20 per cent increase in arable agriculture. This has not only created isolated pockets of prairie but also altered the dimensions of other natural habitats such as riverine forests and woodlands.

As well as the quite distinct changes in land use that can occur in situations such as those discussed above, subtle changes in landscape character and in species composition may occur as management regimes are implemented. The former is exemplified by the increasing frag-mentation of the landscape in the mountains of central Norway. Olsson et al. (2000) have shown that between 1960 and the 1980s forest cover declined as grassland for animals increased. Today, however, the grasslands and heathlands so created are being affected by afforestation programmes. The impact of on-farm manage-ment on the characteristics of Russian grass-lands near Kursk under different management regimes is illustrated by the work of Mikhailova et al. 2000). Table 4.4 gives details of the botanical characteristics. It shows that at the plant species level, biodiversity is greatest where management is most intense, i.e. four years of grazing followed by one year fallow;

| | Native grassland | Grazed/hay field (grazed for four years followed by one year of rest) | Yearly cut grazed/hay field |
|---|---|---|---|
| No. of grasses | 8 | 5 | 11 |
| No. of legumes | 4 | 9 | 5 |
| No. of sedges | 1 | 2 | 1 |
| No. of forbs (herbs) | 28 | 61 | 51 |
| Total no. of families | 19 | 29 | 24 |
| Total no. of species | 41 | 87 | 68 |

**TABLE 4.4** The botanical characteristics of three Russian grasslands in the Kursk region under different management strategies. (Adapted from Mikhailova *et al.*, 2000)

yearly cut/grazed fields were more biodiverse than native grassland. The study shows that some management encourages an increase in biodiversity, especially in relation to herbs (forbs); in this case land use encourages a more biodiverse land cover. Additional data provided by Mikhailova *et al.* show that the level of management employed on these grasslands over the last 50 years has not impaired soil characteristics or forage quality. These results are significant in terms of conservation and sustainable land use. The cessation of grazing activity may also lead to land-cover alteration, as has been reported by Pettit and Froend (2001), who have monitored vegetation change in the Collie River catchment in the southwest of western Australia. Here, widespread clearance for pastoral and arable agriculture has occurred since 1900 (Beresford, 2001) leaving only remnants of the original vegetation cover of jarrah (*Eucalyptus marginata*) woodlands. Many of these remnants have also been used for livestock grazing, but where this is excluded Pettit and Froend have shown that the vegetation can recover to a state similar to that of ungrazed woodlands. The major difference between the grazed and ungrazed plots is the predominance of an understorey of annual herbs in the latter, and perennial shrubs and herbs in the former.

Pastoral activities are also the focus of financial subsidies in the EU, through the Common Agricultural Policy (CAP). Subsidies directed

at upland regions have encouraged increased stocking, especially of sheep, in many parts of the EU, where land cover has altered as a result. This is exemplified by changes that have occurred in parts of upland England where subsidies have been available under the England Rural Development Programme (ERDP) in less favoured areas. As explained by MAFF (Ministry of Agriculture, Fisheries and Food, UK, 2000) subsidies were made available in less favoured areas, notably the English uplands, to encourage the maintenance of rural life through financial support for agriculture and to facilitate land management. However, the original Hill Livestock Compensatory Allowance payments (HLCA) were based on production units, i.e. a subsidy per animal, and this has led to overstocking and resulting land-cover change.

As Routledge (1999) has remarked, combined CAP production subsidies could be as much as £38 per upland breeding ewe. He also states that 'Three [additional] ewes per hectare together with five lambs can attract subsidy and, with the sale of two lambs, produce a financial return greater than that of the ESA (Environmentally Sensitive Area) per hectare payment.' This latter is £145 per hectare for hay meadows in ESAs that have been designated to protect hay meadows and moorlands. Thus one subsidy is counteracted by another. The financial returns have also encouraged altered grazing practices; many sheep are no longer brought off the uplands in winter but are fed *in situ* with hay/silage from the lowlands. The overall result has been extensive overgrazing of both pastures and moorlands, and increased soil and peat erosion. It is likely that as much as 75 per cent of the hay meadows in the UK's Peak District National Park have disappeared or deteriorated since 1990. Populations of numerous birds have also been adversely affected as their habitats have declined or been impaired. MAFF (Ministry of Agriculture, Fisheries and Food, UK, 2000) anticipates that some of these problems may be overcome by switching to the Hill Farm Allowance (HFA) scheme, which offers subsidies based on hectarage rather than numbers of animals.

## 4.5 Arable agriculture and its alteration of land cover

Table 4.1 gives data on the amount of arable land in the developed world (see also Figures 4.1 and 4.2). It also shows that the dominant staple crop is wheat, followed by maize and potatoes. Other crops include roots, legumes, oil crops such as rape, vines, tobacco, and fibre crops such as cotton. The latter is the world's most important fibre crop and in the developed world the major producers are the USA and the Central Asian republics. Most of the arable crops grown in the developed world are grown in intensive agricultural systems that have high inputs of fossil fuels through the use of mechanization, crop-protection chemicals and artificial fertilizers. All occupy land that was once covered in natural vegetation; much of this would have been forests/woodlands or grasslands. As in the case of pasture, the amount of land characterized by this land use has varied spatially and temporally; the expansion, contraction and alteration of arable land use has resulted from socio-economic factors such as population change, market forces, political policies relating to agriculture and technological innovations.

In the European Union and the USA, for example, the last decade has been characterized by agricultural policies designed to reduce productivity. On both sides of the Atlantic such policies have been manifest as so-called set-aside; this involves the removal of land from agricultural production, and either allowing it to revert to natural or semi-natural vegetation or finding alternative land uses such as forestry and/or tourism facilities. Overall, such policies have resulted in a reduction in arable land and thus a change in land use and land cover. Table 4.5 gives data on the extent of change in the area of arable land in EU countries, including the UK, and the USA; it shows an overall decline in the extent of arable land, which is especially marked in the USA. Examples of land-cover alteration as a result of arable set-aside in the UK have been reported by Critchley and Fowbert (2000). Their survey of 50 arable farms, mainly in eastern England, involved sites on which there had been natural regeneration of vegetation and sites on which a cover had been sown; some sites were on land that had been set aside as much as nine years before, i.e. in 1989/90, when the policy was first introduced. The survey shows that perennial grasses predominated on all sites. In addition, the first and second years witnessed an increase in perennial species as annuals declined, as did overall species richness; thereafter species richness increased, including species that were not characteristic of the pre-existing arable condition. Critchley and Fowbert also show that species composition could be influenced if management involving cutting was introduced; this encouraged grass establishment where previously arable fields were sown; the species in the seed mixture predominated and thus helped to suppress species that were weeds of the former arable crop.

| Area of arable land ha × $10^3$ | | |
|---|---|---|
| | 1991 | 1995 | 1998 |
| EU | 71,136 | 68,751 | 69,229 |
| UK | 6554 | 5928 | 6267 |
| USA | 185,742 | 176,950 | 176,950 |

**TABLE 4.5** Changes in the extent of arable land in the EU, UK and USA (Food and Agriculture Organisation, 2001b)

| Year 1–2 | Ruderal weedy species predominant, e.g. *Polygonum aviculare* (knotgrass) and *Elytrigia repens* (quackgrass) |
| Year 3–5 | *Jasione montana*$_2$ (sheep's-bit scabious) and *Rumex acetosell* (sheep's sorrel) became dominant, and other perennials such as *Hieracium pilosella* (hawkweed) and *Deschampsia flexuosa* (wavy hair grass) colonized slowly |
| Year 6–12 | Colonization by ericaceous dwarf shrubs: *Calluna vulgaris* (ling heather) and, later, *Empetrum nigrum* (crowberry) colonize and increase as other perennials decline. |
| Year 12–22 | *Calluna vulgaris* (ling heather) and *Festuca ovina* (sheep's fescue) increased; today, the former predominates, representing the re-establishment of heathland |

**TABLE 4.6** The characteristics of succession on an abandoned arable field in Jutland, Denmark. (Adapted from Degn, 2001)

Degn (2001) has examined the course of vegetation succession on abandoned arable land in Jutland, Denmark, over a longer timescale than that referred to above for sites in eastern England. In the Jutland case, the land was abandoned in 1975 and was monitored until 1997. Table 4.6 gives details of the changes that occurred and their timing. It shows that, within six years of field abandonment, heathland species had colonized from nearby *Calluna-Empetrum* heath with *Calluna* predominating, i.e. developing a cover value of 60 per cent or more by 1991. This represents a reversion to the vegetation cover, albeit a human-induced heathland cover, which characterized the region prior to 1945 when it was customary to cultivate fields on a rotational basis and allow heathland fallow to return periodically.

A similar exercise to that described above has been documented by Hietala-Koivu (1999) for a 39-year period in the Yläne region of southwest Finland. Here, there have been a variety of landscape changes associated with post-war agricultural intensification. Alterations of the

ecological aspects of the landscape have occurred along with other attributes such as an increase in farm size and the area of farmyards. Some of Hietala-Koivu's data are given in Table 4.7, which shows that there has been a substantial decline in area of forest as cereal cultivation has increased since 1958; patches of forest within the agricultural area have also declined in extent. At the same time, the proportion of fields and field sizes have increased to accommodate mechanization and other aspects of agricultural intensification. This is reflected in the marked increase in the size of farmyards, which must now house machinery, increased grain storage facilities etc. Of particular note is the substantial decrease in open ditches resulting from their replacement with subsurface drainage. Hietala-Koivu remarks that the diversity of landscape units has decreased as overall homogeneity has increased.

In Australia, the replacement of natural vegetation communities by arable agriculture has been recorded in many regions over long and short timescales. For example, Hill *et al.* (2000)

| % of total area | 1958 | 1977 | 1993 | 1997 |
|---|---|---|---|---|
| Field size | 47.97 | 52.69 | 55.65 | 55.83 |
| Farmyard | 0.07 | 0.27 | 0.30 | 0.47 |
| Area of cleared forest | 0.22 | 0.33 | 0.46 | N/A |
| Forest patch within the area | 8.87 | 8.81 | 7.46 | 7.43 |
| Ditch margin | 4.15 | 1.43 | 0.65 | 0.54 |

**TABLE 4.7** Data on selected landscape elements in the Yläne region of southwest Finland. (Adapted from Hietala-Koivu, 1999)

used surveyors' plans from 1890, along with aerial photography from 1945 and 1991, to record land-cover alteration in the Mossman district of the coastal wet tropics of north Queensland. Here, rain forest has expanded since 1945. This has generally been attributed to the abandonment of firing practices by Aboriginal people but the historical documents show the recovery to be due to the decline of sugar cane cultivation, which began in the early 1900s. In another example, Fensham and Fairfax (1997) used land survey charts from the late 1800s and early 1900s to reconstruct the original vegetation of the Darling Downs in southeastern Queensland. The results show that the three major natural vegetation units, notably grasslands on alluvial plains, poplar gum (*Eucalyptus populnea*) woodlands on alluvial and terrace clay loam soils, and mountain coolibah (*Eucalyptus orgadophila*) woodlands on ridges and slopes (other than those of the Great Dividing Range), have all been altered considerably by agriculture since 1860. On a shorter timescale, between 1954 and 1989, Fisher and Harris (1999) recorded the loss of tree cover in an agricultural region near Bathurst, west of Sydney, New South Wales. Inevitably, forest cover declined substantially as agriculture and afforestation with pine (*Pinus radiata*) have increased. (This example is discussed in more detail in Section 4.6 because the major land use is mixed farming.)

Once arable agriculture replaces natural vegetation, it does not necessarily mean that the land use remains constant. Technology as well as economic and political factors will result in switches to different crops. This has occurred throughout history; such factors have also influenced the presence or absence of hedgerows, field size and shape, and farm size and shape. One example involves changing cropping practices in Ontario, Canada, in the 1900s. According to Smithers and Blay-Palmer (2001), soybean became a significant crop in the 1970s and by 1995 had replaced maize as the major arable crop; in 1970 some 350,000 acres (141,643 ha) were planted but by 1997 this had increased to 2,250,000 acres (910,563 ha). Smithers and Blay-Palmer attribute this increase to good market prices for soybeans

when compared with wheat or maize, and to various technological innovations to improve soybean productivity. These include the development of cold-tolerant species by introducing genetic material from Swedish soybean strains and genetic material from other 'exotic' species to improve plant establishment, vigour and disease resistance. Non-genetic applications include the improved timing of planting and crop rotation to contribute to pest control.

## 4.6 Mixed agriculture and its alteration of land cover

The conversion of areas of natural vegetation into agricultural land supporting mixed farming in the developed world has been as widespread as the conversion to pasture (Section 4.4) or arable land (Section 4.5). Mixed farming involves the integration of crop and livestock production; it comprises food and fodder crop production, the balance between which is highly variable. In the developed world, most mixed farming systems are characterized by a high technology input and, thus, a high fossil fuel input. In the Old World of Europe especially, modern agricultural systems, mixed or otherwise, have grown out of the mixed farming practices of the historic period.

In Britain, for example, mixed agriculture dominated the English lowlands in the post-Roman period (post-AD 410). According to Higham (1992), sheep were the most important livestock while wheat, barley and rye were the major cereals. Wheat was not grown as extensively as in Roman times, probably because the demand for bread and wheat for export declined, but barley production increased to produce grain-based alcohol. By the thirteenth century, Power and Campbell (1992) have suggested that agriculture in the demesnes of lowland England was characterized by eight basic types. As Table 4.8 shows, all eight had a livestock component. The subsequent rise to prominence of wool for export encouraged livestock farming, including its integration with arable cropping. This occurred in many parts of Europe, as discussed in Section 3.3.

| System type | Dominant crops | Dominant livestock | Location |
|---|---|---|---|
| **A** Intensive mixed farming | Wheat in preference to rye or barley, grown in rotation with legumes to supply nitrogen and animal fodder. | High proportion of livestock due to fodder production. Horses more important than oxen. Cattle used for dairying. Sheep and swine also kept. | East Norfolk, East Kent, area near Peterborough. All had access to markets via rivers and the coast. |
| **B** Light-land intensive | Rye and barley dominate. | Horses outnumbered oxen. Dairy cattle were prominent. Sheep also kept. | Norfolk and Suffolk, especially on light soils. |
| **C** Mixed-farming with cattle (less intensive than **A** or **B**) | Three-course system of cropping winter corn, spring corn (mainly wheat) and fallow; some rye, barley, oats. Legumes less important than in **A** or **B**. | Mostly working animals: oxen provided most power. Cattle were kept for dairying. | Warwickshire, south Somerset, Home Counties. |
| **D** Arable husbandry with swine (a minority type) | Legumes particularly important and were used as fodder for animals. | Horses more important than oxen as working animals. Swine very important. | East Midlands, East Anglia, southeast UK around areas with woodland. |
| **E** Sheep-corn husbandry | Pasture very important. Assorted patterns of cropping: oats, barley, legumes. | Oxen used for draught. Sheep dominate livestock. | Not locally or regionally dominant. Access to grassland was essential. |
| **F** Extensive mixed-farming (see **A** and **C**) | Wheat, oats, some barley and legumes. | High stocking densities but cattle/sheep were most important. | Vale of York to Somerset. |
| **G** Extensive arable husbandry | Arable dominated: wheat, oats, little barley and legumes grown extensively. | Very few working animals. Oxen used for draught. | Geographically most widespread, especially in north, west and southwest England. |
| **H** Oats and cattle (smallest number of demesnes) | Oats dominated; some rye and barley, but few legumes and no wheat. | Cattle and oxen dominated livestock; few horses, sheep and swine. | Near major urban centres; areas where other crops were unsuitable. |

**TABLE 4.8** Demesne-farming systems in England, 1250–1349. (Adapted from Power and Campbell, 1992)

Today, mixed farming concerns are experiencing set-aside in the same way as their arable counterparts (Section 4.5). The impact of set-aside policy on a range of mixed farms in western England has been examined by Critchley and Fowbert (2000). Their data on changes in botanical composition following set-aside show that plant communities similar to permanent grassland predominated, with far fewer of the herbs that are associated with arable fields. This contrasts with the results for arable regions discussed in Section 4.5. Overall, set-aside increases the biodiversity of a previously farmed area even if only rotational or short-term rather than long-term set-aside is practised.

Farm abandonment has a similar affect on land cover as set-aside in the context of biodiversity and landscape homogeneity. This is reflected in the patterns of land-use/ land-cover change that have occurred in two contrasting agricultural regions in Norway (Fjellstad and Dramstad, 1999) in the last 50 years. A comparison of trends in a mixed farming region in Østfold county in lowland southeast Norway with those from the mountainous region of Telemark in the south-west have shown that agricultural intensifica-tion has taken place in the former whilst land abandonment has characterized the latter. Homogenization has occurred in Østfold to accommodate intensification; increased mecha-nization, for example, has caused a loss of field boundaries, woodland ditches, fences etc. and a reduction in pasture as arable cultivation has expanded. In contrast, the tree cover and numbers of buildings, mainly dwellings, has increased in Telemark as tree populations have expanded from formerly isolated woodland patches and as people have chosen to live in this region following road improvement.

Similar processes have characterized the region of Alentejo in southern Portugal. Here the Montado is the name given to the tradition-ally dominant agro-silvo pastoral system, which has high biological, scenic and recre-ational value. According to Pinto-Correia (2000), various processes have conspired to alter the region sufficiently to jeopardize the integrity of the Montado since the 1970s. Amongst the underpinning causes have been a decline in crop prices for cereals, so that their cultivation in the region declined, and rural depopulation reducing the labour force for traditional management tasks such as the hand-clearance of shrubs. As a result the intensity of land use declined and in extreme cases aban-donment occurred in a landscape already faced with innovation as a result of the spread of African swine fever, which caused the demise of traditional pig-raising in woodlands. The subsequent shifts to alternative livestock production, and increased cork production from trees caused sufficient disequilibrium to alter the landscape and the *genre de vie*. As Pinto-Correia points out, European-funded aid schemes have helped to maintain the Montado in some parts of southern Portugal but she

suggests that only a management approach based on the holism of landscape ecology and its maintenance will restore the Montado in the twenty-first century. The traditional agricul-tural system of Castile-La Mancho, in central Spain, has experienced similar problems. According to Caballero (1999), the once-inte-grated cereal production and sheep farming system is threatened for a variety of socio-economic reasons, including the perceived unattractiveness to young people of harsh working conditions.

It is also likely that many other parts of Europe will experience land-cover change in the next few decades as a result of EU policies to encourage diversification and to reduce productivity. This likelihood has been exam-ined by Lütz and Bastian (2002) in relation to Saxony, Germany. They suggest that diversifi-cation etc., involving hedgerow planting and the establishment of herb-rich field margins, could be achieved by withdrawing *c.* 6 per cent of the area under agriculture without any loss of farm income. Political intervention such as this is, thus, a significant cause of land-cover change. Its importance is illustrated by trends in eastern European countries in the last 10–12 years since the demise of controls on agricul-ture within the Former Soviet Union. All types of agriculture have been affected by this polit-ical shift to a free-market economy and the associated changes in land tenure. The processes that have occurred in Hungary are discussed by Burger (2001), and those in Germany by Mathijs and Swinnen (2001). Overall, the decline of collective large-scale farms, as land was re-allocated to individuals and families, led to reduced productivity. This was partly because small farms are less amenable to mechanization, partly because some land was simply abandoned by new owners with no farming tradition, and partly a result of the decline in collective marketing, transport etc.

In Australia, the rapid growth in population and spread of settlement in the area of Sydney in the early nineteenth century led to the trans-formation of the central tablelands and espe-cially of its plains and woodlands. Further intensification occurred after 1850 in response to the discovery of gold, which attracted people into Sydney's hinterland and increased demand for crop and grazing land. Fisher and Harris (1999), quoting Goldney and Bowie (1990), suggest that by 1901 some 25 per cent of the tablelands had been cleared; following a period of relative stability, clearance increased from the 1950s so that by 1981 some 50 per cent of the area had been cleared. The changes that occurred in this 30-year period have been captured in a series of aerial photographs. Analysis of these photographs has shown that grazing and mixed farming were the major cause of tree loss in the west of the region, whereas afforestation and the creation of nature reserves were the major factors in the east. Similarly extensive clearance has occurred in other parts of Australia, not all of it is due to agriculture as logging and mining have also exacted significant tolls (see Chapters 6 and 7).

## 4.7 Conclusion

The wide-ranging examples given above attest to the dynamic nature of the agricultural systems of the developed world. Pastoral, arable and mixed farming systems are equally dynamic. Many have long histories that go back to prehistoric times, whilst those of the Americas and Australia have been profoundly influenced by European practices. All of the agricultural systems referred to above are constrained by the characteristics of the physical environment in which they take place; all are equally affected by socio-economic factors such as market forces, population change and standards of living. Other significant influences include scientific and technological innovations such as mechanization. Governmental influence has also been significant. In Europe, for example, the CAP has had a profound influence on agriculture and, in turn, on rural landscapes in both uplands and lowlands.

On a temporal basis, the major trends in agriculture in the developed world in the post-1750 period have involved the replacement of natural ecosystems with agricultural/rural landscapes that are relatively diverse. These landscapes comprise three major elements: remaining natural or semi-natural habitats; the agriculture itself, which has often involved integrated crop and livestock elements; and the settlement patterns. Much diversity has, however, been lost in the post-Second World War period as agriculture has undergone intensification. Traditional farming practices have been subdued, or even abandoned, so that the *genres de vie* have become increasingly less viable. In the last decade, overproduction in parts of the developed world has been recognized and appropriate changes to agricultural policies made (e.g. set-aside policies). In Europe this has resulted in different processes in upland and/or marginal areas as compared with lowland areas. In general, the former have experienced extensification with some land abandonment, whilst the latter have experienced intensification and increased productivity. Both biodiversity and landscape diversity have declined as a result of this intensification, but they have increased in regions where extensification has occurred. These processes will continue in the coming decades.

## Further reading

**Brouwer, F.** (ed.) 2000: *CAP Regimes and the European Countryside*. CABI, Wallingford.

**Buller, H. and Hoggart, K.** (eds) 2001: *Agricultural Transformation, Food and Environment*. Ashgate, London.

**Krönert, R., Boudry, J., Bowler, I.R. and Reenberg, A.** (eds) 1999: *Land-use Changes and their Environmental Impact in Rural Areas in Europe*. Parthenon, New York.

**OOPEC** 1999: *Agriculture, Environment, Rural Development: Facts and Figures: A Challenge for Agriculture*. OOPEC, Luxembourg.

**Squires, V.R.** (honorary theme ed.) 2002: *The Role of Food. Agriculture, Forestry and Fisheries and the Use of Natural Resources. Encyclopaedia of Life Support Systems*, Vol. 5. UNESCO, Paris.

# Land use related to agriculture: the developing world

## 5.1 Introduction

Agriculture in the developing world – mainly countries in low latitudes – shares many features in common with that of the developed world. In both cases agriculture is constrained by the conditions imposed by the physical environment and at the same time it is a major cause of change in those conditions, especially in relation to land cover. In the past, the agricultural systems of the developing world have been influenced by several centres of innovation, notably the Near East, Far East, Mesoamerica and Sub-Saharan African. In many countries historical events led to the superimposition of European agricultural systems on traditional systems with alterations to crop and animal complexes. The developing and developed worlds are also characterized by a wide variety of agricultural systems, all of which are dynamic, especially in relation to driving forces that are mainly socio-economic factors. Specifically, however, the major factors that affect agricultural systems in the developing world are quite different from those playing a

role in the agricultural systems of the developed world. One of the most important of these is the rate of population growth. In general, higher rates of population growth prevail in the developing world; in most cases, annual growth rates of between 1 per cent, e.g. in Indonesia and Thailand, and 3 per cent, e.g. in Congo and Ethiopia, prevail. Inevitably, such rapid growth rates bring immense pressure to bear on a nation's agricultural systems. Coupled with other immediacies, such as the need to generate income through the export of agricultural produce, especially in nations where the only other possibilities for income generation involve the exploitation of primary resources, agriculture is not only an economic mainstay but also a significant cause of land-cover alteration.

As these pressures continue to intensify in developing countries, agriculture will remain a major agent of land-cover change. Most importantly, the spread of agriculture currently and in the next decade will reduce still further the remaining areas of natural ecosystems. The intensification of production in existing agricultural systems will also alter land use and impair existing natural ecosystems (see comments in Pingali, 2001). Mechanization, the increased use of crop-protection chemicals, artificial fertilizer use and the spread of genetically modified crops will all contribute to the changing patterns of land use. Similarly, the existence of markets, especially in the developed world, for specific products encourages land-use change. For example, in several South American and African countries there has been a major increase in the production of flowers for the growing European and American markets. Plantation production of tea, coffee, cocoa and rubber is also driven to a large extent by requirements in the developed world.

## 5.2 Types of agriculture and their distribution

Figure 5.1 shows the agricultural systems that predominate in the developing world. This is a generalized depiction of a more heterogeneous reality; in particular it does not afford scope to include the myriad forms of shifting cultivation that occur in the developing world and that are mostly absent from the developed world. In

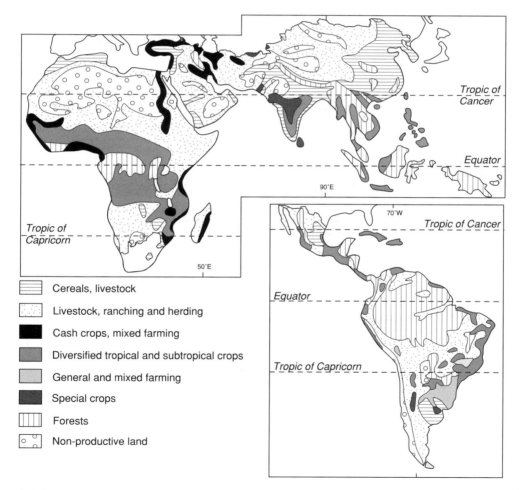

**FIGURE 5.1** Agricultural systems of the developing world. (Adapted from the *Oxford Hammond Atlas of the World*, 1993)

addition, the developing world is characterized by nomadic herding in arid and semi-arid regions and, together with shifting cultivation, these are transitory rather than permanent systems, which influence land cover in varying degrees of intensity. These agricultural systems are also low-input systems and, as discussed in Sections 5.3 and 5.4, represent ways of manipulating environments that are unsuitable for permanent crop or livestock production because of environmental limitations.

A possible classification of agricultural systems is given in Figure 5.2. It shows the significance of four major factors: crop type, production type, crop numbers and energy inputs. All are constrained by the physical environment and all are influenced by socio-economic and political factors. All of these types of agriculture exist in both the developed and developing worlds. As well as the relative importance of transitory systems in the latter in comparison with the former, another major difference concerns inputs. In general, the agricultural systems of the developing world are characterized by fewer inputs and limited technology. Thus, artificial fertilizers and crop-protection chemicals, for example, are less important in the developing world than in the developed world; this also means that less fossil fuel is used in agriculture. The production systems in the developing world that bear the greatest similarity to those of the developed world are those based on plantations. These production systems have completely replaced the pre-existing natural ecosystems; they are monocultural and enjoy a high level of inputs, especially crop-protection chemicals and artificial fertilizers, whilst often still maintaining a high labour requirement because of its low cost.

Another significant difference between the agricultural systems of the developing world and those of the developed world relates to production type (see Figure 5.2). In the former there is a much greater extent of subsistence agriculture than in the latter; this includes agricultural systems that are permanent arable, pastoral and mixed, and not just shifting cultivation or nomadic herding. Agriculture for local consumption is often quite different from that geared towards export markets, even

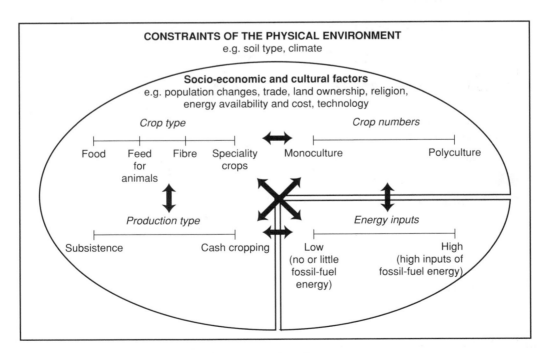

**FIGURE 5.2** A possible scheme for classifying agricultural systems

where both are practised in the same area. In Peru, for example, the production of asparagus for export is quite different, and obviously a more commercial activity than maize production for local consumption. Similarly, in Zimbabwe, mange-tout are grown intensively and commercially for exports to Europe whilst maize (*mealie*) is a subsistence crop.

All of the agricultural systems referred to in Figure 5.2 have contributed to land-cover change. In view of projected population growth rates (see Chapter 10) in many developing countries and other socio-economic factors, the expansion and intensification of agricultural systems will make a major contribution to land-cover change in the future. Such factors, along with developments in technology, will undoubtedly contribute to land-use change as well.

## 5.3 Nomadic pastoralism and its alteration of land cover

Nomadic pastoralism in the developing world occurs mostly in the arid and semi-arid regions of Asia and Africa; it has a long history, extending back at least $7 \times 10^3$ years BP (Smith, 1992) and probably grew out of hunting strategies. In essence, this is an agricultural system that manipulates a naturally occurring land cover rather than one that alters it substantially. With injudicious management, however, especially in relation to carrying capacity, major alterations of land cover may ensue; apart from changes in the species composition of rangelands, desertification may occur, which not only impairs productivity but may also lead to land abandonment. As in the cases of reindeer and saiga herding (see Section 4.3), animals are moved from range to range to take advantage of seasonally produced forage. The animals herded are mainly cattle and camels, with lesser numbers of goats and sheep. These nomadic pastoralist systems rely on mobility, flexibility and good management in relation to range use and carrying capacity; these factors in turn, relate primarily to rainfall distribution since this is the major control on forage production. Consequently, there are inherent land-cover alterations within a year: forage growth following rain; a reduction in forage plants by grazing; and possibly a relatively 'dormant' period before the rains recommence. Whilst the primary (i.e. meat and hides) and secondary (i.e. milk and blood) products are important for both direct consumption and bartering, nomadic pastoralists may relate to crop farmers; sometimes the relationship may be symbiotic, yet at other times it may be antagonistic. This is also a significant stimulus to land-cover and/or land-use change.

There are many different types of nomadic pastoralism in the developing world, as the examples given below reflect. However, all have features in common, notably land ownership, which is collective. This provides essential access to wide areas where there are varying degrees of predictability and reliability in terms of rainfall amount and distribution, and pasture quality and extent. Judicious use of these ranges underpins the sustainability of such systems. In contrast, herd ownership is usually family or individually based. Herd composition is also an important factor in determining land cover because individual animal species are selective ruminants; for example, goats and camels browse leaves and thorny shrubs, whilst cattle and sheep mainly graze grass. There is some evidence that herd composition is determined by the gender of its owner. Turner (1999) has presented data showing that in western Nigeria, and probably the Sahel generally, there has been a shift towards small stock (i.e. sheep and goats) and away from large stock (i.e. cattle and camels). The chief cause of this shift has not been changes in price or livestock productivity but changes in livestock ownership, notably a shift towards smallholders and women. This has occurred because of regional problems such as drought that have threatened subsistence *genres de vie* and increased vulnerability at the household level. This, in turn, has generated local changes because households reflect conjugal and gender roles and relationships, which are also influenced by social factors such as division of labour, level of wealth and religious conventions. For all members of society, livestock represent assets that also play significant roles in society, notably status in terms of wealth, social status as reflected in 'bridewealth' or loan capacity, and the provision of existentialities such as milk, meat, fibre, animal labour and manure. Sheep and goats tend to be the provenance of the poor whilst cattle are owned by the wealthier members of society. According to Turner (1999), national data for cattle, sheep and goat populations for the last 25 years in Senegal, Mali, Burkina Faso, Chad and Niger show substantial declines in the former and increases in the latter two. This has also been a period of recurrent drought. This affects cattle, through high mortality rates, more than it affects sheep and goats, and causes the displacement of cattle-rearing peoples. Persistent drought exacerbates these developments. Thus, as wealth declines, sheep and goats become increasingly prevalent.

The shift towards small livestock is also likely to be a response to religious tenets. For example, Islamic law requires that it is the husband's responsibility to support the family and so it is his stock, mainly cattle, that should be sold first in times of need. In addition, men who leave the home to work elsewhere often have disaffected wives; divorce rates are also high, leaving economically vulnerable females, often with children to rear. Herds of goats and sheep thus provide a valuable resource for women. Consequently, many factors, including complex social relationships, contribute to stock changes that, in turn, alter patterns of land use. Turner states that:

The shift also affects patterns of land use in the Sahel, for there are significant inter-species differences in long-distant mobility, daily grazing radii, grazing/browsing preferences, watering frequency, management demands and manure quality. As a result, a shift in species composition in livestock will affect the pattern of grazing pressure at the landscape, village, patch, and individual plant scales . . .

Complex interactions between socio-economic factors and environmental characteristics are also responsible for land-cover change in the northern Ivory Coast in the last 45 years. Bassett and Koli Bi (1999) point out that, contrary to overall perceptions, this region has experienced increased savanna woodland growth rather than deforestation and/or desertification. Based on field and socio-economic data from two regions, Tagbanga and Katiali, Bassett and Koli Bi have shown that there have been significant shifts in the agricultural systems, with associated land-cover changes that are evident in plant communities. The region is characterized by nomadic cattle herding as well as sedentary cattle rearing and crop agriculture, all of which have been affected by socio-economic factors. In particular, there has been a major increase in the number of grazing animals, especially in the 1970s and 1980s. The prime cause of this has been the movement of transhumant Fulbe

| Niofouin Subprefecture | | | Katiali Village | | |
|---|---|---|---|---|---|
| 1985 | 11,939 | Transhumant cattle | 1982 | 3310 | Fulbe cattle |
| 1990 | 55,618 | Transhumant cattle | 1992 | 7033 | Fulbe cattle |
| 1985 | 13,315 | Sedentary cattle | 1982 | 578 | Peasant-owned sedentary cattle |
| 1990 | 18,484 | Sedentary cattle | 1992 | 1417 | Peasant-owned sedentary cattle |
| 1985 | 15 km² | km² cattle density | 1975 | 4 | Oxen |
| 1990 | 45 km² | km² cattle density | 1992 | 381 | Oxen |

**TABLE 5.1** Data on cattle populations in Katiali village and Niofouin Subprefecture, Ivory Coast. (Adapted from data quoted in Bassett and Koli Bi, 1999)

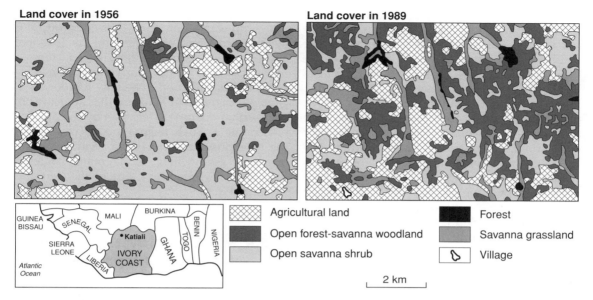

**FIGURE 5.3** Land-cover changes between 1956 and 1989 around Katiali village, Niofouin subprefecture, Ivory Coast. (Adapted from Bassett and Koli Bi, 1999)

pastoralists from Mali and Burkina Faso into the Ivory Coast because of conflicts with crop farmers who sought compensation for crop damage. The magnitude of this change in cattle density in the village of Katiali and its subprefecture of Niofouin is reflected in Table 5.1. In the region overall there has been a threefold increase in cattle populations in just five years, but numbers of transhumant cattle have increased fivefold.

Nevertheless, field-data collection, aerial photographs and GIS mapping all indicate that tree/woodland cover has increased rather than decreased in the wake of this change. Interviews with local people confirm this development. The results of a comparison between conditions in 1956 (a year for which aerial photographs are available) and 1989 are given in Figure 5.3. The most significant change is the increase in savanna woodland at the expense of open savanna types; the former increased from 4 per cent to 31 per cent, whilst open savanna types declined from 46 per cent to 12 per cent. Figure 5.3 shows that cropland has also increased. The reasons for the increase

in savanna woodland are changes in grazing pressure and fire regimes. Both influence the species composition of savanna vegetation; heavy grazing and early fires have discouraged the growth of palatable grass species, and trees and shrubs have invaded. Overall, the value of the savanna for livestock grazing has declined. Such findings, as Bassett and Koli Bi point out, have major implications for development plans that focus on livestock rearing.

Competition for land, and the conflicts that ensue, also generates land-use change. This issue has been explored by Campbell *et al.* (2000) whose work is based on the developments in the Kajiado District in the foothills of Mount Kilimanjaro, Kenya, over the past 30 years. Here, conflicts of interest arise between herding, farming and wildlife; the latter involves the creation of National Parks, a major objective of which is the promotion of tourism, whilst there has been an expansion of rain-fed agriculture since the 1930s. As in the studies referred to above, there are numerous stimuli that have caused these conflicts, including the role of water resources, and have thus generated land-use change. Between 1977 and 1996 there was an increase from 8 per cent to 24 per cent in the number of herders reporting conflicts with farmers. This is partly because the arable area has expanded and because increasing numbers of herders have become involved with cultivation as well. Herders in particular were disadvantaged because land adjacent to water courses has been favoured for cultivation. This restricts access by cattle herds to water and results in increased damage to crops. The situation is exacerbated by high rates of population increase in both herding and farming communities as well as migration into the region. Similarly, there has been an increase in the number of farmers complaining about herders, mostly because of damage to crops, but including the theft of crops. Reports of conflicts with wildlife have also increased; both farmers and herders reported competition from wildlife, including crop damage and kills of domesticated animals.

The various factors that underpin these conflicts and that have also underpinned land-use change are given in Table 5.2. This shows the significance of three major groups of factors: economic; political and institutional; and social and cultural. All of these factors are interrelated, and their relative importance has changed spatially and especially temporally to generate patterns of land use that alter year by year within the broad limits set by the physical environment. Such factors are, of course, not the sole province of Kajiado District but are operative everywhere, to a greater or lesser degree, as is evident in many of the examples presented in this book.

## 5.4 Shifting cultivation and its alteration of land cover

Shifting cultivation involves the clearance of natural or semi-natural vegetation from an area of land so that crops can be produced. After a period of cultivation that may last from between two to five years, the land is

| Driving forces | |
|---|---|
| **Economic** | Increase in international tourism |
| | Increase in rain-fed cultivation |
| | Increase in irrigated horticulture for urban and external markets |
| | Diversification of many herders into farming to reduce the impact of drought |
| **Political and institutional** | Conservation of biodiversity through global initiatives |
| | National initiatives for arid and semi-arid lands |
| | Structural Adjustment Programmes, e.g. privatization, promotion of exports |
| | Local institutional support for these national and international policies etc. |
| | Legal enforcement of policies etc., which reflects local and national power relations |
| | Land tenure and leasing arrangements and their enforcement |
| **Social and cultural** | Colonial influence and policies |
| | Continuance of colonial policies and attitudes |
| | Population increase |
| | Immigration also increases population pressure |
| | Changes in ethnicity |
| | Land rights and legal title to land |
| | Changes in age-related decision-making processes |
| | Gender relationships |

**TABLE 5.2** The major stimuli that contribute to land-use conflicts and land-use change in Kajiado District, Kenya. (Adapted from Campbell *et al.*, 2000)

abandoned because of declining fertility/productivity; farmers move on to cultivate a new area of natural or semi-natural vegetation. Shifting cultivation is today mostly confined to the tropics and subtropics and is considered to be a major cause of land-cover change. According to the World Commission on Forests and Sustainable Development (1999) there are between 200 and 300 million people who are landless shifting cultivators and who rely heavily on forest regions, mostly for the production of staple crops at subsistence level. The clearance required for shifting cultivation not only influences biodiversity but also affects the global carbon cycle (Section 1.5) as carbon is released from the biomass and from the organic matter in the litter and soil.

Table 5.3 gives data on tropical and subtropical forests; it shows that the greatest demise of natural forest cover has occurred in Asia and Africa, whilst the forests of Oceania and Latin America have been least, but still substantially, disturbed. The World Commission on Forests and Sustainable Development (1999) also reports that current rates of deforestation remain high in all of these regions. Shifting cultivation is considered to be a major cause of this deforestation; it converts the forest to land use that may involve the production of between two and five major crops, and as many as fifty crops in total. Some of these will be used for medicinal and construction purposes, or for fibre. Once cultivation has ceased, and if the fallow period is sufficiently long, secondary forest will recolonize the abandoned fields. If the fallow period is short, forest will not regrow and a shrub cover will develop. Thus the entire cycle of forest clearance back to forest recovery involves a dynamic land cover, as the examples given below illustrate.

In Honduras, Central America, deforestation is occurring at a rate of *c.* 1.7 per cent per year (quoted in Kammerbauer and Ardon, 1999), mostly due to an expanding rural population that is engaged in agriculture. Moreover, because much of Honduras is mountainous, with 75 per cent of the land with slopes greater that 25 per cent, and because of land-tenure inequity, most of these subsistent farmers exploit hillsides and small valleys, whilst the plains farmers engage in commercial production. Kammerbauer and Ardon (1999) have examined the land-cover/land-use changes that have characterized the La Lima watershed, part of the Yeguare River basin, in Central Honduras using aerial photographs and GIS for 1955, 1975 and 1995; these changes are similar to those elsewhere in the Yeguare basin as well as in Honduras and in Central America in general.

Figure 5.4 gives details of the alterations that have occurred over this 40-year period. The most obvious difference is the substantial reduction in forest cover and its replacement with agricultural land use. Not all of this land is

| | Percentage of potentially forested land with forest cover remaining | Percentage of potentially forested land with primary forest | % Rate of deforestation |
|---|---|---|---|
| Asia | 32 | 6 | 0.6–1.0 |
| Oceania | 60 | 22 | 0.6–1.0 |
| Africa | 50 | 8 | 0.6–1.0 |
| Latin America Caribbean | 62 | 40 | 0.6–1.0 but >1.0 in Central America and Caribbean |

**TABLE 5.3** Data on tropical and subtropical forests. (Adapted from World Commission on Forests and Sustainable Development, 1999)

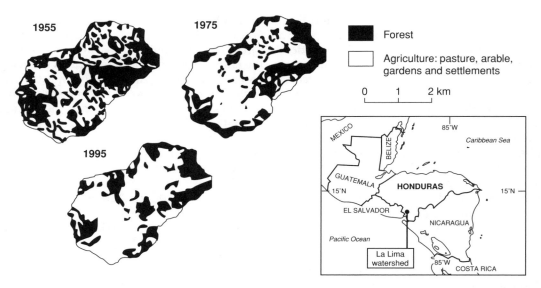

**FIGURE 5.4** Land-cover change between 1955 and 1995 in the La Lima watershed of central Honduras, Central America. Reprinted from *Agriculture, Ecosystems and Environment,* Vol 75, Kammerbauer and Ardon, Land use dynamics and landscape change pattern, 93–100 © 1999, with permission from Elsevier Science.

given over to shifting cultivation and although this is the traditional land use in the region, it has also undergone transformation. Subsistent maize production has intensified and agricultural activities have diversified. Today, the La Lima community comprises 62 family units in a nucleated settlement and a further 119 family units dispersed in the surrounding area. Maize production for local consumption remains the major activity but the study by Kammerbauer and Ardon (1999) shows that *c.* 60 per cent of the agricultural plots, most between 0.1 and 1.5 ha, are now permanently cultivated. Other crops, some for sale in the nearby town of Tegucigalpa, include beans, potatoes, tomatoes, onions and garlic. This activity has been prompted by the construction of a road in 1985 linking La Lima with Tegucigalpa. Between 1955 and 1975 there was a 25 per cent reduction in total forest area, partly caused by the immigration of farmers from outside La Lima; this resulted in a deforestation rate of 1.2 per cent per year. After 1975 this declined to 0.6 per cent per year as agriculture intensified and diversified through the use of pesticides, artificial fertilizers and improved irrigation.

Kammerbauer and Ardon also point out that, not only was the forest cover depleted but the remaining forest was also altered. In particular, forest density has declined due, mainly, to forest exploitation for fuelwood for domestic use (some 80 per cent of family units require fuelwood for cooking).

The conditions referred to above reflect general changes in the La Lima catchment as a whole but Kammerbauer and Ardon also give details of local variations. For example, areas once cultivated and then abandoned as part of the shifting cultivation cycle remain as shrub-dominated plots with degraded soils. Still used, these plots have varied land-cover characteristics, depending on the stage of succession, i.e. various grasses, a shrub vegetation and secondary forest (see Paniagua *et al.,* 1999, for details). Most remaining forest is on land with steep angles of slope but even on land with slope angles of 50 per cent maize-cropping plots can be found despite the problems that exist with soil erosion and landslides. Nevertheless, it is likely that slope angle will be a major factor contributing to the preservation of the remaining forest.

| | Time after abandonment | | | |
| | 1 month | 4 months | 1 year | 3 years |
| --- | --- | --- | --- | --- |
| Annual herbs | Galinsoga parviflora | Galinsoga parviflora Crassocephalium crepidiodies | Conyza sumatrensis Crassocephalium crepidiodies | |
| Perennial grasses | | | | Imperata cylindrica |
| Shrubs | | | | Eupatorium odoratum Trema orientalis |

**TABLE 5.4** The dominant plant species in abandoned agricultural plots near Kundasan, Sabah, north-east Borneo. (Adapted from Ohtsuka, 1999)

All areas in which shifting cultivation is practised experience land-cover change when cleared plots are abandoned. The rate of vegetation recovery will depend on length of abandonment and the sources from which plant species can invade. An example of vegetation regrowth has been presented by Ohtsuka (1999) who has documented community structure and floristic change after plot abandonment near the city of Kundasan, in Sabah, northeast Borneo. The sites examined had been abandoned at different times and thus represent different stages in the succession. The results are summarized in Table 5.4. In the first year, three stages of succession can be identified on the basis of distinct plant communities comprising short-lived annual herbs. The three species referred to in Table 5.4 are all annual species (therophytes); they produce small seeds that are wind dispersed. According to Ohtsuka, species with these characteristics dominate the early stages of succession in areas where shifting cultivation is practised elsewhere in the tropics. Within three years, perennial grass and shrub species had begun to invade along with pioneer tree species, with secondary forest beginning to emerge some ten years after abandonment.

Secondary forest growth is very important in so far as it provides services and goods similar to primary forest. Most importantly, it is a bigger store of carbon than either the arable crops produced when the land is cultivated or the successional stages that preceded it. Its other major roles include the recovery of biogeochemical cycles, especially the link between the soil and vegetation, protection for the soil and the provision of fruits, nuts, spices etc., which can generate income. In some regions where shifting cultivation is practised, secondary forest may be extensive. For example, Smith *et al.* (1999) have shown that in the Peruvian Amazon, areas that have been settled may comprise *c.* 33 per cent secondary forest whilst the remaining land is under cultivation. As agriculture becomes increasingly sedentary, however, the extent of forest cover, either primary or secondary, declines. This is illustrated by Imbernon (1999) who has documented land-cover/land-use change in North Lampung, Sumatra. Here, maps and satellite imagery show that in 1930, when shifting cultivation was dominant, forest cover was extensive at 77.5 per cent. Between 1969 and 1985 the forest area was reduced to only 10 per cent as a transmigration programme was initiated and forest clearance ensued to accommodate small farms for settlers and agro-industrial plantations. By 1996 the remaining forest had disappeared, leaving an agricultural landscape of small and large holdings. A trend towards sedenterization has also developed in Indonesia's other outer islands, where there has been a gradual move away from shifting cultivation for food production to tree-crop production. According to Sunderlin *et al.* (2001), this process has altered recently in response to Asia's economic recession, which has meant that recouped funds from exported food crops could not offset increased production costs. To compensate, and to avert future financial crises, small farmers have shifted their emphasis from food crops for export to forest crops for export. These include not only wood and rattan but also rubber, forest clearance for which has occurred. Further deforestation has occurred because some farmers wanted to establish tenure rights and so have land available for future enterprise. A similar process of change has occurred in Acre, Brazilian Amazonia. According to Peralta and Mather (2000), Landsat satellite imagery for 1975, 1985 and 1989 shows that, prior to 1975, small plots of forest were cleared by family groups involved in natural resource extraction, e.g. rubber tapping. Thereafter the patches cleared grew larger with some coalescence, so that by 1989 the area was characterized by a settled agricultural economy.

Another example of land-cover alteration due to shifting cultivation is that of the Mikea Forest area of southwest Madagascar. According to Seddon *et al.* (2000), this is a tropical dry forest with high plant and animal biodiversity but little protection in terms of conservation. Satellite imagery has facilitated the calculation of rates of deforestation, which increased from 0.35 per cent per year in the period 1962–94 to 0.93 per cent per year between 1994 and 1999. Seddon *et al.* indicate that the major cause of deforestation is *hatsake*, a form of slash-and-burn maize cultivation practised by the Mikea and Masikoro people, who began this activity in the 1950s. Recent immigrants from the south, the Tandroy people, also practise *hatsake*. They, however, cultivate plots of *c.* 10 ha as compared with plots of 1–3 ha used by the Mikea and Masikoro people. Maize production is for local consumption, though some is exported to mainland Africa and the nearby islands of Mauritius, the Seychelles and Réunion. Since 1990 the demand for maize has intensified, and commercial activities have increased. This, together with increasing demand for charcoal and wood for domestic use, has resulted in accelerated rates of deforestation.

## 5.5 Ranching/pastoral agriculture and its alteration of land cover

Pastoral agriculture is especially important in the world's semi-arid regions where rangelands comprise grassland and savanna vegetation, and in cool desert or semi-desert such as Patagonia in southern South America. In many regions this type of agricultural activity is associated with a nomadic lifestyle, as discussed in Section 5.3. However, many agropastoral systems are associated with permanently settled groups, though in times of crisis (such as prolonged drought) the distinction between the two is unclear as settled agropastoralists adopt a nomadic lifestyle to survive. It is also important to note that pastoral activities have become increasingly important in the humid tropics in the past 40 years or so. Attempts to develop specific regions, notably in Central and South America, have resulted in the establishment of US-style ranching. Along with shifting cultivation (Section 5.4), this is considered to be a major cause of deforestation. Elsewhere, the rearing of cattle for milk has become important around large cities in the developing world where there is a growing demand for milk and dairy products. Similarly, as individual wealth has grown in those nations that are industrializing, notably in Southeast Asia and China, the demand for dairy products and meat has stimulated pastoral agriculture. All of these developments, but especially those involving deforestation and pasture/rangeland degradation, affect the global carbon cycle by releasing carbon from above-ground biomass. As the examples given below illustrate, the conversion of forest or savanna to pasture represents a major shift from land cover to land use, but less overt alterations can occur in pastures such as replanting, fertilizer use, firing and conservation measures etc. Changes may also occur in response to land tenure or lack of it, and government policies.

The latter is illustrated by the situation in Botswana where, since independence in 1966, ranching has been encouraged in the Kalahari region under the auspices of private rather than communal enterprise. According to Dougill *et al.* (1999) the two major policies aimed at expanding the livestock industry are the Livestock Development Project (LDPI) of 1970, which was incorporated into the second, broader, Tribal Grazing Lands Policy (TGLP) of 1975. This remains in operation today but was reinforced by the 1991 National Policy on Agricultural Development, which continues to advocate the expansion of privatized ranches. Dougill *et al.* have examined the environmental/land-cover changes that have occurred as a result of these policies in Botswana's Kalahari region. Field data were collected from the Makoba Ranches that were first established under the TGLP in 1975 and botanical surveys reflect significant changes in vegetation communities, which are linked to grazing patterns. These patterns are, in turn, associated with water provision from the boreholes. In essence, the higher the grazing density, the greater the cover of bush species over grass species (i.e. the bush:grass ratio increases with increasing grazing intensity). Thus, ranches on which only one borehole was provided per 8 km² have limited bush encroachment and have maintained extensive grass-dominated savanna. In contrast, where further boreholes have been provided to increase grazing capacity, bush encroachment has been more intensive. According to Dougill *et al.*, this bush encroachment has become sufficiently significant to be apparent on satellite imagery. Whilst it constitutes a change in land cover, bush encroachment need not, however, cause degradation in the longer term, possibly because of the preference of certain palatable grass species for habitats beneath low-growing *Acacia* spp. In so far as these grass species are protected from grazing by the acacias, the seeds they produce become incorporated in the soil seed bank so that when disturbance of bush through grazing

and/or burning occurs, a grass cover of palatable species can regenerate. This case study not only reflects the stimulus of ecological change through water provision and resulting increased stocking, but also illustrates the complex interrelationships between savanna species. Overall, the TGLP has not improved the lot of indigenous peoples nor generated a vastly improved livestock industry, but neither has it led to desertification or appreciable land degradation (Thomas *et al.*, 2000).

Where rangelands have become degraded, for whatever reason, several methods for rehabilitation have been devised. These methods give rise to altered species composition, biodiversity and plant cover. For example, van der Merwe and Kellner (1999) have documented rangeland characteristics following two methods of mechanical rehabilitation in 26 sites in arid and semi-arid areas of South Africa. Both 'dyker' plough cultivation and 'ripper' cultivation techniques were assessed for their effectiveness in reducing the area and frequency of bare patches. Ripping the soil breaks up the surface crust, and the creation of deep and shallow furrows improves the infiltration rate of water and makes it available to subsurface soil horizons at depths up to 380 mm for plant roots. The dyker plough, in contrast, makes hollows in the soil so that the crust is broken but water is retained within 100 mm of the surface. The results of the survey indicate that dyker cultivation results in an improved vegetation cover in clay-rich soil, whilst ripper cultivation is particularly appropriate in sandy soils. The regularity of renovation as well as the methods of renovation can influence pasture characteristics, as has been pointed out by the work of Vera *et al.* (1998) on pastures in the savanna environment of eastern Colombia, South America. Here, the most important techniques include the application of phosphorus fertilizer and mechanical treatment with a disk. Vera *et al.* show that pasture quality, and hence land cover, varies between flat and dissected land because of differences in technique and frequency of treatment. For example, the favoured flat lands were treated with both phosphorus and disking at average intervals of three years whilst dissected lands were often only treated mechanically at intervals of up to 10 years.

Rangeland degradation is also a serious problem in China where land for agriculture is precious and where agriculture is intensive. As Ho (2000) has discussed, China's pastoral areas have experienced growing environmental problems following the major economic reforms of 1978 when state control of agriculture was relaxed. The reasons for this accelerated degradation, including desertification, focus on the tenure system, or lack of one, in the wake of the demise of the commune and the embrace of a free-market economy. The control of the commune over rangeland was replaced with a family-based contract system; farmers were rewarded for what they produced. However, specific land units were never allocated, nor was ownership defined. Confusion has resulted in 'eating from the common rice pot', i.e. a tragedy of the commons, whereby responsibility or incentive for pasture management and maintenance is not clear-cut, so discouraging investment and improvement. To overcome some of the environmental problems, Ho points out that government intervention to restore rangeland quality has involved the sowing of seed by plane and improvements in the water supply through the provision of wells. Nevertheless such measures are treating only the symptoms of degradation and not the underlying, apparently legal and institutional, causes. Moreover, in some of China's pastoral lands, notably in the central province of Ningxia, the problem of degradation is compounded by people digging up liquorice root. Whilst this is a useful plant because it is good fodder, it is also important in Chinese medicine. Collectors can, thus, earn cash from its sale but, environmentally, this is aggravating degradation because its deep roots mean deep holes must be dug, leaving land – including below-surface horizons – susceptible to wind erosion. There is also the social impact of this conflict of interests, with confrontations occurring between the liquorice-root diggers and the farmers.

Many parts of Latin America have experienced the spread of cattle ranching in the last 100 years as frontiers have been opened up and as governments have fostered development. Several types of natural vegetation communities have been affected as a result. For example, widespread tropical forest destruction has taken place in Costa Rica and Brazil while both savannas and grasslands of Argentina, Bolivia and Paraguay have been substantially modified. In all of these regions ranch abandonments, natural and anthropogenic fire, periods of drought, and overgrazing have caused landcover change following the original disturbance. There is a wealth of literature that attests to such changes, e.g. Bucher and Huszar (1999) have renewed the history of, and suggested improved management practices for, the Gran Chaco savanna/grassland region of Argentina, Bolivia and Paraguay where encroachment by unpalatable woody shrubs is reducing productivity. This is also a problem in the rangelands of southwestern Mendoza, Argentina (Guevara *et al.*, 1999), where management by fire can redress the balance between grassland and woody species. Moreover, in Patagonia, there was a major expansion of sheep farming between 1880 and 1952, which has also resulted in land-cover alteration. According to Aagesen (2000), the number of sheep peaked at $22 \times 10^6$ in 1952, since when there has been a decline to $13 \times 10^6$ today. The most significant changes in natural ecosystems occurred at the onset due to the rapid introduction of sheep for a burgeoning wool market in Europe. Range quality was swift to decline and uncontrolled grazing exacerbated vegetation loss in many parts of Patagonia. This accelerated soil erosion in areas with steep slopes, and encouraged the spread of unpalatable shrub and herb species at the expense of grasses. Aagesen opines that vast areas are substantially degraded, with an urgent need for conservation and rehabilitation measures. These examples reflect conditions similar to those in other regions where pastoral activities predominate, e.g. Botswana (see above), South Africa and other parts of South America.

The removal of tropical forests for ranching is of major concern, not least because of the resulting loss of biodiversity and the release of carbon dioxide to the atmosphere where it contributes to global warming. For example, some 40 per cent of the land in Costa Rica's

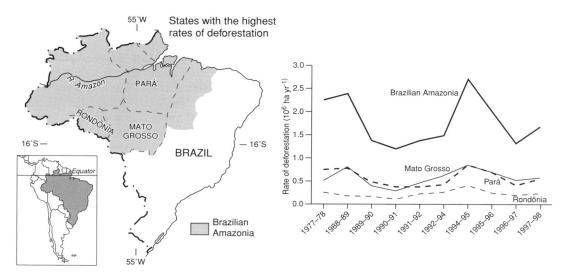

**FIGURE 5.5**  Rates of deforestation in Brazilian Amazonia since 1978. (Adapted from data from INPE (the Brazilian Space Agency), quoted in Houghton *et al.*, 2000)

Northern Atlantic Zone has been converted to pasture since the late 1800s, though approximately 70 per cent of this can be considered as degraded because of the presence of weed species (data quoted in Bulte *et al.*, 2000). Two reasons account for this degradation: overgrazing and lack of nitrogen. The latter is partly caused by the reluctance of farmers to use artificial fertilizers. However, what is especially worrying is the conclusion of Bulte *et al.* that, economically, pasture degradation and even abandonment is more efficient, given current meat prices etc, than better management. Thus the incentive, i.e. financial gain, for forest destruction initially, has not been sustained and the economics of forest destruction, in this case, have not brought rewards. Ecologically, however, if forest regrowth can occur, there will be environmental benefits such as increasing carbon storage and reduced rates of erosion.

A similar abandonment of cattle ranches has occurred in Brazil where at least 50 per cent of deforestation is attributed to ranch establishment and most of the rest to shifting cultivation and logging. Data on rates of deforestation are given in Figure 5.5, which shows that in the late 1980s and 1990s between one and two million hectares were removed annually. Government policies have underpinned much of this deforestation, and especially the expansion of ranching, through development programmes that encourage migration into frontier zones, infrastructure provision such as roads and, for large enterprises, tax incentives (see discussion in Walker *et al.*, 2000). Apart from the obvious changes in land cover that ranching brings, there is a major change in the distribution of carbon between the soil, vegetation and atmosphere. Logging (see Chapter 6) also contributes to carbon emissions through the exposure of soil and litter, whilst the abandonment of pasture on which secondary forest growth occurs redresses this, to some extent, by sequestering carbon. Recently, Houghton *et al.* (2000) have estimated the annual fluxes of carbon brought about by forest removal and forest regrowth for the 1989 to 1998 period using three different models. Their data are given in Figure 5.5; they indicate that there is almost a balance between sinks and sources, i.e. that the amount of carbon released through deforestation, abandonment and logging, i.e. 0.1–0.4 PgC yr$^{-1}$, is nearly balanced by the carbon uptake of natural forests. Consequently, the anthropogenic impact is offset by natural ecosystems although, had no anthropogenic disturbance occurred, the Amazonian region would have remained a net sink of carbon. The implication is that as further deforestation occurs, Amazonia will become a net producer of carbon.

## 5.6 Permanent arable agriculture and its alteration of land cover

Substantial areas of natural vegetation in the developing world have been replaced with arable crops that are grown on a permanent basis, either for subsistence or cash cropping. A wide range of crops are cultivated, including fibre crops, fodder crops, food crops for direct consumption by humans, and speciality crops such as tea, coffee, herbs and spices. These crops are grown in agricultural systems that employ little technology at one extreme, to those that rely on a huge technological input involving fossil fuels (see Figure 5.2) at the other. The most important crops, in terms of the area they occupy and the role that they play in the diets of most people in the developing world, are however the staples of maize, wheat and rice. Data on the production of these crops are given in Table 5.5, which shows that rice is the most important staple crop. It occupies more land than any other staple and much more of it is produced in comparison with maize and wheat, which are produced in similar quantities on similar land areas. Apart

from the fact that the $747,828 \times 10^3$ ha of land currently occupied by arable crops represents a significant land-cover change from the original vegetation cover, arable agricultural systems are as dynamic as the other agricultural systems described in Chapter 4 and above. They react to environmental stimuli and to socio-economic factors, as well as to technological change (see Figure 5.2).

One such stimulus is urbanization. Market forces created by agglomerations of people requiring food products cause alterations in agricultural systems, as Aweto (2001) has disclosed in relation to contrasting types of agriculture in the vicinity of the city of Ibadan, Nigeria. Here the natural vegetation of moist, semi-deciduous rain forest has been removed for permanent and shifting cultivation. The major crops are cassava, maize and yams, which are intercropped on permanent or shifting plots; other crops include vegetables such as pumpkin, and fruits such as tomatoes. Pressure on land due to population increase and the expansion of Ibadan has resulted in a reduction in the length of fallow in the shifting cultivation system, itself a cause of land-cover/land-use change. Moreover Aweto's comparison of land-cover characteristics shows that the density of trees is substantially different in the two agricultural systems. Tree density is higher on land under permanent cultivation at the city edge than it is in areas nearby, under shifting cultivation (i.e. 511

|  | 1996 | 1998 | 2000 |
|---|---|---|---|
| Maize production Mt $\times$ $10^3$ | 268,902 | 281,211 | 250,911 |
| Area of maize production ha $\times$ $10^3$ | 93,169 | 91,170 | 90,722 |
| Rice production Mt $\times$ $10^3$ | 544,072 | 552,472 | 567,407 |
| Area of rice production ha $\times$ $10^3$ | 146,695 | 147,853 | 149,407 |
| Wheat production Mt $\times$ $10^3$ | 274,113 | 278,737 | 264,846 |
| Area of wheat production ha $\times$ $10^3$ | 106,356 | 106,345 | 99,895 |
| Total amount of arable land in developing countries ha $\times$ $10^3$ | 748,994 | 747,828 | N/A |

**TABLE 5.5** Data on staple crop production in the developing world (Food and Agriculture Organisation, 2001a)

trees ha$^{-1}$ compared with 143 trees ha$^{-1}$). Moreover, there is also a significant difference in terms of the species present. In particular, the permanently cultivated areas had high densities of *Gliricidia sepium*, an exotic tree legume that, because of its nitrogen-fixing role, helps maintain soil fertility. Recent land-cover changes in the Narok district of Kenya also reflect the operation of different stimuli and their impact on a variety of arable systems in the same area. Serneels and Lambin (2001) have shown that the most important factor affecting large-scale mechanized cereal production is distance to market, i.e. costs of transport, as well as climate; the expansion of smallholder arable agriculture is limited by the availability of water, land suitability, proximity to tourist markets, and access to facilities such as clinics and schools. Large-scale farming has also resulted in cultivation of *c.* 45 × 10$^3$ ha for wheat. Overall, *c.* 8 per cent of the area has lost its natural vegetation of savanna because of agriculture.

Land-cover/land-use change in the Sudan/ Sahel countries has been examined by Reenberg (2001), and Stéphenne and Lambin (2001), who demonstrate that a myriad of factors influence the nature and direction of change. The cultivation of sorghum and millet is the main activity in Reenberg's (2001) study areas in Burkina Faso where population growth is a major stimulus to land-cover/land-use change. In the process of increasing food production, farmers are influenced by many factors including labour availability, climate and climatic variability, tenure rights, cultural factors such as ethnicity (which influences choice of soil type), field size (which is related to social hierarchies for historic reasons), and gender relationships and responsibilities. Whilst Reenberg gives the detail of land-cover/land-use change, Stéphenne and Lambin (2001) attempt to develop a model to predict land-cover change at the national scale. They too highlight population growth as the under-pinning factor and have used data from Burkina Faso to test their model. Their results show that change occurs at two time frequencies. High frequency change, i.e. every few years, is driven by climatic variability; during periods of drought there is a small expansion of cropland and pasture into areas used for fuel-wood collection and also into unused land. Low- frequency change, e.g. at decadal scales, is associated with population trends; examples of this relationship include increases in cropland and pasture between 1970 and 1985, which occurred at the expense of fuelwood extraction areas, and a subsequent phase of increase in cultivated and fallow areas in response to urbanization.

Arable agriculture is of primary importance in China because of its large population and scarcity of suitable land. However, Yang and Li (2000) have shown that the total amount of cultivated land has declined by 4.73 × 10$^6$ ha, or 4.45 per cent, since 1978. This loss has not occurred evenly, with some 50 per cent of the decrease occurring between 1983 and 1987. Nor has it been even throughout the country; the greatest declines have been in coastal provinces with significant declines in the central provinces and some increases in the western border provinces. Data on changes in each province are given in Figure 5.6. These are general figures that mask internal changes of land use such as a shift from cultivated land to land used for horticulture (this is considered to be a separate land-use category). The data also mask any additions to the amount of cultivated land through drainage, reclamation and the conversion of pasture, horticultural or afforested land to arable land. Conversely, some causes of losses are obvious, e.g. urban spread, the construction of rural housing, and loss of land due to disasters such as flooding, mass movement, desertification etc. Yang and Li's analysis of detailed statistics does, however, show that there has been a real decrease in cultivated land, notably through a reduction in the southeast provinces. As Figure 5.6 shows, the greatest decreases have been in Shanghai, Zhejiang and Guangdong provinces. This is because these provinces have been particularly well placed to benefit from China's economic growth following the economic reforms of the late 1970s (see also Section 5.5). With the exception of Guangxi Province in the coastal region, all other provinces in both the coastal and central regions experienced a loss of cultivated land, mainly due to its conversion to horticulture

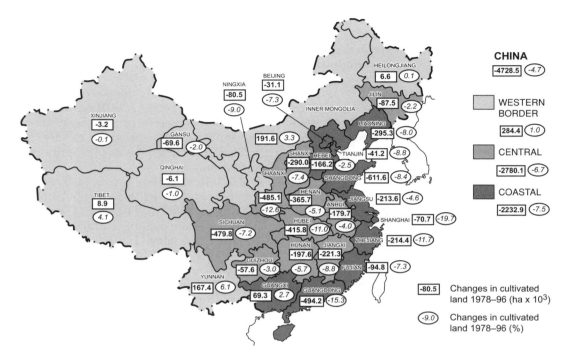

**FIGURE 5.6** Changes in the area of cultivated land in China, 1978–96. (Adapted from Yang and Li, 2000)

(including orchards) and fish ponds. This shift in land use reflects rising incomes and standards of living, which have increased demand for horticultural products, such as fruits, and aquatic products. Clearly, this conversion could be reversed and in some respects Yang and Li point out that horticulture and aquaculture increase land values because they are producing higher-value goods than staple cereals, and this in turn prompts improved land management and soil conservation.

However, Figure 5.6 shows that the situation is quite different in the western-border provinces where significant gains in cultivated area have occurred, especially in Yunnan and Inner Mongolia. The resulting predominant land use involved forestry and pasture, rather than horticulture or fish ponds. Much of the addition has been brought about through reclamation schemes, including the Three-North Forest Protection Belt Programme. Nevertheless the abandonment of damaged land has continued; declining grain output from arable

land may lead to abandonment, a process that may be reported as conversion to pasture or forest, and thus distort values for given land-use categories. At the same time, ready markets for arable produce have stimulated conversion of pasture and forest to arable. Consequently, the situation in the western-border provinces is particularly dynamic in terms of land-use change. Another analysis of land use in China by Verburg *et al.* (2000) reflects a similarly complex pattern of change. They also show that in parts of the western provinces the efficiency of grain production could be increased to compensate for losses of cultivated land.

In relation to future land-use change in China, Verburg and van der Gon (2001) estimate that there will be further losses in cultivated land and in the area under rice, whilst animal husbandry will increase. The possibilities are summarized in Table 5.6. The increased demand for meat and meat products is a result of rising incomes (see above in relation to fruit and aquaculture products). Apart from creating land-use

| | Value in 1991 | Projected change 1991–2010 | Methane production 1991–2010 | |
|---|---|---|---|---|
| Cultivated land | 132 × 10⁶ha | −8% | 20.2 | 18.4 |
| Area under rice | 45.8 × 132 × 10⁶ha | −13% | | |
| Large animals* (cattle, buffalo) | 132 × 132 × 10⁶ head | +21% | | |
| Pigs | 369 × 132 × 10⁶ head | +12% | 8.8 | 13.6 |
| Sheep and goats* | 199 × 132 × 10⁶ head | +61% | | |
| Poultry | 2.6 × 132 × 10⁶ head | +70% | | |

* Will require additional grazing land

**TABLE 5.6** Projected changes in China's agriculture by 2010. (Adapted from Verburg and van der Gon, 2001)

change, alterations in livestock numbers give rise to changes in methane production; projections for such changes are given in Table 5.6. Overall, by 2001 an increase of 11 per cent is anticipated. This is a net value as declines in emissions from rice are offset by increasing emissions from livestock. Most importantly, methane emissions to the atmosphere are significant in relation to global climatic change.

Rising standards of living and the demand for an increasing variety of food often have repercussions in regions well beyond those in which such improvements occur. For example, Murray (2000) considers that such processes are part of globalization and cites the case of recent agricultural change in the Pacific islands. In particular, Murray points out that so-called staple exports, e.g. sugar, copra and taro, are now being despatched to new markets while niche products, e.g. kava (a species of pepper, *Piper methysticum*, the root and stem of which is used to produce a narcotic drink), are being produced in increasing quantities for markets in North America, Europe and Asia. Similarly, in some South American countries 'luxury' vegetables, such as mange-tout and asparagus, and flowers are produced in Peru and Ecuador respectively, for foreign markets while indigenous people survive on subsistence crops sometimes augmented by food aid.

Mountainous regions, especially those where population growth is high, have also witnessed land-cover and land-use change in the last few decades. Such changes are exemplified by the

work of Tiwari (2000), Jain *et al.* (2000), and Rao and Pant (2001) on regions of the Himalaya. The former highlights the significance of population growth as a major agent of land-cover/land-use change in the Himalayan Kumaon region of India. Using field data from three catchments, Tiwari (2000) has shown that cultivated land and pasture has expanded, mostly at the expense of forest. These data are summarized in Table 5.7. The significant loss of forest area has altered hydrological regimes and is considered by Tiwari to be the cause of recurrent floods in the Indo-Gangetic plains. In the nearby Sikkim Himalaya, Jain *et al.* (2000) have documented land-cover/land-use change in the Khecheopalri Lake watershed, where similar losses of forest cover have occurred as cultivation has increased. Jain *et al.*, whose results are summarized in Table 5.8, also show that not only has forest cover declined but so has the character of the remaining forest, which has experienced considerable degradation. In addition, soil erosion and nutrient loss was highest from the cultivated land; as some of this material is deposited in the lake itself, it has encouraged vegetation growth so that the area of open water has almost halved. Rao and Pant (2001) have documented land-cover/land-use changes in the Sadiyagad watershed near Almora in Uttar Pradesh State, India. Here, there are 47 settlements mostly engaged in grain production, especially wheat and finger millet, horticulture and milk production. As in the other studies referred to above, remotely

|                                         | Sheil Gad | Balia |
|-----------------------------------------|-----------|-------|
| **A** Land under grazing % total 1965   | 48.98     | 12.30 |
| + miscellaneous 1995                    | 42.70     | 15.81 |
| % change                                | −6.28     | 3.51  |
| **B** Forest % total 1965               | 24.70     | 65.84 |
| Forest % total 1995                     | 19.62     | 53.94 |
| % change                                | −5.08     | −11.90|
| **C** Cultivated land % total 1965      | 26.32     | 21.86 |
| Cultivated land % total 1995            | 37.68     | 30.25 |
| % change                                | 11.36     | 8.39  |
| **D** Total land area (ha)              | 1014.00   | 74.00 |

**TABLE 5.7** Data on land-cover and land-use change in two catchments in the Kumaon Himalaya of northern India. (Adapted from Tiwari, 2000)

|                                            | 1963    | 1997   |
|--------------------------------------------|---------|--------|
|                                            | ha      | ha     |
| Area of dense mixed forest                 | 1160.00 | 272.20 |
| Open mixed forest                          | –       | 492.01 |
| (includes cardamom agroforestry)           |         |        |
| Degraded forest                            | –       | 254.00 |
| Settlement and cultivated area             | 38.20   | 162.50 |
| Rock                                       | –       | 17.50  |
| Lake area                                  | 7.40    | 3.79   |
| Bog area                                   | 3.40    | 7.01   |

**TABLE 5.8** Data on land-cover and land-use change in the Khecheopalri Lake watershed, Sikkim Himalaya, India. (Adapted from Jain *et al.*, 2000)

sensed data show that deforestation has occurred since 1963 and that grain production has become increasingly industrialized as the diversity of crops cultivated has declined.

## 5.7 Mixed agriculture and agroforestry, and their alteration of land cover

Mixed agriculture involves the integration of crop and livestock production and is widespread throughout the world, except in the extreme climatic zones. It is highly variable in terms of its components and in relation to energy subsidy (see Figure 5.2). For example, only a few animals may be kept to provide the farmer with essential animal products such as milk and manure for use as a crop fertilizer, in other cases the animals themselves may be the major product, e.g. intensive pig rearing. Similarly, energy subsidies may range from very low to very high. Agroforestry systems may also involve livestock rearing or they may be concerned only with crop production. It is an integrated land-use type in which trees are retained or introduced for the protection and stability of the soil and, in many cases, to provide a product; beneath the trees, other crops such as cereals or legumes are grown or a pasture is encouraged for grazing animals that may also graze fallow vegetation. Mixed farming is a significant agent of land-cover and land-use change in developing countries, where it is often a subsistent type of agriculture. Agroforestry is also widespread in the developing world. It is valued for its capacity to provide environmental protection, to produce economic products and to store carbon and, like mixed farming, its components can be many and varied.

In a historical context, Viglizzo *et al.* (2001) have documented the development of agriculture in the Argentinian pampas during the past 100 years. Here, cattle and crop production are combined in various proportions in agricultural systems that have transformed the natural, mainly grassland, vegetation, and which themselves have been altered through intensification. The major products today are sunflowers, soybean, peanuts, wheat, maize and sorghum as well as beef. The analysis of Viglizzo *et al.*, based on natural census data and historical maps, shows that significant changes have occurred, especially in the extent of cultivated land, since the 1880s. At this time, the extent of cultivated land was low; only *c*. 5 per cent of the $52 \times 10^6$ ha was planted with annual crops, whilst natural grassland remained elsewhere and was used for cattle grazing. By the 1940s, cultivation had spread and intensified; all parts of the pampas experienced some ploughing, even in the marginal western regions, and in the north-central region of rolling pampas more than 60 per cent of the land was cultivated. Further transformation had occurred by the 1980s, notably in the rolling central and southern pampas where between 40 and 60 per cent of the land had been brought under cultivation. In addition, many remaining grasslands are no longer in their natural state as the degree of management has intensified and fragmentation has occurred.

The complex interaction of various factors that stimulate land-cover alteration are illustrated by the study of Nielsen and Zöbisch (2001) on the village of Im Mial in northwest Syria. Here, mixed farming is the mainstay of the local economy. In recent years productivity has declined and grazing areas have become degraded. A major factor underpinning these changes is a reduction in the length of the

fallow periods; this in turn has been caused by population increase and technological innovations, which include the use of tractors and combine harvesters in place of donkey ploughs and harvesting by hand. Not only have areas of arable agriculture been affected but so too has land used for pasture. Cropping has encroached into the latter, so reducing the amount of land for grazing in a culture that measures its wealth by size of flock. On a positive note, however, greater integration of crop and livestock farming is occurring in response. For example, animals are increasingly being fed on crop residues and through stubble grazing. This is reducing pressure on degraded pastures, which should recover. In addition, trees have been planted in the upper parts of fields; these not only help to conserve soil and water, but may also produce an economic crop, e.g. olives. Nielsen and Zöbisch also observe that the trend of villagers finding off-farm work will need to continue in order to protect the environment. In this respect educational and training facilities require improvement in order to provide the necessary skills.

Mixed farming is especially characteristic of the middle hills of Nepal (Figure 5.7) where crop and livestock production are integrated, and often linked with forest grazing and forestry. The middle hills of Nepal support a large and rapidly growing population, which is stimulating agricultural expansion in a region that requires careful land management in order to avoid degradation. According to Pilbeam *et al.* (2000), the success of many subsistence systems is due to a net movement of nutrients, especially nitrogen, from the forest to the arable fields. This occurs because forest litter is used for animal bedding and because forage is grazed directly or is collected from the forest for animals; the animals then manure the arable fields. Even with this addition, plus crop rotations that include legumes and the use of flood waters with high concentrations of nutrients, crop productivity is low. On the one hand, the construction of terraces for arable cultivation has altered the land-cover characteristics and on the other hand any loss of forest has repercussions not only for nutrient inputs into arable agriculture, but also for the provision of fuelwood.

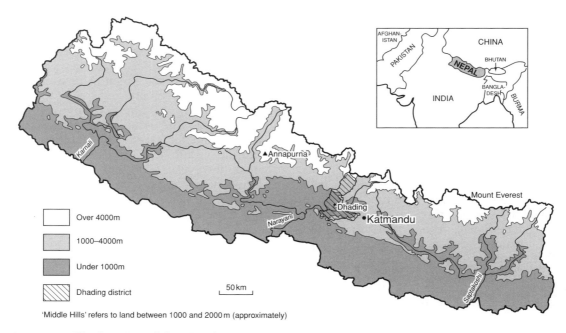

'Middle Hills' refers to land between 1000 and 2000 m (approximately)

**FIGURE 5.7** The location of the Nepalese regions mentioned in the text

Agroforestry may be one way to improve productivity as well as a means of reducing soil erosion. Its potential impact has been discussed by Neupane and Thapa (2001) based on their research in the Dhading district of the middle hills of Nepal. Their survey of 223 households revealed that problems of declining crop productivity were linked with increasing livestock populations and overgrazing in the forests, deforestation and soil erosion, all of which contribute to land-cover alteration as well as declining farm incomes. To overcome some of these problems, the Nepal Agroforestry Foundation (NAF) has been promoting various forms of agroforestry at farm level, including fodder and grass species as well as trees. Neupane and Thapa have determined the degree of success enjoyed by these projects through the collection of environmental and socio-economic data. The latter provided most evidence of improvement with gross income increasing by *c.* 45 per cent on farms that employed agroforestry. Even more remarkable is the 140 per cent increase for those farms that had introduced mulberry trees for silk production. Indeed, such trees are multi-

functional: they provide protection for the soil; nutrients for the soil through the production of leaf litter; fruit; and food for silkworms. Inevitably, the widespread adoption of agroforestry brings about land-cover change and an increasingly biodiverse landscape.

In some regions agroforestry is being introduced and/or adopted to replace shifting agriculture (see Section 5.4). One such region where this is occurring is the forest zone of southwest Cameroon. According to Adesina *et al.* (2000), the alteration of food-production systems is necessary in this region because of increasing population pressure and resulting land-use intensity, the major impact of which has been to reduce the length of fallow periods and thus to reduce productivity. Farmers' responses to questionnaires following the adoption of agroforestry indicate that considerable improvements in soil fertility, weed suppression, need for fallow and crop yield had occurred. In this region the form of agroforestry adopted is alley cropping, which involves the cultivation of food crops, such as maize, between hedges of shrub or tree species which are leguminous. The latter provide protection for the soil, a regular input of organic matter and, through their association with nitrogen-fixing bacteria, an increase in nitrate in the soil. In southwest Cameroon alley cropping is practised alongside shifting agriculture, and both represent a change in land cover from the original forest. Agroforestry may also have a role in mitigating carbon dioxide emissions because woody species store more carbon than crops or pasture. Adesina *et al.* (1999), for example, have calculated that, in Nigeria, a possible 1530 Mt of carbon could be withdrawn from the atmosphere by 2030 if some 76 per cent of the *c.* $39.5 \times 10^6$ ha of farmland adopted agroforestry systems. This would more than counteract emissions through deforestation at the rate of 2.6 per cent per year.

## 5.8 Conclusion

Vast areas of natural ecosystems in the tropics and subtropics have been transformed into agricultural systems. The many and varied types of agriculture have, to a greater or lesser extent, altered natural land-cover types. Least transformation has occurred in the semi-arid regions where extensive, often nomadic, pastoralism is predominant. As the examples given above attest, there are many different types of grazing regimes and many different ways of manipulating species composition. Shifting cultivation is another extensive system of food production; it relies on fallow periods to restore soil fertility and is a low-energy input system. It is increasingly coming under pressure, mainly because of population increase and because of a growing number of landless poor, and alternative systems of food production are being introduced (e.g. agroforestry). Both nomadic pastoralism and shifting cultivation involve cycles of intensive and relaxed land use, and regions in which they predominate are thus characterized by internal dynamics that are reflected in land cover and land use.

Permanent pasture, arable cultivation and mixed farming are equally significant; in terms of land-cover alteration the agroecosystems have substantially altered or completely replaced natural vegetation communities. All three types of agriculture are influenced significantly by exter-nal as well as internal factors (Figure 5.2), not least of which are population growth, market forces, development policies, gender-based decisions and religious laws concerning family relationships. Of prime importance are ready markets in the developed world for many types of agricultural commodities, e.g. tea, coffee, oils, speciality crops, meat and ornamental plants. These factors are as dynamic as the agricultural systems they influence.

A combination of socio-economic pressures and, in many cases, environmental degradation, causing declines in agricultural productivity, has forced innovation. The multiplicity of agroforestry systems provides various ways to address both pressures. In general, they help mitigate environmental problems, such as soil and water conservation, and can contribute to economic success by providing an increased variety of crops as well as fuelwood. In a wider context, the increased cultivation of shrub and tree species is important for sequestering carbon to offset emissions due to deforestation and fossil-fuel consumption.

As discussed in Chapter 10, however, agriculture in the developing world is set to expand and intensify considerably as population increases of at least 25 per cent occur in the next two decades.

## Further reading

**Azam-Ali, S.N.** 2001: *Principle of Tropical Agronomy.* CABI, Wallingford.

**Buck, L.E., Lassoie, J.P. and Farnandes, E.C.M.** (eds) 1999: *Agroforestry in Sustainable Agricultural Systems.* CRC Press, Boca Raton.

**Lefroy, E.C.** (ed.) 1999: *Agriculture as a Mimic of Natural Ecosystems.* Kluwer, Dordrecht.

**Mannion, A.M.** 1995: *Agriculture and Environmental Change. Temporal and Spatial Dimensions.* John Wiley and Sons, Chichester.

**Mørch, H.** (ed.) 1999: *Agriculture, Systems, Landscapes, Resources.* Royal Danish Geographical Society, Copenhagen.

**Morse, S., McNamara, N., Acholo, M. and Okwoli, B.** (eds) 2000: *Visions of Sustainability: Stakeholders, Change and Indicators.* Ashgate, Aldershot.

# 6

# Land cover and land use related to forestry

## 6.1 Introduction

Forests are one of the world's greatest natural resources. Not only do they provide a wide range of products, they also supply vital ecosystem services. The major products are construction materials, pulp and paper, whilst the ecosystem services involve the regulation of biogeochemical cycling with implications, especially through the global carbon cycle, for the regulation of global climates (see Section 1.5). In the latter context the world's forests comprise the largest terrestrial store of carbon in the biomass and leaf litter. Any alteration of the forest cover, especially when it is replaced with agriculture or other non-forest cover, increases the amount of carbon in the atmosphere and thus contributes to global climatic change.

As shown in Figure 1.1, forests have been a very important component of the Earth's vegetation cover in the present interglacial (the Holocene). Only where average annual temperatures are too low, at high latitudes and altitudes, and where water is a major limitation on growth, as in arid lands, do trees fail to grow. However, since the advent of agriculture *c.* 10,000 years ago (see Section 2.6), the Earth's forest cover has been altered; some of it has been fragmented or removed completely, as in large parts of Europe, and some of it has been modified as it has been manipulated for economic gain. Agriculture is not the only cause of forest change as logging and mineral extraction are also important. This exploitation has resulted in forests that, today, are often markedly different in terms of distribution and composition from their predecessors. The current distribution of forest is reflected in Figure 1.2, which is itself a generalization of reality, and is discussed in more detail below (see Figures 6.1 and 6.4). Forest destruction and removal is currently occurring at an alarming rate, though the emphasis has shifted from Europe and the temperate world to the boreal zone of Asia and the tropics.

Overall, there are many factors that have contributed, and that will continue to contribute, to forest demise. As Mather *et al.* (1998) have discussed, population growth is particularly significant; they conclude that approximately half of the deforestation that has occurred throughout human history can be attributed to population growth. However, they also point out that this relationship is not simplistic or linear but complex; it can involve factors such as land tenure, poverty and lack of education which may, especially collectively, be just as significant driving forces as population increase. Moreover, in many countries a so-called forest transition stage (Mather, 1992) has occurred. For example, in most European countries the population/deforestation relationship no longer obtains. In the late 1800s and early 1900s, forests began a period of expansion, often as a response to the political recognition of a vital resource in decline and the development of alternative products to those of forests. It could be argued that most countries where deforestation rates are high currently have not yet reached this crucial threshold. Those countries in which the transition has occurred often have aggressive afforestation policies and forest management strategies, whilst those that have not reached the threshold have weak afforestation and management policies as well as relatively ineffective measures to protect remaining forest areas or to develop sustainable forest use.

Although the subdivision is not distinct, most developed countries have experienced the forest transition, in contrast to most developing countries. This subdivision is used in this chapter, which focuses on forest exploitation, i.e. logging or timber extraction and forest management strategies, as well as afforestation in terms of its implementation and subsequent management. All of these activities cause land-cover and land-use change. Note that urban forests are discussed in Chapter 8.

## 6.2 Forest exploitation in the developed world

Figure 6.1 is a generalized map of forest distribution in the developed world. It shows that there are many different types of forest, ranging from needleleaved boreal species in high latitudes, mainly in the Northern Hemisphere, to Mediterranean-type woodlands dominated by both deciduous and evergreen species adapted to low moisture in the summer and low, but not below-zero, temperatures in wet winters (see Archibold, 1995, for a description of various biomes). Almost all of these forests are subject to some form of management or exploitation. Indeed, forest exploitation, including deforestation, has been occurring since the inception of agriculture 10,000 years ago (see Section 2.6). In some countries the forest resource underpins well-developed lumber and pulp and paper industries, as in Canada, USA, Scandinavia and Russia. Table 6.1 gives some basic data on wood and wood-product output in selected developed countries. These are nations where a substantial proportion of the land area is occupied by forest and where there are well-developed forest-related industries. In all cases the huge volumes of wood extracted for roundwood, pulp etc. (see Table 6.1) cause land-cover change whether the areas of forest are replaced with a completely different land use, are replanted or abandoned to successional processes. Moreover, the use and abuse of these extensive forests goes relatively unreported in comparison with the well-publicized plight of tropical forests. This is despite the importance of boreal forests for wealth generation and the provision of ecosystem services.

Forest industries are particularly well developed in North America where there are extensive areas of both boreal and temperate forests. For example, in Canada, the world's leading exporter of timber products, 11 categories of forests have been distinguished, as shown in Figure 6.2. According to Global Forest Watch: Canada (2001), forests occupy some 418 $\times$ $10^6$ ha with the majority being boreal forests, and 94 per cent are publicly owned; 78 per cent of this forest has been affected by human activity, with the remaining intact forest being situated in the northernmost areas of forest; the forest industry generated US$47 $\times$ $10^9$ in 1996. This has not been achieved without considerable land-cover and land-use alteration. The timber industry is organized on the basis of a tenure system whereby individual provinces grant licences to companies or individuals for up to 25 years. During this time each company has responsibility for harvesting, management

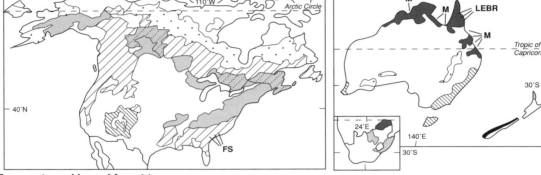

**Temperate and boreal forest types**

```
/////  Evergreen needleleaf forest
       >30% canopy cover; canopy is >75% needleleaf and evergreen

°  °   Deciduous needleleaf forest
°  °   >30% canopy cover; canopy is >75% needleleaf and deciduous

       Mixed broadleaf/needleleaf forest
       >30% canopy cover; canopy is between 50:50% and 25:75%
       needleleaf and broadleaf

■      Broadleaf evergreen forest
       >30% canopy cover; canopy is >75% evergreen and broadleaf

       Deciduous broadleaf forest
       >30% canopy cover; canopy is >75% evergreen and broadleaf

       Sclerophyllous dry forest
       >30% canopy cover; canopy is mainly sclerophyllous
       broadleaves and is >75% evergreen

·  ·   Sparse trees and parkland
·  ·   Between 10–30% canopy cover (e.g. steppe regions); trees of
       any type
```

```
       Plantations
       Planted, intensively managed forests with <30% canopy cover

FS     Freshwater swamp forest
       <30% canopy cover; trees of any mixture of leaf type and
       seasonality; soils are waterlogged
```

**Tropical forest types**

```
       Sparse trees and parkland, and thorn forest
       Mainly 10–30% canopy cover (e.g. savanna regions); trees of any
       type, but deciduous trees with thorns and succulent
       phanerophytes with thorns predominate locally, and have >30%
       canopy cover

M      Mangroves
       <30% canopy cover; mangrove trees, along coasts in or near
       brackish or salt water

LEBR   Lowland evergreen broadleaf forest
       <30% canopy cover; canopy is >75% evergreen broadleaf;
       altitude below 1200m, little or no seasonality is displayed
```

**FIGURE 6.1** Forest distribution in the developed world. (Adapted from UNEP World Conservation Monitoring Centre, 2001)

etc., with licence agreements varying according to intensity of use and provision for replanting etc. The logging rate is controlled centrally through provincial governments; it is reviewed every five to ten years and takes into account a wide range of factors including employment capacity and revenues, as well as environ-mental factors. According to Global Forest Watch: Canada, 64 per cent of Canada's forests are tenured and have experienced a significant increase in the rate of logging in the past 100 years, but especially during the period 1975 to 1988, from $c.\ 600 \times 10^3\ \text{ha yr}^{-1}$ in 1975 to almost $800 \times 10^3\ \text{ha yr}^{-1}$. The majority of the harvest is

| | Forest area 1995 ha $\times 10^3$ | % of total land area | Annual change 1990–95 ha $\times 10^3$ | Industrial roundwood M$^2 \times 10^3$ | | Sawnwood M$^2 \times 10^3$ | | Pulp for paper M$^2 \times 10^3$ | | Paper and paperboard M$^2 \times 10^3$ | |
| --- | --- | --- | --- | --- | --- | --- | --- | --- | --- | --- | --- |
| | | | | Production | Consumption | Production | Consumption | Production | Consumption | Production | Consumption |
| Canada | 244,571 | 26.5 | 175 | 183,113 | 187,528 | 62,829 | 14,068 | 183,113 | 187,528 | 62,829 | 14,068 |
| Finland | 20,029 | 65.8 | –17 | 42,503 | 49,077 | 9270 | 2378 | 9,676 | 8198 | 10,441 | 2098 |
| Japan | 25,146 | 66.8 | –13 | 22,897 | 70,750 | 24,493 | 36,011 | 11,065 | 14,394 | 30,014 | 30,595 |
| New Zealand | 7,884 | 29.4 | 43 | 16,999 | 10,727 | 3052 | 2130 | | | | |
| Norway | 8,073 | 26.3 | 27 | 7,695 | 10,288 | 2405 | 2495 | 2269 | 1983 | 2096 | 783 |
| Russian Fed. | 763,500 | 45.2 | N/A | 67,000 | 51,652 | 21,600 | 16,927 | 3725 | 2758 | 3212 | 1788 |
| Sweden | 24,425 | 59.3 | –2 | 52,600 | 56,406 | 14,370 | 2919 | 52,600 | 56,406 | 14,370 | 2919 |
| USA | 212,515 | 23.2 | 589 | 406,595 | 387,884 | 109,654 | 146,338 | 406,595 | 387,884 | 109,654 | 146,338 |
| World | 3,454,382 | 26.6 | –11,269 | 1,489,530 | 1,492,939 | 429,645 | 426,233 | 178,543 | 177,962 | 284,383 | 278,767 |

*Note:* Data for production and consumption relate to 1996

**TABLE 6.1** Data on forest area and the production of forest-based commodities (Food and Agriculture Organisation, 2001a)

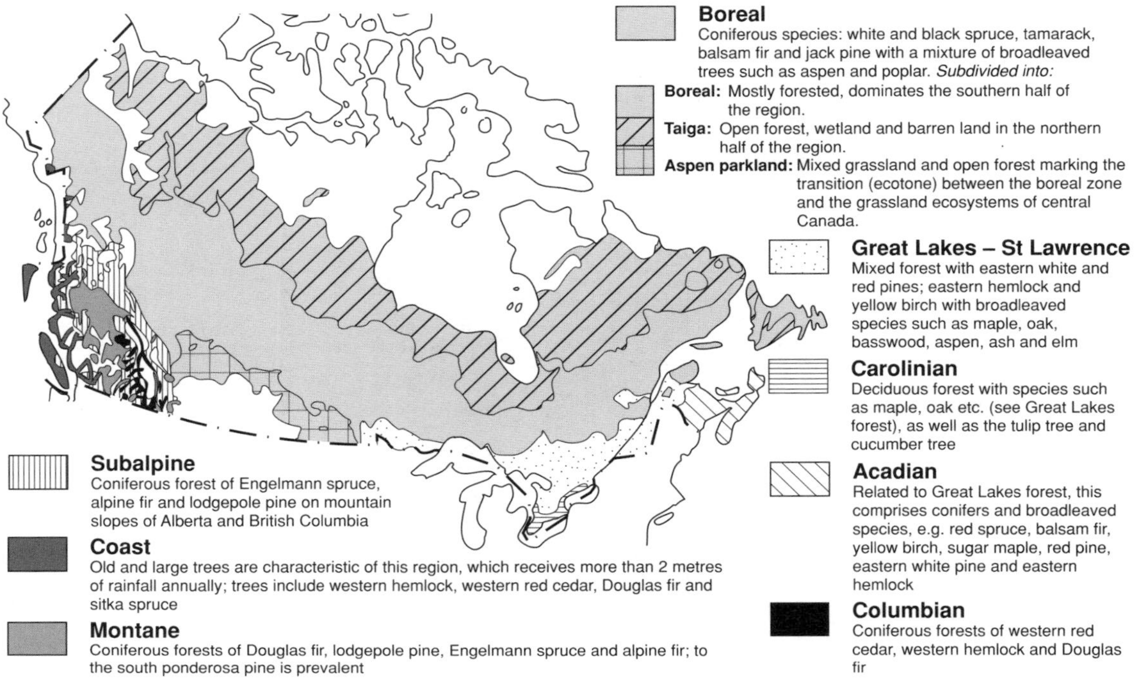

**Boreal**
Coniferous species: white and black spruce, tamarack, balsam fir and jack pine with a mixture of broadleaved trees such as aspen and poplar. *Subdivided into:*
**Boreal:** Mostly forested, dominates the southern half of the region.
**Taiga:** Open forest, wetland and barren land in the northern half of the region.
**Aspen parkland:** Mixed grassland and open forest marking the transition (ecotone) between the boreal zone and the grassland ecosystems of central Canada.

**Great Lakes – St Lawrence**
Mixed forest with eastern white and red pines; eastern hemlock and yellow birch with broadleaved species such as maple, oak, basswood, aspen, ash and elm

**Carolinian**
Deciduous forest with species such as maple, oak etc. (see Great Lakes forest), as well as the tulip tree and cucumber tree

**Subalpine**
Coniferous forest of Engelmann spruce, alpine fir and lodgepole pine on mountain slopes of Alberta and British Columbia

**Acadian**
Related to Great Lakes forest, this comprises conifers and broadleaved species, e.g. red spruce, balsam fir, yellow birch, sugar maple, red pine, eastern white pine and eastern hemlock

**Coast**
Old and large trees are characteristic of this region, which receives more than 2 metres of rainfall annually; trees include western hemlock, western red cedar, Douglas fir and sitka spruce

**Columbian**
Coniferous forests of western red cedar, western hemlock and Douglas fir

**Montane**
Coniferous forests of Douglas fir, lodgepole pine, Engelmann spruce and alpine fir; to the south ponderosa pine is prevalent

**FIGURE 6.2** The distribution of forests and their characteristics in Canada. (Adapted from Global Forest Watch: Canada, 2000; categories are those of Rowe, 1977)

reaped through clear cutting or clear felling, with most activity occurring in Quebec, Ontario and British Columbia (see Table 6.2 for data).

The use of clear felling is a drastic and rapid means of altering land cover. As discussed by the World Commission on Forests and Sustainable Development (1999) it is a controversial issue in forest exploitation. This is because it involves the wholesale removal of tracts of forest of varying size and its impact on environmental factors, such as soil drainage, hydrological regimes and animal life, is often devastating. It is, clearly, a rapid means of altering land cover, is often followed by replanting with one tree species and is thus a means of transforming natural or semi-natural forest into monocultural plantations. Where

| | **Area**<br>ha $\times$ 10$^3$ | **1997**<br>**% of total**<br>area | **Volume**<br>M$^3$ $\times$ 10$^6$ | **% of total**<br>volume |
|---|---|---|---|---|
| **Quebec** | 364 | 36 | 41 | 22 |
| **Ontario** | 198 | 19 | 25 | 13 |
| **British Columbia** | 176 | 17 | 69 | 38 |

**TABLE 6.2** Data on logging in 1997 in the three Canadian provinces with the most extensive forest industries. (Adapted from Canadian Forest Service, quoted in Global Forest Watch: Canada, 2001)

natural succession is allowed to occur, a key factor in forest regeneration following disturbance is the availability of seed from remnant stands. For example, Asselin *et al.* (2001) have shown that in a boreal forest of southwestern Quebec, regeneration of species such as balsam fir (*Abies balsamea*), white spruce (*Picea glauca*) and white cedar (*Thuja occidentalis*), depends more on the availability of seed sources than on method of clearance, which in the sites investigated comprised burning and logging. Such findings have implications for felling strategies in areas left to revegetate naturally, i.e. mature sources of seed need to be preserved within the area harvested.

Fire is also an important factor in land-cover alteration in Canada's boreal forests, as it is throughout the boreal zone. Fire can occur naturally, and may be necessary for regeneration by releasing seeds from cones, or it may be used as a management technique. These issues have been discussed by Weber and Stocks (1998) who report that, on average, some $2.5 \times 10^6$ ha of Canada's boreal forest are burnt annually, though in extreme years the figure may rise to $c. 7 \times 10^6$ ha. The likelihood that the incidence of fire has increased since the pre-1960s, from $c.$ 6000 fires to almost 10,000 fires in the 1990s, may reflect a response to global warming. Consequently, further increases in fire incidents may occur. This will accelerate land-cover alteration and could impair regeneration if the numbers of remnant stands are reduced. Such events may result in the considerable alteration of the boreal biome, not only in Canada but also in Eurasia.

Forest extraction and management are just as significant causes of land-cover change in the Russian Federation as they are in Canada. Figure 6.3 shows the distribution of forests in the Russian Federation, which houses between 750 $\times 10^6$ ha of forests; this is the most extensive area of forest in any one nation which includes some 70 per cent of the Earth's boreal forest (World Resources Institute, 1997). As Table 6.1 shows, the Russian Federation is a major producer of industrial roundwood and sawnwood, at least 10 per cent of which is exported. There is much concern about the state of these forests and their management (World Resources Institute, 1996;

World Commission on Forests and Sustainable Development, 1999), partly because of paucity of information and the known interest of multinational companies in the region. The World Resources Institute (1996) observes that available data indicate slight increases in forest area but that these mask the state of the forest, much of which has been altered significantly in the last century. The major causes of this degradation are pollution, mainly the production of sulphurous and nitrous oxides from metal smelters, and the fallout from the Chernobyl explosion, and logging. The results of these and other activities are reflected in Figure 6.3, which shows that approximately 50 per cent of these northern forests are categorized as modified, fragmented or planted forests. Much of this degradation occurred during the 1930–50 period under the Soviet regime and its development programmes; it focused on the European/Ural zone initially and subsequently spread into Siberia. Today, economic decline means that increasing emphasis is placed on the Russian Federation's primary resources, including its forests, to generate income. There is considerable interest, especially from Japanese, US and South Korean companies, in Siberian forests as the demand for wood and wood products continues to rise. According to the World Resources Institute (1996) an agreement with South Korea's Hyundai Corporation has generated significant clear felling in the Primorsky region. As Figure 6.3 shows, the areas of forest under medium or high threat of alteration are situated in the east, which reflects the interest of Pacific Rim countries.

Apart from the loss of biodiversity that is associated with forest degradation and/or demise, the capacity of forests to absorb carbon is an important issue because of the role of forests in the regulation of global climates (Section 1.6). The role of 'Eurosiberian' boreal forests in this context has been examined by Schulze *et al.* (1999). Their data show that there is considerable variation between the productivity of Siberian and European boreal forests with the former achieving $c.$ 27 per cent of the latter. This is mainly because of climatic constraints, especially the reduced growing season in continental Eurasia. In both regions,

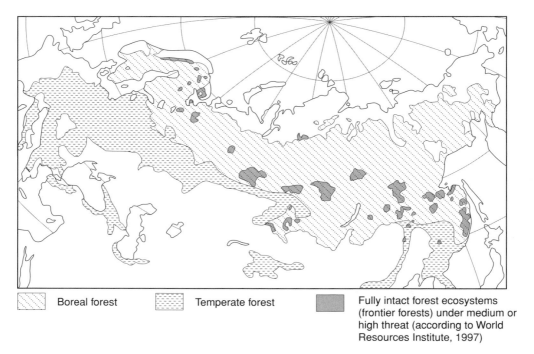

Boreal forest     Temperate forest     Fully intact forest ecosystems (frontier forests) under medium or high threat (according to World Resources Institute, 1997)

**FIGURE 6.3** The general distribution of forests in Eurasia. (Adapted from World Commission on Forests and Sustainable Development, 1999)

disturbances caused by logging, wind and fire influence the carbon cycle. Moreover, in Siberia Schulze *et al.* conclude that '. . . maintaining old growth pristine forest contributes more to the global net carbon sink than reforestation after logging'. It can only be deduced that large-scale logging enterprises in Eurasia will not only alter land cover but will also have a significant effect on the global carbon cycle.

In many European forests, management is changing in an attempt to reconcile forestry and conservation as well as to accommodate recreational activities. Toivonen (2000), for example, has examined various aspects of such attempts in Sweden and Finland where objectives involve a network of nature conservation areas juxtaposed with ecologically sustainable forestry. These new objectives also relate to European Union legislation designed to preserve biodiversity and encourage conservation, e.g. the EU Habitat Directive of 1998 and the selection of sites for inclusion in the Natura 2000 Network. The latter are sites considered typical of the boreal zone. The establishment of nature reserves of various sizes, often linked to, or within, forest areas that are otherwise exploited, is a key element in conservation, and access by the public may be restricted. Instead, designated recreational areas may be provided with visitor centres, car parks etc. As Toivonen points out, forestry practices in conservation and recreational areas must be suited to the objectives and will be necessary not only for management but also for restoration. The nature and scale of such practices, notably burning, selective felling etc., will vary according to the size of reserve and whether or not multi-use is important, as in National Parks. The implementation of management strategies, which are, of necessity, underpinned by legislation, will gradually bring about landcover alteration. This approach to forest exploitation is in stark contrast to that of eastern Siberia, as discussed above.

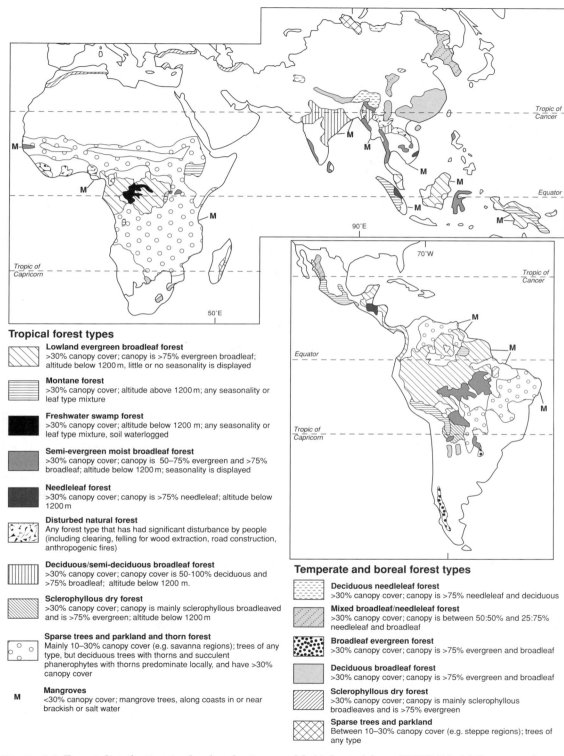

**Tropical forest types**

**Lowland evergreen broadleaf forest**
>30% canopy cover; canopy is >75% evergreen broadleaf; altitude below 1200 m, little or no seasonality is displayed

**Montane forest**
>30% canopy cover; altitude above 1200 m; any seasonality or leaf type mixture

**Freshwater swamp forest**
>30% canopy cover; altitude below 1200 m; any seasonality or leaf type mixture, soil waterlogged

**Semi-evergreen moist broadleaf forest**
>30% canopy cover; canopy is 50–75% evergreen and >75% broadleaf; altitude below 1200 m; seasonality is displayed

**Needleleaf forest**
>30% canopy cover; canopy is >75% needleleaf; altitude below 1200 m

**Disturbed natural forest**
Any forest type that has had significant disturbance by people (including clearing, felling for wood extraction, road construction, anthropogenic fires)

**Deciduous/semi-deciduous broadleaf forest**
>30% canopy cover; canopy cover is 50-100% deciduous and >75% broadleaf; altitude below 1200 m.

**Sclerophyllous dry forest**
>30% canopy cover; canopy is mainly sclerophyllous broadleaved and is >75% evergreen; altitude below 1200 m

**Sparse trees and parkland and thorn forest**
Mainly 10–30% canopy cover (e.g. savanna regions); trees of any type, but deciduous trees with thorns and succulent phanerophytes with thorns predominate locally, and have >30% canopy cover

**M    Mangroves**
<30% canopy cover; mangrove trees, along coasts in or near brackish or salt water

**Temperate and boreal forest types**

**Deciduous needleleaf forest**
>30% canopy cover; canopy is >75% needleleaf and deciduous

**Mixed broadleaf/needleleaf forest**
>30% canopy cover; canopy is between 50:50% and 25:75% needleleaf and broadleaf

**Broadleaf evergreen forest**
>30% canopy cover; canopy is >75% evergreen and broadleaf

**Deciduous broadleaf forest**
>30% canopy cover; canopy is >75% evergreen and broadleaf

**Sclerophyllous dry forest**
>30% canopy cover; canopy is mainly sclerophyllous broadleaves and is >75% evergreen

**Sparse trees and parkland**
Between 10–30% canopy cover (e.g. steppe regions); trees of any type

**FIGURE 6.4** Forest distribution in the developing world. (Adapted from UNEP World Conservation Monitoring Centre, 2001)

| | Land area (LA) ha × 10³ | Natural forest area ha × 10³ | Total forest area (TFA) ha × 10³ | TFA as % of LA |
|---|---|---|---|---|
| **A Africa** | 2,936,960 | 515,455 | 520,237 | 17.7 |
| Tropical Africa | 2,236,262 | 502,743 | 504,901 | 22.6 |
| Central Africa | 423,361 | 204,352 | 204,677 | 48.3 |
| **B Asia** | 3,073,436 | N/A | 503,001 | 16.4 |
| Insular Southwest Asia | 244,417 | 126,038 | 132,466 | 54.2 |
| Tropical Asia | 847,459 | 255,751 | 279,766 | 33.0 |
| **C South America** | 1,751,708 | 863,315 | 870,594 | 49.7 |
| Tropical South America | 1,385,678 | 822,385 | 827,946 | 59.8 |
| Brazil | 845,651 | 546,239 | 551,139 | 65.2 |
| **D Central America** | 241,942 | 74,824 | 75,018 | 31.0 |
| Costa Rica | 5106 | 1220 | 1248 | 24.4 |

Note: If total forest area exceeds natural forest area, the positive difference is due to plantations/afforestation

**TABLE 6.3** Data on the distribution of forests in the developing world. (Adapted from Food and Agriculture Organisation, 2001a)

## 6.3 Forest exploitation in the developing world

Figure 6.4 provides a generalized map of forests in the developing world. According to the Food and Agriculture Organisation (1999) approximately 55 per cent of the world's forests are located in developing countries. Estimates of distribution by continent are given in Table 6.3. In particular, the humid tropical regions of South and Central America, Central Africa, and South and insular Southwest Asia house some of the Earth's most biodiverse communities, which store a huge volume of carbon. The forests of these regions are experiencing major alteration through human agency, especially through the expansion of agriculture (see Chapter 5), logging and mining. Often, such activities are encouraged through development programmes fostered by governments and all three activities are frequently related. In many countries, extractive industries are brutal, with the harvesting of desired species resulting in the demise of many other species; in other countries

accidental extraction is reasonably well managed and replacement schemes are in operation. Nevertheless, there is a net loss of forest cover in the developing world annually as well as a substantial amount of forest degradation such as reduced canopy cover and fragmentation.

According to Whitmore (1997) some $6 \times 10^6$ ha of tropical forest are logged annually and most of this is virgin forest. Logging is a major activity in Brazil, which has a total forested area of $c.\ 550 \times 10^6$ ha (see Table 6.3). Much of this forest is in the Amazon basin, but forests are also present along the Atlantic coast and have been subject to modification since European colonization. Whilst logging involves the harvesting of selected tree species, clear felling involves the removal of all trees from large areas (see Section 6.2). Methods of logging can also cause deforestation and damage, which often goes unrecorded (see discussion in Putz *et al.*, 2000). For example, damage to other trees and plants is widespread from vehicular use, and fires can occur, especially after drought periods, reducing the forest cover further and altering soil carbon stocks. Such impacts contribute to degradation and fragmentation but, as Laurance (2000) has highlighted, the most significant impact of forest exploitation is the opening up of forest areas through the provision of roads and tracks, as discussed in Section 5.4. These provide access for shifting cultivators, ranchers and hunters, often culminating in complete forest removal (see below).

The issue of direct logging impact has been addressed by Putz *et al.* (2000; 2001) who have detailed the characteristics of various logging techniques and intensities. First, they point out that it is difficult to propose general management policies because of the wide variation that occurs in logging intensity, notably from less than $1\,m^3\,ha^{-1}$ to more than $100\,m^3\,ha^{-1}$. Nevertheless, a literature survey of logging intensities shows that the higher the intensity the greater the impact. Putz *et al.* also report that the proportions of soil and residual trees damaged by logging range from 5 per cent to 50 per cent. The actual methods employed in logging include chainsaw cutting and log removal in a process known as yarding. This latter may cause considerable immediate damage, especially where bulldozers are used, and especially in comparison with manual removal or removal by draft animals. The impact of yarding is particularly bad on land with moderate to steep slopes or on flooded land and, even where helicopter retrieval is used in difficult terrain, road or tracks are still essential. Other activities associated with logging that influence land cover include restrictions on tree removal such as the setting of minimum tree diameters, the retention of seed trees, and seedbed preparation such as weed clearance, herbicide treatments etc. In view of the adverse impact of many logging practices, efforts to encourage reduced-impact logging have been fostered. In particular, a substantial reduction in damage due to yarding can be achieved through considered design of roads and their maintenance. This, plus appropriate silvicultural methods to suit individual species and environments, requires adequate planning and good management, and both are often lacking. Moreover, road provision of any description provides access for forest invasion

and the impact that it generates. For example, Wilkie *et al.* (2000) have collected data on bushmeat acquisition in the Democratic Republic of Congo's (DRC's) tropical forests following road construction by logging concessions. Where roads provide access to animals and to markets, bushmeat consumption increases substantially. Wilkie *et al.* also discuss future prospects for the DRC's road network since most of it is in need of repair. This has implications both for future economic revival based on forest products and minerals after a civil war, and for forest invasion/destruction. Consequently, there is a need for a fully planned approach to road reconstruction. Wilkie *et al.* state that:

> Using GIS-based scenarios that combine information on road reconstruction and repair costs with market access, land use and land cover, and areas of ecological sensitivity, DRC's policymakers can avoid *ad hoc* or opportunistic road development, devising instead a strategic approach that prioritizes investment in revitalization of the road system.

A further example of the impact of road construction is given by Maki *et al.* (2001), who describe the land-cover changes associated with road construction between Iquitos and Nauta in the northwestern Peruvian Amazon. Over the construction period, between 1970 and 1995, land-cover changes included the expansion and establishment of settlement patterns, deforestation and changing agriculture to provide for new settlers.

Some of these issues have been addressed by Nepstad *et al.* (1999), who have shown that in Brazilian Amazonia, logging results in damage to $10,000\,km^2$ to $15,000\,km^2$ per year, which is not considered in deforestation inventories. They also suggest that, in drought years especially, damage by fires may be considerable so that, overall, estimates of annual deforestation may be in error by as much as 50 per cent. These findings are the result of a survey of wood mill operators in 75 logging centres in Amazonia, which account for more than 90 per cent of Amazonia timber production. The data of Nepstad are summarized in Figure 6.5, which shows that logging is particularly intense in the

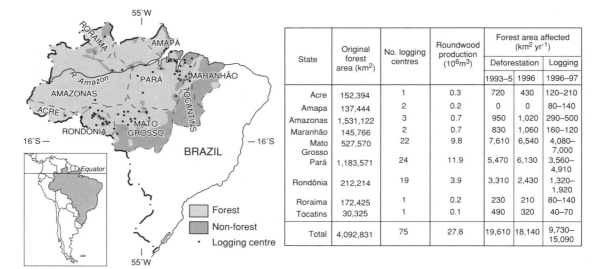

| State | Original forest area (km²) | No. logging centres | Roundwood production (10⁶m³) | Forest area affected (km² yr⁻¹) | | |
|---|---|---|---|---|---|---|
| | | | | Deforestation | | Logging |
| | | | | 1993–5 | 1996 | 1996–97 |
| Acre | 152,394 | 1 | 0.3 | 720 | 430 | 120–210 |
| Amapa | 137,444 | 2 | 0.2 | 0 | 0 | 80–140 |
| Amazonas | 1,531,122 | 3 | 0.7 | 950 | 1,020 | 290–500 |
| Maranhão | 145,766 | 2 | 0.7 | 830 | 1,060 | 160–120 |
| Mato Grosso | 527,570 | 22 | 9.8 | 7,610 | 6,540 | 4,080–7,000 |
| Pará | 1,183,571 | 24 | 11.9 | 5,470 | 6,130 | 3,560–4,910 |
| Rondônia | 212,214 | 19 | 3.9 | 3,310 | 2,430 | 1,320–1,920 |
| Roraima | 172,425 | 1 | 0.2 | 230 | 210 | 80–140 |
| Tocatins | 30,325 | 1 | 0.1 | 490 | 320 | 40–70 |
| Total | 4,092,831 | 75 | 27.8 | 19,610 | 18,140 | 9,730–15,090 |

**FIGURE 6.5** Data on logging activities in the Brazilian Amazon. (Adapted from Nepstad *et al.*, 1999)

states of Mato Grosso, Para and Rondonia, where rates of deforestation are highest. Using data provided by landowners on areas experiencing fire, coupled with data on fire damage from satellite imagery, Nepstad *et al.* determined the area logged (see Figure 6.5). When compared with data on deforestation (from Brazil's Space Agency, see reference in Nepstad *et al.*) logging affects a further area of between *c.* 50 per cent and *c.* 90 per cent of that deforested. Moreover, much of this impact is in unlogged areas. Thus rates of forest damage are far higher than those revealed by straightforward deforestation data. Nepstad *et al.* also suggest that, because of this deficiency, estimates of carbon release from Amazonia are likely to be underestimates. They calculate that in 1996 logging released between 4 and 7 per cent of the net annual carbon release due to deforestation in Brazilian Amazonia, i.e. *c.* $0.3 \times 10^9 \, \text{Mg yr}^{-1}$. This figure could increase considerably if the incidence of surface forest fires increased. This research suggests that not only is deforestation/forest damage more extensive than is formally recorded but that extinction and biodiversity loss are more serious than hitherto considered. Recent research by Siegert *et al.* (2001) adds weight to the suggestion that

logged forests experience further damage by fires. Their work on East Kalimantan, Borneo, shows that $2.6 \times 10^6$ ha of forest was burned in the vast fires of 1997–98, which accompanied the drought associated with El Niño/Southern Oscillation (ENSO). Using satellite imagery and ground survey, Siegert found that almost twice as much damage occurred in recently logged forests than in old logged forests, and that least damage occurred in pristine forests.

That rates of deforestation increased in the Brazilian Amazon in the late 1990s is shown by data from Brazil's Space Agency (INPE) as quoted in Laurance *et al.* (2000). The latter also state that many logging industries operating in the region are not Brazilian, and that since 1996 companies from Malaysia, Indonesia, Taiwan and others have invested more than US$500 × 10⁶; these companies control 12 × 10⁶ ha of Amazon forest and there are another 400 Brazilian timber companies. Thus, and in common with forest extraction in Siberia, pressures for land-cover change are generated within the host counties and from outside them. On the one hand, host countries are attempting to export their primary resources to earn foreign currency and, on the other, the growing demand for wood and wood products worldwide provides a ready market, especially in countries with little forest cover or where forests are protected. The world demand for agricultural products such as soybeans is also prompting deforestation in countries such as Bolivia (see Steininger *et al.*, 2001) and Brazil (Fearnside, 2001) though, in the latter, savanna lands have been affected more than rain forest areas.

The problems described above are inherent in all tropical forest regions. Even very selective logging can have a substantial impact on land cover, especially in the long term. This is illustrated by a study in the moist evergreen forest of the Western Ghats, South India. Pelissier *et al.* (1998), for example, have monitored vegetation change in two large plots, one of which was selectively felled in 1979–80, whilst the other was in its natural state. In terms of the species dynamics, there were notable differences between the plots: in the logged plot lower canopy and intermediate stratum species had much higher mortality rates than in the unlogged plot. This will affect the species composition and physiognomy of the eventual forest covers. In addition, damage and fragmentation are likely to have an impact on animal communities, and these may affect tree regeneration through seed predation and dispersal. These relationships can be influenced

further by the degree of protection afforded to forest areas, as is illustrated by the work of Guariguata *et al.* (2000) on two rain forest sites near the Caribbean lowlands of Costa Rica. They examined seed removal by terrestrial mammals at La Selva, a logged but protected site, and at Tirimbina, a logged but unprotected site. Their results show that hunting reduced animal populations, especially rodents, in the latter to such an extent that the seeds of many timber species were not well dispersed; in the protected area the seeds of timber species were much more widely dispersed and thus more likely to contribute to forest generation.

Similarly, the loss of biomass from forests in other parts of the tropical world is a significant contributor to global carbon emissions. Reference has already been made to carbon emissions from the Brazilian Amazon (see above), whilst emissions from tropical Asia have been discussed by Houghton and Hackler (1999). The latter's data are given in Table 6.4, which shows that, between 1850 and 1995, major adjustments in carbon storage occurred in parallel with land-use change. In particular, the area of forest declined by 37 per cent during this period, causing substantial declines in carbon storage in the biomass and soils. Much of this forest has been replaced with permanent croplands and this transformation has resulted

| | Pre-disturbance | 1850 | 1995 | % change 1850–1995 |
|---|---|---|---|---|
| **A** Area of forest ha $\times$ $10^6$ | 546 | 467 | 294 | 37.0 |
| Carbon in vegetation Pg | 100 | 76 | 32 | 57.9 |
| Carbon in Forest Pg | 100 | 73 | 27 | 63.0 |
| Carbon in soil Pg | 51 | 44 | 27 | 38.6 |
| **B** Emissions due to wood harvest and land-use change 1850–1995 | 43.5 PgC | | | |
| % of world total carbon emissions due to land-use change | c. 33.3 | | | |
| % of World total carbon emissions | c. 3.0, i.e. 7.7 PgC | Of total of 253 PgC | | |

**TABLE 6.4** Data on carbon storage in and emissions from forestry and land-use change in tropical Asia. (Adapted from Houghton and Hackler, 1999)

in the release of 33.5 PgC, i.e. about 75 per cent of total emissions. A further 11.5 PgC was released from the remaining forests due to shifting cultivation, logging and fuelwood collection, whilst the establishment of plantations caused 1.5 Pg C to be withdrawn from the atmosphere. In recent years, Houghton and Hackler state that deforestation rates in tropical Asia have declined, but that emissions of carbon from fossil-fuel use have increased. Whilst logging is not a major direct cause of carbon emissions in Asia it contributes its share, and as discussed above, logging activities are often the first stage in opening up pristine forests for other, more destructive and widespread, activities.

## 6.4 Afforestation in the developed world

Afforestation programmes are widespread throughout the world because wood and wood products are important everywhere, and because the overall ecological and environmental benefits are widely recognized despite certain drawbacks. According to the World Commission on Forests and Sustainable Development (1999), plantation forests now occupy $c$. $135 \times 10^6$ ha worldwide; about 75 per cent are in the temperate regions (mostly developed nations) and $c$. 25 per cent are in tropical and subtropical regions (mostly developing nations). The latter are addressed in Section 6.5.

| | Area of plantation ha $\times$ $10^6$ |
|---|---|
| Europe | 12 |
| Russian Federation | 17 |
| USA | 13 |
| Japan | 10 |
| New Zealand | 1.5 |
| Australia | 1 |

**TABLE 6.5** The area of plantation forests in selected developed countries. (Adapted from Food and Agriculture Organisation, 2001a)

In contrast, the Food and Agriculture Organisation (1999) suggests that there are only $60 \times 10^6$ ha of plantation forests worldwide, the distribution of which is given in Table 6.5. The difference in values probably reflects the fact that it is often difficult to distinguish between natural forests that are intensively managed, including replanting programmes, and forests that can be described as true plantations. As the Food and Agriculture Organisation (1999) points out, the distinction is especially difficult to make in many parts of the developed world, notably in Europe, where boreal forests in particular have been subject to intense management for at least two centuries. In Australia and New Zealand, however, and to a large extent in Japan, there is a clear distinction between natural forests and plantations. In addition, land-cover characteristics have been altered through widespread small-scale tree planting designed to provide fuelwood and shelterbelt habitats for wildlife, and as components of agroforestry systems (Section 5.7).

Afforestation programmes in the developed world have been undertaken to provide wood and wood pulp etc., whilst others have been designed for different or dual objectives, e.g. for soil and water conservation and recreation. Management strategies, and hence land cover, vary according to the objective. In many cases afforestation has been undertaken using non-native species; in the UK, for example, Sitka spruce (*Picea sitchensis*), Norway spruce (*Picea abies*) and Corsican pine (*Pinus nigra*) are commonly used, and none are native. Similarly, in New Zealand, non-native species that have been widely planted include Monterey pine (*Pinus radiata*), poplars (*Populus spp.*) and willows (*Salix spp.*). Often, monocultures predominate and this in itself represents a major change from natural land cover even if this was forest. In addition, many afforestation programmes have been encouraged by government action through the creation of national forestry services, e.g. the Forestry Commission in the UK. However, this does not mean that afforestation is confined to public land as policies linked with tax incentives, grants or subsidies also encourage afforestation on private land. Moreover, small-scale forestry activities are being encouraged on farmland in many parts of Europe (for example, to provide alternative incomes in place of crop and animal production). The situation in Denmark has been reviewed by Madsen (2002), who points out that an afforestation programme was introduced in 1989 in response to EU directives and that, since 1991, farmers have been able to apply for grants to finance afforestation programmes. On a larger scale, in Hawaii, many thousands of hectares have been planted with Sydney blue gum (*Eucalyptus saligna*) to replace sugar cane, which has become uneconomic (Binkley and Resh, 1999), and thus to provide an alternative income. Another example is the planting of mallee eucalypts on wheat farms in southwest Australia; this has the dual purpose of reducing soil salinization caused by vegetation clearance and the provision of an economic return through the production of eucalyptus oil (S.J. Bell *et al.*, 2001).

The fact that plantations have a higher productivity rate than natural forests is due to intense management, which is essential to guarantee a crop in a reasonable space of time. Management will vary according to the species of tree involved and is important at all stages between the establishment of seedlings and the harvest. A major objective of plantation management is to remove competition and to improve conditions for growth, as is the case in agriculture, and various silvicultural techniques to improve productivity, as described in Table 6.6. All of these influence land cover to a greater or lesser degree, though harvesting methods are particularly important. For example, clear cutting, selective cutting and shelter-wood cutting give rise to quite different land-cover characteristics (see Table 6.6). Whether forestry is sustainable or not, it promotes land-cover alteration.

Forestry is an important component of New

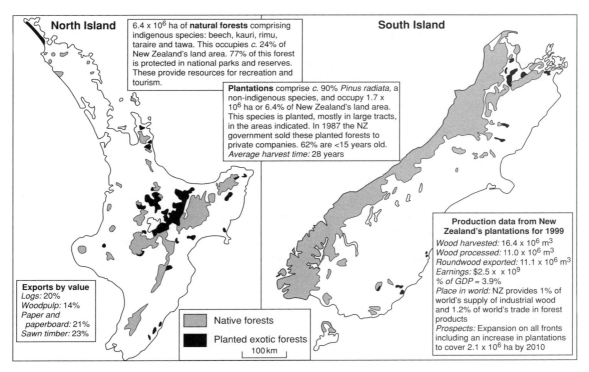

**FIGURE 6.6** The distribution of forests in New Zealand, with data on forest industries. (Adapted from New Zealand Forest Service, 1977; data from Ministry of Agriculture and Forestry (MAF), New Zealand, 2001)

**1  INTERMEDIATE TECHNIQUES BETWEEN ESTABLISHMENT OF SEEDLINGS AND THE HARVEST**

A  *The regulation of species composition and growth rates*

(a)  Release cuttings: the removal of larger trees of competing, undesirable species in stands of saplings or seedlings; this may involve the use of herbicides.

(b)  Improvement cuts: the removal of diseased or poor-performing species at the post-sapling stage to improve quality.

(c)  Thinning: the density of the stand may be reduced in order to promote growth of the remaining trees; this can be achieved by low/high or mechanical thinning, respectively through cutting lower/higher canopy species or the clearance of strips; thinning may be undertaken periodically, e.g. every ten years.

(d)  Fertilizing: this is important to promote growth productivity where some nutrients are deficient, e.g. nitrogen.

**2  PREDICTION OF FOREST GROWTH**

Prediction of growth helps determine the volume of the annual cut. Forest-growth models are widely used and are based on observable relationships between growth, age, site, quality and competition.

**3  REGENERATION**

A  *Natural regeneration*

Natural regeneration occurs in a variety of ways, including advance regeneration. This includes seedlings and saplings that were established before harvesting, sprouting from stumps and regeneration from the seed bank. How effective this is, and how rapidly it occurs, depends on a variety of factors, including the quantity and regularity of seed production, germination success, local microclimate, competition and predation by insects and mammals.

B  *Artificial regeneration*

Artificial regeneration is more manageable and predictable than reliance on the natural process; it also ensures the regeneration of desired species in the desired mix.

(a)  Direct seeding by hand or mechanically: this is especially appropriate where it is necessary to seed large areas rapidly and is most successful where logging is thin or absent

(b)  Planning: this is more effective than direct seeding but is more expensive because of the extra costs involved in the nursery production of seedlings.

**4  HARVESTING METHODS DEPEND ON THE NATURE OF THE STANDS**

A  *Even-aged stands*

(a)  Clear-cutting: all trees are removed from a given area in a relatively short time but it is no longer a favoured technique because of adverse aesthetic and environmental impacts such as the acceleration of soil erosion.

(b)  Seed-tree methods: these involve the maintenance of scattered mature trees to act as a seed source.

(c)  Shelterwood: trees are maintained to provide seed and shelter for new seedlings; once the seedlings are established, the older trees can be removed.

(d)  Coppicing: this involves the removal of the upper portions of mature trees so that subsequent sprouts from the old bole can be harvested; it is not widely used.

Using these methods in even-aged stands, it is only possible to approach sustained yield by cutting a small proportion of a forest annually; over a period of 50–100 years the entire forest is cut.

B  *Uneven-aged stands*

Uneven-aged stands are managed by selective cutting, which involves the harvesting of scattered trees or groups of trees. This approach ensures that the stands are maintained and that disturbance to the ecosystem is minimal.

*Source:* based on Lorimer (1982) and Smith (1986)

**TABLE 6.6** Silvicultural techniques (Mannion, 1997)

Zealand's economy. According to MAF in New Zealand (Ministry of Agriculture and Forestry, New Zealand, 2001), forestry employed 21,000 people and generated 3.9 per cent of New Zealand's Gross Domestic Product (GDP) in 1999. Data on New Zealand's forests and forest industries are given in Figure 6.6, which shows that New Zealand enjoys a substantial forest cover of *c*. 8.1 × $10^6$ ha or 30.4 per cent of the total land area. The native forests comprising indigenous species occupy upland areas and the west coast of South Island has a cover of temperate rain forest. For the most part this does not support an extractive forest industry, but is important for recreation and tourism. In contrast, planted forests occupy 1.7 × $10^6$ ha, or 6.4 per cent of New Zealand's land area, mostly in the central part of North Island. Planting has occurred on a large scale since the 1950s and it is from these forests, which comprise *c*. 90 per cent *Pinus radiata*, that wood and wood products are generated. This species is not native to New Zealand and has thus been controversial. Economically, however, it is desirable because it is fast growing, with the average time between planting and harvesting only *c*. 28 years. As *c*. 61 per cent is 15 years old or less, there will be a major increase in wood production during the next decade (Ministry of Agriculture and Forestry, New Zealand, 2001). The planting of this 1.7 × $10^6$ ha represents in itself a significant land-cover change since the 1950s, whilst management has also altered land-cover characteristics even if these concern only different stages in forest growth.

New Zealand has also experienced land-cover changes involving the planting of a range of woody species in various configurations for various purposes. As Wilkinson (1999) has discussed, this has involved the planting of non-indigenous poplars (*Populus spp.*) and willows (*Salix spp.*) for riverbank stability, shade, windbreaks and small woodlots. Such planting programmes, encouraged by government subsidies, were widespread in the 1960s and 1970s, when *c*. 2 × $10^6$ poplars were planted in programmes designed to control soil erosion. Wilkinson argues that 1 × $10^6$ ha of New Zealand's pastoral lands are in need of similar tree planting to combat soil erosion.

This will also constitute a significant land-cover change should it occur.

There are some similarities between New Zealand's forestry enterprises and those of Japan. First, both have a substantial proportion of plantation forest; second, most plantation forest comprises coniferous species; and, third, both have developed forestry industries based on plantation forests and have protected their remaining natural forests. One major difference, however, is that Japan is a major importer of wood and wood products from other parts of the world, including New Zealand. As shown in Table 6.1, Japan consumes more than three times the industrial roundwood it produces and half as much again of sawnwood. Overall, Japan has 25 × $10^6$ ha of forest, which covers *c*. 66 per cent of its land area, as shown in Figure 6.7. This is a very large forest cover for a temperate country that is heavily populated. Of this, approximately 73 per cent is planted, most of which comprises coniferous species including cypress (*Chamaecyparis obtusa*) and cryptomeria (*Cryptomeria japonica*), known as *hinoki* and *sugi* respectively. This is the basis of Japan's forestry industry.

However, the Japanese Ministry of Agriculture, Forestry and Food (MAFF) also provides data on the distribution of forest for purposes other than extraction. These data are given in Table 6.7, which shows that forests have been retained or planted for a variety of purposes relating to hydrology, soil conservation, climatic amelioration and, especially, recreation. Planting has been especially prevalent since the 1950s in Japan's mountainous regions to help control floods, landslides and river siltation, as well as to provide a forest industry. Nevertheless, Japan, in common with many European countries, has experienced rural depopulation and this has led to a decline in the forest industry in recent years. As Knight (2000) has discussed, this decline and neglect of forest lands has given rise to new types of forest land use, including forest recreation and tourism. These activities have been encouraged by the government, and the significance of recreation forests is reflected in the data given in Table 6.7. This new use of forests has, however, led to management programmes

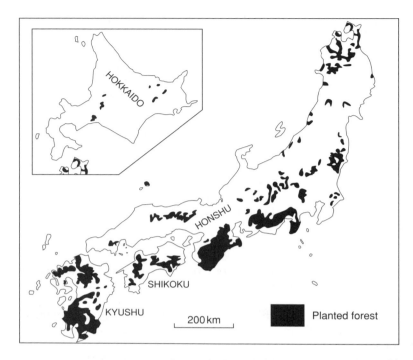

**Figure 6.7** Distribution of planted forests in Japan
*Source:* based on Fujita, 1985; reproduced in Himiyama, 1992

**Extent of forest**

| | Area of needleleaved ha × 10³ | | Area of broadleaved ha × 10³ | |
| --- | --- | --- | --- | --- |
| | Natural | Planted | Natural | Planted |
| 1980 | 9435 | 24,702 | 149 | 2500 |
| 1990 | 10,050 | 23,771 | 203 | 2401 |

**A  Non-extractive forest types by use (area = ha)**

| | |
| --- | --- |
| Headwater conservation forest | 6,548,595 |
| Soil loss prevention forest | 1,846,129 |
| Crumbling soil control forest | 43,308 |
| Shifting sand control forest | 16,244 |
| Windbreak forest | 54,596 |
| Flood control forest | 759 |
| Tide water control forest | 12,032 |
| Drought disaster control forests | 38,126 |
| Fog prevention forest | 51,277 |
| Recreation forest | 549,609 |
| Scenic beauty protection forest | 16,078 |
| Others | 45,223 |
| Total | 8,223,976 |

**Table 6.7** Data on the area and uses of Japan's forests (from Ministry of Agriculture, Forestry and Food, Japan, 2001)

designed for this purpose. Gradually, alterations are being made to primary and planted forests to provide facilities and this gives rise to land-cover alteration. Knight also observes that the national reforestation programme is still in operation, although its emphasis has changed. Tree planting is now directed at multipurpose forests, and there is an anticipation that forestry will re-emerge as a major industry. The trees being planted now include broadleaved species, e.g. varieties of cherries (*Prunus spp.*) and maples (*Acer spp.*) as well as conifers. Consequently, Japan's extensive land-cover of trees is in a state of constant flux, as are plantations and forests everywhere.

## 6.5 Afforestation in the developing world

According to the World Commission on Forests and Sustainable Development (1999), *c.* 25 per cent of the world's plantations are in the tropics and subtropics. As Table 6.8 shows, the largest area of plantations is in Asia, with $21 \times 10^6$ ha in China and $20 \times 10^6$ ha in India. Plantations have been established for a variety of purposes; industrial production of wood and wood products is the major purpose but, as in Japan (see Table 6.7), other objectives include soil and water conservation. In addition, plantations have been established to exploit specific tree crops, notably rubber, coconut and oil palm, and many millions of hectares are occupied by shrub species with commercial value, e.g. tea and coffee. The Food and Agriculture Organisation (1999) states that *c.* 57 per cent and 43 per cent of the plantation area for industrial and non-industrial uses (see Table 6.8) comprises hardwood species and softwood species respectively; eucalypts occupy *c.* 30 per cent of the area planted with hardwoods and pines occupy *c.* 61 per cent of the area planted with conifers. The area under plantation for both forest and non-forest species is increasing annually, though not at a rate that compensates for the deforestation and degradation of natural forests. Plantations do, however, compensate to a certain extent for the loss of carbon from biomass due to deforestation and, in some countries, plantations supply most of the required industrial wood and so take pressure off remaining natural forests. This is not the case for the developing world as a whole, as is illustrated in Section 6.3. It must

**A   Plantation areas**

|  | Industrial ha × 10³ | Non-industrial ha × 10³ | Total ha × 10³ | Annual rate of planting ha × 10³ |
|---|---|---|---|---|
| Africa | 3787 | 3025 | 5861 | 288 |
| Asia and Oceania | 31,781 | 21,216 | 52,997 | 2330 |
| Latin America | 7826 | 2134 | 9960 | 401 |
| Total | 43,394 | 26,375 | 69,769 | 3019 |

**B   Plantation areas of non-forest species**

|  | Rubber ha × 10³ | Coconut ha × 10³ | Oil-palm ha × 10³ | Total ha × 10³ |
|---|---|---|---|---|
| Africa | 529 | 461 | 922 | 1912 |
| Asia and Oceania | 8718 | 10,546 | 4587 | 23,851 |
| Latin America | 238 | 269 | 265 | 772 |
| Total | 9485 | 11,276 | 5774 | 26,535 |

**TABLE 6.8** Plantation areas in the developing world in 1995. (Adapted from Food and Agriculture. Organisation, 2001a)

also be noted that afforestation may be undertaken for reasons other than those related to economics. Many Andean slopes, for example, have been clothed with eucalypts to provide stability and to reduce soil erosion. This is also common in many African countries and in Madagascar.

Plantation forests have received much criticism for a variety of reasons. First, in some countries, native forests have been destroyed and replaced with plantations. Second, most plantations involve the monoculture of specific tree species because this is economic; management can be applied uniformly and harvesting is relatively easy. Third, most plantations comprise large blocks of even-aged stands with regularly spaced trees. None of these characteristics encourages a diversity of wildlife, and instead produce landscapes that are monotonous; the trees may be particularly susceptible to fire and outbreaks of insect pests, viral and fungal diseases that could be seriously damaging. On the other hand, plantations provide employment and an economic return as well as a means of sequestering carbon. As the Food and Agriculture Organisation (1999) indicates, there have been attempts to diversify into mixed-species plantations and, where large blocks of land are now scarce, as in much of Asia, farmers have been encouraged by governments via 'outgrower' schemes to plant trees in blocks for eventual sale to wood-processing industries. This could be described as another form of agroforestry (see Section 5.7), and can be beneficial in terms both of income diversification and environmental protection. Moreover, plantation forestry is likely to become influenced by biotechnology, especially genetic engineering, which will allow the production of trees with specific characteristics that are economically desirable. For example, the acceleration of the growth of specific hardwood species, oil-producing or fruit-producing trees could change the character of plantations in the next few decades (see discussion in Chapter 10).

Another reason why plantations are likely to increase in extent in the future is because of their capacity to sequester carbon. This may be politically and economically motivated in the wake of international agreements such as the Kyoto Protocol. This was established under the Framework Convention on Climate Change (FCCC) and was adopted by signatories in December 1997. This protocol recognizes that measures are required to limit and control carbon emissions to the atmosphere, and international and national targets have been set. In particular, the industrialized nations have agreed to reduce emissions by 5 per cent of those in 1990. As well as the trading of emissions between nations, the expansion of forests and plantations provides possibilities for emission mitigation. As the World Commission on Forests and Sustainable Development (1999) states:

> ... with respect to both carbon sequestration and how the accounting is done for the role forests play as sources and sinks of carbon, through mechanisms such as the clean development mechanism and joint implementation, the Kyoto Protocol is of great consequence both for forests and biodiversity use and their conservation.

The combination of carbon emission mitigation and genetic engineering could mean that, in the twenty-first century, considerable land-cover and land-use change will occur as plantations expand.

The prospect of carbon sequestration in tropical regions in the context of the reforestation of abandoned agricultural land has been examined by Silver *et al.* (2000). They consider the impetus for this as the creation of so-called carbon credits (an outcome of the Kyoto Protocol), which are allocated to units at scales from individual companies to countries. If such units emit more than their allocation of carbon the excess can be nullified if they create a carbon sink on the landscape and, of course, forests are just such a sink. Thus, the dynamics of forest regeneration and plantations in terms of their capacity and patterns of carbon sequestration, as well as the most appropriate management techniques to encourage optimum carbon sequestration, are of primary concern. On the basis of published literature, Silver *et al.* demonstrate that there is considerable potential for this, especially for the first 40 to 80 years when carbon sequestration due to growth and soil carbon accumulation are greatest. For periods longer than this, there may be little if any carbon accumulation. However, active management to encourage the early growth stages in favour of old forest would counteract this to some extent. As Fearnside (1999) has suggested, a dynamism could be created if wood from afforested areas and/or plantations were used as a biomass fuel to provide an alternative to fossil fuels.

Plantations for biomass-based energy are already widespread, especially in Africa and Asia. Many have been planted to reduce deforestation caused by domestic fuel needs; this is acute in many rural areas in developing countries, especially the poorest countries, where biomass fuel is essential for almost all domestic heating and cooking. One such example is the Blantyre fuelwood project in southern Malawi, which was initiated by the government in the mid-1980s. According to Kalipeni and Feder (1999), some 8500 ha was planted with exotic eucalyptus trees which, together with the management of local indigenous woodlands, was intended to supply the city of Blantyre with its wood energy needs. Through management, a second objective was to protect native woodlands. From its inception, however, the project experienced difficulties, mainly because of lack of consultation with local people. For example, subsistence arable farmers were removed from their land so that afforestation could take place and pastoral activities were disrupted. This region was already experiencing problems in so far as it was receiving migrants displaced by the civil war in neighbouring Mozambique, and displaced farmers from nearby Thyolo and Mulanje where land was required for tea plantations. Coupled with high population rates these factors compounded an already acute fuelwood situation. Nevertheless the afforestation caused conflict rather than providing relief, a situation attributed by Kalipeni and Feder to a 'top-down' response by government rather than a 'bottom-up' approach involving discussion and participation at the local level.

As Islam *et al.* (1999) point out, deforested, degraded and marginal lands provide suitable sites for these so-called energy plantations, but the degree of success depends on the choice of species, especially in relation to site suitability and biomass production. Moreover, decisions must be made as to whether indigenous or exotic species are most appropriate. Using data from an afforestation venture on the campus of the University of Chittagong, Bangladesh, a particularly poor country with a fuelwood crisis, Islam *et al.* compared the suitability of both indigenous and exotic species. They found that *Cassia siamea* and *Derris robusta*, both indigenous, as well as *Acacia auriculiformis*, an exotic, were particularly successful; as well as being fast growing, these species survived the soil degradation that had occurred following the removal of the original tropical semi-evergreen mainly dipterocarp forest, and the invasion of grasses and shrubs. A major reason for the success of these species is that they all have root nodules that are occupied by nitrogen-fixing bacteria. This counteracts nitrate deficiency in the soil, and the survival of the trees results in litter production and thus enrichment of the soil with organic matter. This capacity for nitrogen enrichment is also a key to the reclamation of land damaged by pollution and activities such as mining, as discussed in Chapter 7.

However, it is also important to note that other factors may influence afforestation, especially in relation to communal activities, in countries like Bangladesh. For example, Khan (1998) has drawn attention to land tenure and land ownership as a deterrent to the development and success of afforestation. Land ownership and tenure place constraints on afforestation etc. in other parts of the developing world. For example, Klooster (1999) has discussed these issues in relation to rural environments in Mexico. Despite the government's recognition of local tenure rights and its devolvement of management to local groups, community forestry is still impaired by numerous factors: corruption, the benefits to local social elites and timber-smuggling operations.

## 6.6 Conclusion

The Earth's forest cover has been altered considerably in the last 5000 years. Most of this alteration is the consequence of human activity, especially since the advent of agriculture. Most removal of the forest cover has occurred in the temperate zone, notably in Europe, which is the most disturbed of all the continents (see Section 3.5). Large areas of forest have been replaced with arable and pastoral agricultural land and, more recently, with urban agglomerations. What remains of Europe's forests are intensively managed to provide an economic return. This is particularly important in Scandinavia. Forest exploitation is also an economic mainstay in other parts of the developed world. In the USA, Canada and Russia, the extraction of wood contributes significantly to the Gross Domestic Product and to export earnings. Both state-owned and private companies are involved in forest management and in the Russian Far East many foreign companies have logging concessions. All types of management cause land-cover alteration but clear felling is particularly drastic. It is often followed by replanting, or natural regeneration is allowed to take place; both are components of the forest dynamic. The forest exploitation in these American and Asian regions is focused mainly on boreal forests and the major species extracted are conifers.

Forest exploitation is prolific in developing countries, in many of which forest products are very important economically. Logging is particularly important and is thus a major agent of land-cover alteration. As in the case of Russia's Far East, many of the logging companies are foreign and the stimulus for forest extraction is mainly economic as wood exports generate income. It can cause considerable damage due to the movement of logs to transport points, but its most significant impact is the provision of roads and tracks, which facilitate the movement of agriculturalists into the forest. Forest regeneration also creates land-use change.

Afforested areas and plantations also represent an altered land cover from the original. Both occur in the developed and developing worlds, but plantations are more abundant in the latter. They have replaced a former land-cover type, which may or may not have been forest, and their objective is the provision of an economic crop, which may be wood or a product such as palm oil or coconuts. Afforested areas and plantations are managed to optimize productivity, but most are mono-cultural, in stark contrast to the previous land cover. Afforested areas may also be managed for purposes other than forest products. The provision of facilities for recreation and tourism, for example, leads to diversified land cover within forests.

All types of forest exploitation and creation influence the global carbon cycle. Deforestation and timber extraction cause carbon to be released into the atmosphere from forests and forest soils. Conversely, forest regeneration, afforestation and plantations sequester carbon from the atmosphere. Globally, human activity results in the release of more carbon from forests than is sequestered through afforestation etc. Moreover, the allocation of carbon quotas under the 1997 Kyoto Protocol is likely to encourage the creation of carbon sinks in order to offset emissions from fossil fuels. Forests are amongst the best carbon sinks in the biosphere, so the Kyoto agreement is likely to foster tree planting on large and small scales.

## Further reading

**Backman, C.A.** 1999: *The Forest Industrial Sector of Russia. Opportunity Awaiting*. Parthenon Publishing, Paris.

**Evans, J.** (ed.) 2001. *The Forest Handbook*. 2 Vols. Blackwell, Oxford.

**Food and Agriculture Organisation** 1999: *State of the World's Forests*. FAO, Rome; also at www.fao.org (accessed 2001).

**Gibson, C.C., McKean, M.A. and Ostrom, E.** (eds) 2000: *People and Forest. Communities, Institutions and Governance*. MIT Press, Cambridge, Massachusetts.

**Goldsmith, B.** (ed.) 1998: *Tropical Rain Forest: A Wider Perspective*. Chapman and Hall, London.

**World Commission on Forests and Sustainable Development** 1999: *Our Forests Our Future*. Cambridge University Press, Cambridge.

**Woodwell, G.M.** 2001: *Forests in a Full World*. Yale Univesity Press, New Haven.

# 7

# Land use and land cover related to mining

## 7.1 Introduction

The retrieval of inorganic substances, which are of practical use and value to society, from the Earth's surface or crust can be achieved in numerous ways. The obvious methods are mining and quarrying. These methods are employed to obtain a wide range of substances, or resources, which can be classified into three groups: energy resources; metalliferous minerals; and non-metalliferous minerals. Whatever the method and whatever the resource, the extraction causes land-cover change at the site of extraction. In some cases (for example, where deep mining to obtain minerals at depth within the Earth's crust is used), land-cover alteration may be relatively small scale; in others (for example, where strip mining is used), land-cover alteration may occur over many hectares.

Land-cover alteration may also occur well beyond the site of extraction through contamination and pollution, as well as the construction of mining camps or towns, and roads and tracks to facilitate transport of the resource. The latter may have a particularly significant impact in forested regions of the developing world because the infrastructures provided by

the mining companies may be used by other groups to infiltrate the forest and cause widespread land-cover change for shifting agriculture and ranching; such activities may also be encouraged if mine workers provide a ready market. Mining may thus have a similar affect to logging, as discussed in Section 6.3. Contamination and pollution may also affect an area far beyond the site of extraction, especially if mineral processing occurs on-site. Dust, acid rain and, where appropriate, heavy metals may cause land-cover alteration; such substances entering drainage systems will cause changes in aquatic systems, riparian vegetation communities and wetlands.

The recovery and reclamation of mine-damaged land may either be a slow or relatively rapid process. Recovery through the natural process of vegetation succession will be slow, taking hundreds of years to recover fully. Where reclamation schemes are instituted, recovery may be relatively rapid; in two decades, for example, some types of mine-damaged land may be reclaimed and have a clear afteruse. Reclamation schemes are, however, expensive and time consuming, and economic returns from the reclaimed land may be limited. Moreover, each type of extraction method, and type of resource, generates specific problems, which include physical and chemical constraints, e.g. instability, high concentrations of toxic substances etc. Restoration of land cover may involve a vegetation type that is different from the original and, in some cases, reclamation may involve the creation of wetlands.

The factors that influence the spread of mining and/or abandonment are many and varied. Economics is a big factor involving supply-and-demand relationships and trade relationships. The quality and quantity of the mineral will dictate the lifespan of the mine and economic returns, whilst technological developments may extend the life of a mine by improving extraction techniques and the viability of low-grade ores. Equally, technological developments may result in mine closure, as has happened in parts of Europe in relation to a decline in coal use, and a switch to natural gas and oil. Such shifts may also be encouraged by political motivation; gas and oil are cleaner than coal in terms of emissions.

Similarly, there are many factors that influence reclamation, not least of which is legislation. In many countries, mining licences involve requirements for reclamation, a proposal for which is often necessary in the application for a licence. Financial penalties are imposed if agreed conditions are not honoured. Where there is no legislation, or if enforcement is weak, there may be no prescribed aftercare; reclamation will proceed slowly and the contaminated land may constitute a hazard. Scientific research is also an essential component of reclamation; for example, the reclamation of spoil heaps will involve suitable plant species and management procedures, such as regular fertilizer applications. In addition, afteruse must be appropriate – for example, grazing land cannot easily be created on land with high concentrations of toxic substances.

## 7.2 Mineral extraction: some basic facts

Tables 7.1, 7.2 and 7.3 give statistics on the quantities of selected minerals that are currently

being extracted annually. As Table 7.1 shows, coal mining is particularly extensive, with China and the USA being the world's largest producers. Coal can be mined in a variety of ways: by deep, opencast, drift or strip mining. The latter three have the greatest impact on land cover since large areas of the land surface are altered. The extraction of oil and natural gas causes less disruption at the point of extraction, but the transport of both commodities to points of use via pipelines can affect large areas, as is the case in northern Canada, Alaska and Siberia. In addition, the processing of oil prior to its use requires the construction of refineries. These usually comprise extensive plant and are often constructed in coastal regions with easy

| **A  Petroleum** $\times$ **10³ tonnes** | |
| --- | --- |
| Saudi Arabia | 441,200 |
| USA | 353,500 |
| Russian Federation | 323,300 |
| Iran | 186,600 |
| Mexico | 172,100 |
| Venezuela | 166,800 |
| China | 162,300 |
| World | 3,589,600 |

| **B  Coal** $\times$ **10³ tonnes oil equivalent (toe)** | |
| --- | --- |
| USA | 570,700 |
| China | 498,000 |
| Australia | 155,600 |
| India | 154,300 |
| South Africa | 118,800 |
| Russian Federation | 115,800 |
| Poland | 68,100 |
| World | 2,103,500 |

| **C  Natural gas m³** $\times$ **10⁶** | |
| --- | --- |
| Russian Federation | 545,000 |
| USA | 555,600 |
| Canada | 167,800 |
| UK | 108,100 |
| Algeria | 89,300 |
| Indonesia | 63,900 |
| Netherlands | 57,300 |
| World | 2,096,800 |

**TABLE 7.1** Energy resources: the production of petroleum, coal and natural gas in selected countries in 2000. (Adapted from BP Amoco, 2001).

| A  Copper × 10³ tonnes (1999) | | B  Bauxite × 10³ tonnes (1998) | |
|---|---|---|---|
| Chile | 4382 | Australia | 44,653 |
| USA | 1600 | Guinea | 17,300 |
| Indonesia | 740 | Jamaica | 12,646 |
| Australia | 735 | Brazil | 14,000 |
| Canada | 614 | China | 9,000 |
| Peru | 536 | Venezuela | 4,826 |
| World | 12,600 | World | 128,000 |

*Source:* US Geological Survey (2001)    *Source:* British Geological Survey (2001)

| C  Iron × 10³ tonnes (1998) | | D  Uranium tonnes (2000) | |
|---|---|---|---|
| China | 260,500 | Canada | 10,682 |
| Brazil | 195,300 | Australia | 7,578 |
| Australia | 161,100 | Niger | 2,895 |
| Russian Fed. | 129,700 | Uzbekistan | 2,350 |
| USA | 62,700 | Russian Fed. | 2000 E |
| Canada | 38,900 | USA | 1,465 |
| World | 2,071,300 | World | 34,746 |

*Note:* E = estimate
*Source:* Euromines (2001)    *Source:* Uranium Information Centre (2001a)

**TABLE 7.2** Copper, bauxite, iron and uranium output from the world's major producers

access to areas of consumption. Similarly, the production of electricity from coal, natural gas or oil requires appropriate plant, the construction of which constitutes land-cover and land-use change.

In relation to the mining of metalliferous ores, it is difficult to generalize because many different types of metals are mined, each requiring specific techniques and each creating specific land-cover changes. Moreover, the mining of

|  | 1999 (× 10³ tons) | 2000 (× 10³ tons) |
|---|---|---|
| USA | 28,900 | 29,500 |
| Paraguay | 10,000 | 10,000 |
| Germany | 7000 | 6800 |
| France | 6500 | 6600 |
| Austria | 6000 | 5800 |
| Spain | 6000 | 6000 |
| UK | 4000 | 4000 |

**TABLE 7.3** Industrial sand and gravel production and reserves. (Data from United States Geological Survey, 2001)

metals is widespread and occurs in almost every country and in all types of ecological context, from the tundra to the dry and humid tropics. In some cases, ore processing occurs on-site, which necessitates plant additional to that for the actual mining. As Table 7.2 shows, the extraction of iron is a particularly significant mining activity when measured in terms of volume extracted. Just over $2000 \times 10^6$ t were produced worldwide in 1998, the major producers being China, Brazil and Australia. As in the case of coal, various types of mining are employed to extract iron ore; inevitably, extraction from near the surface has the greatest impact. This is also the case with copper and bauxite extraction (the latter is the ore from which aluminium is produced), which are both mined in large quantities because they have a wide range of uses. Table 7.2 shows that the major producers of bauxite are Australia and Canada, while the USA, Chile, Indonesia, Poland and the Russian Federation produce most copper.

The impact that the mining of metal ores exacts is determined to a large extent by the grade of ore, i.e. how much metal versus non-metal present. For example, the minimum ore grade, expressed as a percentage of metal present, for aluminium and iron is 25 per cent, but for gold and platinum the minimum ore grades are 0.0008 and 0.0003 per cent respectively (see Mather and Chapman 1995). These data reflect several factors: the concentration of metal produced by geological processes, the minimum concentrations of metal that are necessary before mining is considered worthwhile, and the current technology of mining practices. It is important to note that as science and technology advance, ores of poorer grade can be exploited. However, the lower the concentration of the required substance the more 'waste' material is produced. Disposal of the latter also causes land-cover alteration, with additional transformation if reclamation is effected.

Non-metalliferous minerals are many and varied. They range from rare gemstones such as diamonds and emeralds, to the more ubiquitous aggregates like sand and gravel. As in the case of the minerals referred to above, the extraction methods and resulting wastes are unique to each mineral, as are the environmental impacts and reclamation procedures. As

Figure 7.1 shows, the major producers of diamonds are Australia, Botswana, Zaire, Russia and South Africa. These are regions in which igneous rocks are prevalent, and which have cooled very slowly and under immense pressure. Both gemstone-quality and industrial-quality stones are produced, the difference being the size and quality, i.e. the state of cracking or discoloration of the stones. Diamond-bearing rocks are known as kimberlites and they emerge at the Earth's surface as pipes; deep mining is necessary to extract the diamonds, such as that which occurs in the Kimberley region of South Africa and at the Argyle mine in Australia. Where kimberlite pipes are eroded by rivers, diamonds may occur in secondary situations; these are known as placer deposits. For example, sediments off the coast of Namibia are diamond-rich and recovery is effected through the filtration of sediments by specially adapted ships. Although overall grades are poor, Namibia produces a high volume of gem-quality stones; these, inevitably, provide the best income. In this case, the impact on the terrestrial environment is relatively low because filtered sediment is returned to the oceans.

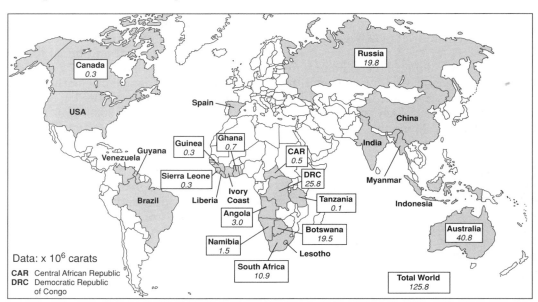

**FIGURE 7.1** The world's diamond-producing regions with data for production in 1998. (Data from the Diamond Registry, 2001)

THE DIRECT IMPACT OF MINING

This is certainly not the case for aggregates, which are extracted in huge quantities for construction industries. As Table 7.3 shows, European countries dominate sand and gravel production along with the USA. Crushed rock is also used as an aggregate in concrete and is especially important for road construction. Commonly, limestone, dolomite and granite are used to produce crushed rock, which is obtained through open quarrying of rock, usually with crushing on-site. Where deposits or rocks such as these are absent, other substrates, e.g. pumice or even coral, are used; the latter is used in some island nations where there are no alternatives. Wherever extraction occurs, there is substantial land-cover alteration; large pits of varying depth are created, often with drainage problems. Reclamation schemes are varied and may involve the creation of wetlands and recreational facilities or infill, sometimes as repositories for domestic refuse, prior to the implacement of a topsoil and an eventual return to agricultural use.

This summary presents a snapshot of only a small proportion of the world's mining activity. It nevertheless indicates the significance of mining as an agent of land-cover and land-use change, more detailed examples of which are given in the sections that follow.

## 7.3 The direct impact of mining

As stated above, the exact type and extent of land-cover alteration that occurs where minerals are extracted depends on the type of mineral and the type of mining. Strip mining is a particularly disruptive method of surface mining because it affects a larger surface area than quarrying or deep mining. There are three types of strip mining: contour stripping involves the cutting of benches into sloping land following the contours; hilltop removal flattens land to produce plateaux; and area stripping involves the removal from the near surface of overburden from a large area on relatively flat land. These techniques have been widely applied in many areas to extract coal, iron, copper and bauxite amongst others.

In the United States, for example, there has been a big increase in the surface mining of coal and a shift to coalfields west of the Mississippi since 1970. According to Craig *et al.* (1996) there are five reasons for this: increased demand for lower-sulphur coals which occur close to the surface in western states (see Figure 7.2); an increased number of power plants near these western deposits; concerns about safety as surface mining is less hazardous than below-surface mining; the increased productivity that is possible in surface mines because of easier use of equipment; and poor worker unionization in the western states. All of these factors are, thus, stimuli to land-cover/land-use change. Both eastern and western coalfield areas have experienced damage due to surface mining. In Appalachia (see Figure 7.2), for example – the earliest region to be exploited for coal resources – contour strip mining has affected c. $1.6 \times 10^6$ ha and created c. 40,000 km of contour benches, about 8 per cent of which are subject to major landslides (data from Toy and Hadley, quoted in Mannion, 1997). This also causes land-cover change. Of the western states, Wyoming is the most important coal producer (see data in Figure 7.2), mostly through opencast mines. Amongst the largest of these are Black Thunder and Coal Creek (Mining Technology, 2001a) in the Powder River basin (Figure 7.2). The former comprises three open pits with two huge storage silos from which rail links transport the low-sulphur subbituminous coal to power plants. The pits plus plant have created very different land use to the original prairie grassland. The USA is also home to the world's largest open-pit mine, notably the Bingham Canyon copper mine near Salt Lake City, Utah. It is more than 700 m deep, 4 km in diameter and 7 km$^2$ in area. Benches have been cut into the bedrock surrounding the site, and these carry railway lines and roads to facilitate the transport of ore. Total ore extraction in 1999 was 56.6 Mt and although copper is the major product, gold, silver and molybdenum are also produced. As well as roads and a railway line to transport ore, there is now a conveyor system linking the mine through 8 km with the Copperton concentrator where the ore is processed. This takes 80 per cent of the ore

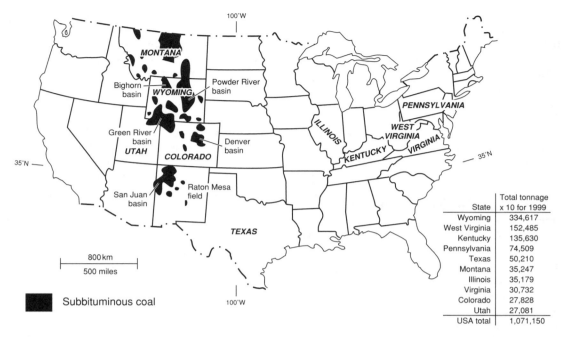

**FIGURE 7.2** The major subbituminous coal fields of the USA. (Adapted from Garbini and Schweinfurth, 1986; the coal production data for 1999 in the major coal-producing states of the USA are from the United States Office of Surface Mining, 2001)

while the remaining 20 per cent is transported by rail to the old North concentrator complex some 24 km away. The resulting concentrate is then transported to a smelter 27 km distant (data from Mining Technology, 2001b).

The extraction of materials for some types of manufacturing industries and the construction industry also has a relatively large surface impact. Such materials include sand and gravel, often the residues of fluvial or fluvioglacial processes in the past, as well as rocks such as limestone. As Table 7.3 shows, the major producers are the developed nations of North America and Europe, notably the USA, Germany, France and the UK. A recent survey of aggregate extraction in England and Wales (British Geological Survey, 2000) has detailed patterns of aggregate sales (a proxy measure of production), flows and uses for 1997. These are summarized here in Figure 7.3, which shows that the major producing regions are the southwest, southeast and East Midlands, whilst the major consuming regions are in the southeast,

East Midlands and northwest. In terms of distribution, the southeast produces the greatest volume of sand and gravel, whilst the East Midlands produces the largest volume of crushed rock. The latter comprises both limestone and igneous rock.

One example of a quarry is that of Merehead in the Mendip Hills near Bristol. This is vast, covering c. 125 ha. Most aggregate material is extracted from open pits or quarries. All such activity alters land cover through pit or quarry excavation and plant construction. Similarly, in the Newcastle region of New South Wales, Australia, the mining of sand, coupled with fluoride pollution from the Tomago aluminium smelter and the firing of forest vegetation has generated land-cover change (Taylor and Fox, 2001). The resulting disturbance of natural vegetation has, in turn, affected local lizard populations by altering the balance between open habitat and closed (forest) habitat plant species.

The extraction of uranium is another cause of land-cover change. According to the Uranium

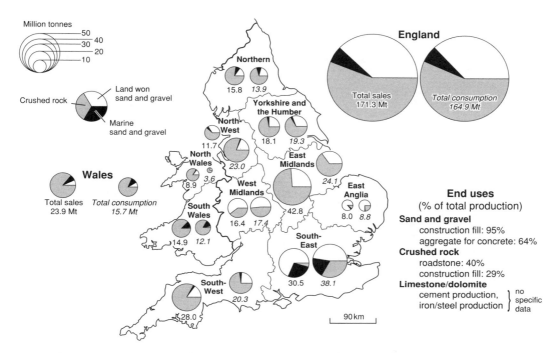

**FIGURE 7.3** Data on aggregate production and consumption in England and Wales in 1997. (Adapted from British Geological Survey, 2000)

Information Centre (2001a), 38 per cent of the world's uranium is obtained using open-pit mines, 33 per cent from underground mines, 17 per cent by *in situ* leaching and 12 per cent as a by-product of other mining activities. Open-pit mines create the greatest extent of land-cover alteration. Moreover, because the uranium content of ore is low, usually between 0.1 per cent and 0.2 per cent, a large amount of waste material is generated. The world's largest uranium mine is that of Rössing, 55 km east of Swakopmund in Namibia (Rössing, 2001). The open pit is 3 km long, 1 km wide and 300 m deep; production began in 1973, reaching full capacity in 1979. Today *c.* 4500 tonnes of uranium oxide are produced annually, making Rössing the fifth largest producer in the world with *c.* 7 per cent of world output (see Table 7.2).

A number of the world's largest uranium-producing mines are listed in Table 7.4. Unfortunately, no data are given by the Uranium Information Centre on the spatial extent of each mine, so it is impossible to give an estimate of their alteration of land cover.

Nevertheless, the production data provide a crude proxy measure of disturbance as level of production can roughly be equated with degree of disturbance, especially in relation to open-pit mines. In Australia there are currently two active uranium mines: Ranger in the Northern Territory and Olympic Dam in South Australia, with a third being commissioned at Beverley, also in South Australia. The locations of these are shown in Figure 7.4. Although Ranger mine is open pit and Olympic Dam is a deep mine (it also produces copper and gold), the same method of uranium extraction is employed. The ore is crushed to a fine powder and made into a slurry with water; excess water is removed in a thickener facility and the slurry is pumped to leach tanks; sulphuric acid is added to dissolve the uranium, and other minerals, and after removal of the solid component, treatment with kerosene strips the uranium, which is precipitated with ammonia; the resulting ammonium diuranate is then converted to uranium oxide in a furnace known as a calciner; the oxide, in powder form, is packed into large

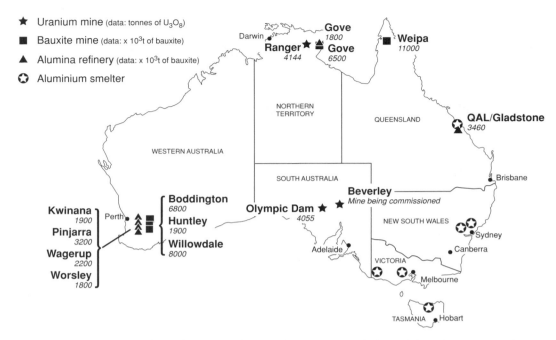

★ Uranium mine (data: tonnes of $U_3O_8$)

■ Bauxite mine (data: x $10^3$t of bauxite)

▲ Alumina refinery (data: x $10^3$t of bauxite)

✪ Aluminium smelter

**FIGURE 7.4** The location and productivity of Australia's uranium (adapted from Uranium Information Centre, 2001b) and bauxite mines (adapted from Anon., 2000); data are for 1999

drums for export. Thus, at both mines, the land surface is disturbed not only by the mining activity but also by the plant used for processing the ore. The Ranger mine, for example, is on a 7860 ha lease and is surrounded by the $1.98 \times 10^6$ ha of Kakadu National Park, a World Heritage Site; approxi-mately 500 ha, or 0.025 per cent, of the total area of Kakadu are directly disturbed by mining and processing (Uranium Institute, 2001). The Olympic Dam mine is more extensive; the mine, smelter etc. occupy c. 840 ha of a total lease of 12,000 ha. In both cases, there are environmental protection strategies through

| Mine | Country | Type | Annual production (tonnes) | % of world production |
|---|---|---|---|---|
| Key Lake | Canada | Open pit | 3715 | 12.0 |
| Ranger | Australia | Open pit | 3266 | 10.5 |
| Olympic Dam | Australia | Underground by product | 2713 | 8.7 |
| Rabbit Lake | Canada | Underground | 2705 | 8.7 |
| Rössing | Namibia | Open pit | 2689 | 8.7 |
| Akouta | Niger | Underground | 1917 | 6.2 |

*Note:* Data are for 1999

**TABLE 7.4** The world's largest uranium—producing mines (Uranium Information Centre, 2001a)

active management and plans for eventual rehabilitation (Uranium Institute, 2001).

Figure 7.4 also gives the location of Australia's major bauxite mines, the largest of which are in Western Australia. All are open-pit mines. Those of Western Australia have resulted in the removal of a natural vegetation cover of jarrah (*Eucalyptus marginata*) forest on lateritic soils. According to Hinds' (2001) review, mining began in 1963 and the primary interest is that of the Aluminium Company of America (Alcoa). The pits are relatively shallow at *c.* 4 m depth and up to 10 ha in area. On an annual basis *c.* 450 ha of jarrah forest is mined and an equal area is rehabilitated (see Section 7.6). Alcoa has also constructed three refineries, which produce alumina from the bauxite, to the south of Perth (see details in Anon., 2000, and Figure 7.4). This land-cover alteration is a major generator of wealth for Alcoa and Australia. Many other deposits of bauxite in Western Australia remain unexploited, but provide opportunities for land-cover alteration in the future.

The extraction of other, equally valuable, substances have also led to the creation of surface pits. A classic example is that of the 'Big Hole' at Kimberley, South Africa. The discovery of diamonds in the area occurred in 1869 and caused an influx of fortune-seekers, who created the town of Kimberley. Opencast diamond mining commenced in 1871 in Kimberley itself as individual miners purchased claims to the land on which a kimberlite diamond pipe extruded at the surface. Eventually all the claims were purchased by the famous Cecil John Rhodes who founded DeBeers Consolidated Mines Limited in 1888, and through which he controlled most of the world's diamond supplies. Eventually, mining at the Big Hole in Kimberley was abandoned as conditions deteriorated, but the mine site itself remains today as a major tourist attraction. It is *c.* 6000 m$^2$, almost 500 m deep and is partially filled with water. The old town of Kimberley, with its saloon, bowling alley, shops etc. is an open-air museum. Indirectly, diamonds thus continue to provide a means of wealth generation through tourism, a not uncommon feature in other regions of the world where mining has generated now-abandoned towns.

In Kimberley, in the centre of South Africa's karoo region, the Big Hole and its museum buildings are in sharp contrast to the original karoo vegetation of shrubs and grasses. Similarly, the so-called ghost town of Kolmanskop, near Luderitz in Namibia, was the result of diamond mining in the 1930s and 1940s. Located in the desert, it was affluent in its heyday when all provisions etc. were imported; today many buildings are partly covered in sand and the town is now a tourist attraction. Many other nations exploit their mining heritage as foci for tourism, which in turn generates further land-cover change as hotels, restaurants etc. are constructed.

On a smaller but nevertheless significant scale, gold mining in the Amazon basin by unregulated prospectors has contributed to deforestation in several countries, including Brazil and Surinam. Impact on the latter has been reported by Peterson and Heemskirk (2001), who have shown that gold miners clear between 48 and 96 km$^2$ of old-growth (pristine) forest annually. This will amount to approximately 750 to 2280 km$^2$ by 2010. Moreover, Peterson and Heemskirk note that forest regeneration is particularly slow when compared to areas affected by agriculture and logging. This is because of severe topsoil disturbance, pit excavation and frequent re-excavation. Abandoned mining camp sites were recolonized more rapidly than mined areas, but even these were slow to recover in comparison with areas abandoned by shifting cultivators.

# 7.4 Land-cover alterations at distance from the site of extraction

Mining activities frequently have an impact that extends well beyond the mined area. As well as the construction of plant for ore processing etc. on-site, as referred to in Section 7.3, the provision of smelters and transport facilities effects land-cover/land-use change. Transport facilities include not only road and rail networks but also pipelines, which are discussed in Section 7.5. The former encourage extensive land-cover alteration when they are constructed in otherwise inaccessible areas such as tropical forests; here they have a similar impact to logging (Section 6.3) by opening up the forests to settlers who then clear land for agriculture. Moreover, the disposal of waste material, i.e. tailings, often in the vicinity of the source mines, extends the impact of land cover/land use well beyond the mine itself. Similarly, the use of minerals such as coal produces large quantities of waste materials (e.g. fly ash) where the coal is consumed, e.g. near power stations, rather than where it is mined. Hydrological systems in the vicinity of mines are also affected, especially where minerals such as iron sulphides are exposed. In contact with water these substances produce acids that acidify drainage water, which may then contaminate surface waters and groundwater. This is a serious problem in coal- and metal-mining areas where sulphides are abundant. Contamination with substances that are toxic when concentrated, e.g. cadmium, mercury and arsenic, can be equally problematic; uranium mining is especially important in this context because of high levels of radioactivity.

The wider impact of mineral extraction and use is well exemplified by the Australian aluminium industry (see Section 7.2). As shown in Figure 7.4, bauxite is mined in Western Australia and the Northern Territory where nearby refineries process the bauxite to produce alumina. This is aluminium oxide, a tonne of which is extracted from 2 to 3 tonnes of bauxite. Aluminium is obtained from this by electrolysis in a smelting process; in Australia the smelters are located in areas with an abundance of cheap power and which are close to the coast for export. The ramifications of bauxite mining are thus far reaching.

An example of multifaceted impacts due to mining in one area is that of the Witbank Coalfield, c. 120 km east of Johannesburg, South Africa. Here, extraction began in 1908 and consisted mostly of underground rather than surface mining. Mining ceased in 1947 but the area still bears evidence of degradation at the site itself and in the drainage systems nearby.

According to F.G. Bell et al. (2001), extensive subsidence, spontaneous combustion, groundwater and surface-water contamination with acid-mine drainage, spoil heaps and vegetation destruction remain a legacy. First, subsidence has been widespread, causing problems for surface land hitherto unaffected by the extractive activities. This began in the 1930s and 1940s because underground pillars were removed (technically, the term is 'robbed') both on final retreat from specific underground passages and due to pillar crumbling after passage abandonment. Crownholes and surface cracks emerged. This has repercussions for spontaneous combustion – its frequency increases as coal workings become exposed to the atmosphere. F.G. Bell et al. state that c. 150–200 ha are affected by burning that is hazardous. Beyond the mine, waste material is present as two large spoil heaps of 56,250 m$^2$ and 66,000 m$^2$, which remain relatively unvegetated because of the mineral composition and spontaneous combustion. Subsidence has also resulted in water accumulation after summer rains. It becomes acidic and percolates into the subsurface passages, which contaminates drainage water and results in aquatic impoverishment. High acidity and high concentrations of aluminium are, as F.G. Bell et al. point out, also responsible for the loss of vegetation over 3 ha along the eastern edge of the coal outcrop.

Such problems are common in coal-mining areas in general. For example, Donnelly et al. (2001) have reported widespread subsidence causing damage to residential areas and agricultural land in the Amaga (Angelopolis and Venecia) Bolombolo coal-mining regions of the Andean region of western Colombia. Here the replacement of shallow room and pillar constructions underground, which are typical of only partially mechanized mining, have been replaced with longwalls as mechanization has increased. This has resulted in poor support for the overburden, which has collapsed into the mine passages. Subsidence is also a problem in the Northern Coalfield of New South Wales, Australia, where disturbed ground, open fissures, craters, seam fires and sediment movements (creeps) have occurred since the early 1900s (McNally, 2000).

Active and abandoned mines can cause contamination of wide areas, as is exemplified by the Mole River mine, also in northern New South Wales, Australia. This mine was actively worked in the 1929s and 1930s for arsenopyrite, from which arsenic chemicals were obtained, and used to produce herbicides for the control of prickly pear. According to Ashley and Lottermoser (1999), abandoned mine workings,

waste dumps and a treatment plant have given rise to high concentrations of arsenic, and other heavy metals such as iron, silver, copper lead and zinc, in the 12 ha mine site itself. Arsenic concentrations as high as 9300 ppm have been recorded in soils, which compare with 8 ppm in background soils. However, contamination has occurred over a much wider area than the 12 ha mine site, as shown in Figure 7.5. Ashley and

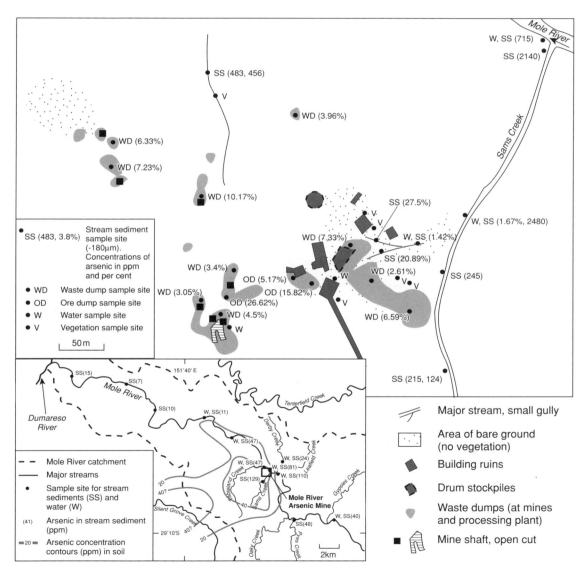

**FIGURE 7.5** Mole River Mine, New South Wales, Australia, and its impact on the region. (Adapted from Ashley and Lottermoser, 1999)

Lottermoser estimate that $c.$ 60 km$^2$ have been affected, with soil arsenic values of $c.$ 55 ppm. Wind and water erosion of arsenic-rich particles has caused this contamination, which is also evident in stream sediments. The latter have an arsenic concentration of 62 ppm in the vicinity of the mine, which contrasts with a background average of 23 ppm. There is also a problem with acid drainage. Arsenic is toxic to most plants and its high concentrations have prevented vegetation growing on mine dumps, though a few species are surviving, e.g. *Cynodon dactylon* (couch grass) and *Cassinia laevis* (cough bush), around the edge of the site.

As stated above, mineral extraction affects not only the land but also drainage systems which spread the mining impact well beyond the point of origin. There are many examples in the literature that reflect this impact. Amongst the most common effects on drainage networks is acid-mine drainage, which results from many different mining activities including coal and metal-ore extraction. Not only is water pH affected but accompanying discharges of heavy metals also have an impact on the aquatic and littoral flora and fauna of water bodies. In addition, tailings from mines are frequently discharged into rivers and lakes, causing increased sediment deposition. This too may contain toxic concentrations of metals, and thus contributes to reducing aquatic and littoral biodiversity. These impacts are illustrated by the Ely Creek Watershed in Lee County, Virginia, USA, the chemical and biological characteristics of which have been compiled by Cherry *et al.* (2001). Table 7.5 summarizes their results and shows that severe degradation of the Ely Creek drainage has occurred. In particular, the low pH reflects the output of acid drainage from the mined area; a pH of 2.73, for example, is extreme even in areas of acid bedrock. The metal concentrations are all well above background concentrations and the biological measures of mining impact, e.g. macroinvertebrate populations and Habitat Index, all point to considerable alteration of the ecosystem. They also highlight the need for restoration treatment.

Another example of aquatic alteration by mining is that of the Ok Tedi/Fly River system, which is affected by the Ok Tedi mine in Papua

| | | |
|---|---|---|
| **A** | **Water pH range** | 2.73 to 5.2 |
| **B** | **Sediment metal concentrations mg kg$^{-1}$** | |
| | Iron | $c.$ 10,000 |
| | Aluminium | $c.$ 1500 |
| | Magnesium | $c.$ 400 |
| | Manganese | $c.$ 150 |
| **C** | **Habitat Index** | (based on 9 parameters concerned with the quality of the physical environment, such as channel scouring, bank-vegetation stability etc., which influence bottom-dwelling fauna) |
| | 10 sample stations: | Partially supported or non-supported |
| | 10 sample stations: | Excellent or supported |
| | Cause of problem: | Mainly sedimentation |
| **D** | **Benthic macroinvertebrate populations** | |
| | 6 sample stations: | No macroinvertebrates |
| | 8 sample stations: | 1–9 organisms |
| | 4 reference stations: | >100 organisms and 13 taxa |
| **E** | **Other toxicity tests** | |
| | These indicated varying degrees of stress at sampling stations with 12 stations being categorized as severely stressed | |

**TABLE 7.5** Chemical and biological characteristics of Ely Creek Watershed, Virginia, USA. (Adapted from Cherry *et al.* 2001)

New Guinea. This mine is a major producer of copper and gold, and is an open-pit mine where production began in 1984. Despite its relatively recent origin, mountain terrain characteristics have prevented waste retention. Consequently, *c*. $66 \times 10^6$ tonnes of residues plus waste rock and overburden are released annually into the Ok Tedi River, a tributary of the River Fly, which discharges into the Gulf of Papua and the Coral Sea (Hettler *et al.*, 1997). The effects of this include alterations to the natural sedimentary processes, including increased deposition in oxbow lakes and creeks, as well as lakes and swamps close to the Fly River. Sediment is carried into the latter by overbank flow, the incidence of which has increased as channel deposition has increased. However, the sediment deposited is enriched in copper and many other metals, as is the drainage water. Hettler *et al.* state that, on average, sediment derived from the mine is enriched in copper by a factor of 15 and that there is much variability in copper enrichment throughout the drainage system; higher concentrations occur in the sediments of oxbow lakes than in floodplain sediments, for example. Where variable redox conditions occur, as in the latter, high copper concentrations may have a negative ecological impact on aquatic and littoral species; in contrast, the less variable and carbon-rich oxbow environment probably creates a reducing environment in which copper is less mobile and therefore less available for organisms. The Ok Tedi situation is not unusual in Papua New Guinea as tailings from many mines enter river systems directly. In the longer term this will affect not only drainage basin morphology, but also human communities within the floodplains through an increase in flooding and an impact on aquatic species, including fish.

## 7.5 The impact of oil and gas extraction and transport

Whilst the provision of road and rail networks for the transport of minerals is an important consequence of mining resulting in land-cover change, these facilities are usually components of multipurpose transport systems. In contrast, the transport of oil and natural gas requires unique provision in the form of pipelines. These have been constructed over thousands of kilometres in order to bring oil and gas from the point of extraction to the point of processing and/or consumption. In addition, a great deal of oil is transported by sea, in ships designed for the purpose. These oil tankers can be the subject of accidents such as inadvertent discharges, collisions and poor navigation causing then to run aground. The resulting oil spills can have extensive and long-lasting affects on land-cover characteristics in coastal regions.

Figure 7.6 gives the distribution of some of the world's major oil and gas fields and their pipelines. Not only do some of these pipelines link the more remote parts of the globe with ports and centres of use, they also transect some of the most politically sensitive regions, notably the Middle East. Apart from the intrinsic problems of constructing and maintaining oil and gas pipelines, they may also be subject to attack in times of war, as indeed may be point sources such as oil wells and oil rigs.

The impact of pipeline construction has been widespread and, from the tundra to the tropics, it has had implications for both land cover and human communities. Some of the most extensive impacts of oil and gas extraction and transportation have occurred in Arctic regions, especially in northern Eurasia. According to Vilchek (2002), the Tyomen Oblast (a territorial unit in the Former Soviet Union) in western Siberia (see Figure 7.5) is the most important region for oil and gas production in the Russian Federation; it incorporates lands occupied by indigenous peoples, notably the Khants and Mansis, and the Nenets. Reference was made to the problems encountered by the latter in relation to hydrocarbon extraction and their nomadic herding activities in Section 4.3. The extraction of hydrocarbons occurs north of 60°N in tundra/taiga underlain by permafrost. Production from the major oilfield, Samotlor, increased throughout the 1970s but declined from the late 1980s and many smaller oilfields were brought into production. Inevitably, a substantial infrastructure has been established

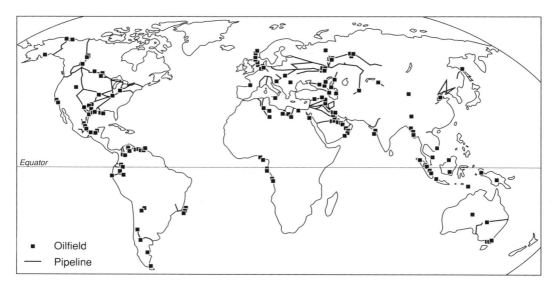

**FIGURE 7.6** Major world oilfields and pipelines. (Adapted from *The Times Concise Atlas of the World*, 1995)

to extract and transport the oil and gas; as new fields have been exploited, the infrastructure has expanded. Poor maintenance and lack of investment in the last decade have, however, accentuated the environmental impact of this industry as increasing numbers of spills and leaks have occurred.

Saiko (2001) has reviewed the impact of development in general and industrialization in particular in Arctic Eurasia. Referring to Yablokov (1996), Saiko states that, 'Three to ten million tonnes of oil is being spilt annually in approximately 300 major and 11,000 minor technical accidents on pipelines and wells in the main oil-producing regions of Russia.' She goes on to say that, in western Siberia, major accidents, involving oil spills of 10,000 tonnes, include a pipeline rupture near the Sos'vinskiy nature reserve, which resulted in the release of 420,000 tonnes of oil in 1993. Moreover, between 1995 and 1997 there were *c.* 40,000 technical accidents causing oil leaks. In total, *c.* 200,000 ha of land in western Siberia is now covered with a layer of oil greater than 5 cm thick. Quoting Wolfson (1994), Saiko also reports that only half of the gas produced for export reaches its customers because the other half is lost through leaks, accidents and the high volume of gas needed for pumping. Clearly, such losses of hydrocarbons have a substantial impact on the flora and fauna of Siberia, including the lichen-rich grazing grounds of reindeer (Section 4.3) and, because of the harsh environment recovery is likely to take centuries.

Hydrocarbon extraction and transport in temperate and tropical regions has had an equally significant impact, not only on land cover but also on human communities. For example, hydrocarbon extraction around the Caspian basin has caused a considerable amount of pollution of both the land and the sea (Efendiyeva, 2000; see also papers in Kalyuzhnova, 2002). Particularly severe problems occur on the Apsheron Peninsula of Azerbaijan as a legacy of 100 years of oil production. Severe contamination of soils in this area is commonplace as are standing oil ponds and shorelines covered with oil residue along Baku Bay. Similar problems occur in the Caspian basin of Kazakhstan, especially as a result of oil transport from the Tenghiz oilfield. Pollution of the sea itself is also severe, especially by heavy metals and hydrocarbon products such as phenols. These and other pollutants have had an impact on the

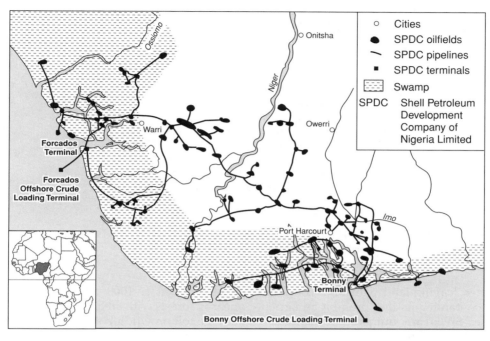

**FIGURE 7.7** The oilfields and pipelines of Shell in the Niger Delta, Nigeria. (Adapted from Shell, 2001)

health of human populations around the Caspian coast.

Other countries that have experienced the impact of hydrocarbon extraction include Ecuador and Nigeria. In both cases, controversy has arisen because environmental and cultural impacts have been dramatic. Oil extraction in Ecuador's Oriente region (eastern Ecuador or Amazonian Ecuador) began in the early 1970s and has had a significant impact on the region's tropical forests through accidental spills, the dumping of drilling by-products and chemicals into rivers, and the opening up of the forest to invaders. According to Rainforest Action Network (2001), an oil spill of 275,000 gallons in 1992 contaminated large areas of land as well as rivers, and since *c.* 1980 pipeline cracks have resulted in the spillage of *c.* 16.8 $\times$ $10^6$ gallons of crude oil into Ecuador's rivers. Moreover, Rain Forest Action Network ascribes a large proportion of Ecuador's annual deforestation rate of *c.* 54 $\times$ $10^3$ ha, i.e. 2.3 per cent, to the direct and indirect effects of oil development. In a cultural context, tribal lands have

been appropriated and pollution has caused health problems, as well as a loss of plant and animal species used by native people.

Similar problems have occurred in Nigeria, where oil extraction began in the 1950s. Much of the oil derives from the Niger Delta, home to the Ogoni people. Pipelines connecting numerous oilfields in the wet and dry parts of the delta, as shown in Figure 7.7, have generated land-cover change and in the areas designated as oil bearing, oil-extracting plant has replaced mangroves, swamp forests, lowland rain forests and agricultural land. According to Shell (2001), some 400 km² of land are used for its activities in this area. However, there are indirect impacts. For example, oil extraction has also caused an influx of people from outside the delta. Consequently, the urban centres of the upland have expanded, as have the dispersed settlements of the coastal swamplands. Oil extraction and its transport have caused other land transformations. Oil spills, leaks, blowouts, the dumping of drilling fluids and refinery effluents have all contributed to land-cover alter-

| Ship name | Year | Location | Oil Lost (tonnes) |
|---|---|---|---|
| *Atlantic Empress* | 1979 | off Tobago, West Indies | 287,000 |
| *Castillo de Bellver* | 1983 | off Saldanha Bay, South Africa | 252,000 |
| *Amoco Cadiz* | 1978 | off Brittany, France | 223,000 |
| *Haven* | 1991 | Genoa, Italy | 144,000 |
| *Torrey Canyon* | 1967 | Scilly Isles, UK | 119,000 |
| *Independenta* | 1979 | Bosphorus, Turkey | 95,000 |
| *Braer* | 1993 | Shetland Islands, UK | 85,000 |
| *Sea Empress* | 1996 | Milford Haven, UK | 72,000 |
| *Metula* | 1974 | Magellan Straits, Chile | 50,000 |
| *Exxon Valdex* | 1989 | Prince William Sound, Alaska | 37,000 |

**TABLE 7.6** Some of the world's largest oil spills, which have contaminated coastlines. (International Tanker Owners Pollution Federation Ltd, 2000)

ation, often leaving land with high concentrations of contaminants that prohibit an agricultural use. Spills are numerous; Shell (2001) admits to 340 reported oil spills in 2000, which amounts to 30,751 barrels. These were caused by corrosion, equipment failure and human error, though sabotage accounted for *c.* 40 per cent of the spills and *c.* 57 per cent of the oil spilt. This is an interesting twist to the story of oil in Nigeria. Shell provides no analysis of why the sabotage occurs, though some of it at least could relate to the Ogoni claim that they and their lands are exploited by Shell without due care and recompense.

It is also important to recognize that areas far beyond sites of hydrocarbon production have been severely affected by oil spills. These are mostly coastal regions that have been in receipt of oil spills from large tankers. Table 7.6 gives details of some of the world's largest oil spills that have affected coastlines (there are other equally large or larger spills not listed because they occurred in the open ocean and had no effect on land cover). The impact of these coastal oil spills is usually well publicized by the popular media because of the devastating effect they have on the environment, and especially on wildlife. For example, the *Exxon Valdez* tanker ran aground in Prince William Sound, Alaska, and *c.* 37,000 tonnes of oil were released. According to Bregman (2000) this contaminated *c.* 1000 miles of coastline and nearby forests with an oil film, and also exacted a huge toll on wildlife such as sea otters, seals, whales and birds. In the years since this accident, the shoreline has made a remarkable though not complete recovery, as have the populations of many but not all marine mammals and birds. The recovery began with human inputs of detergents etc., but the action of water movement and sunlight probably contributed most of the effort. Other coastlines badly affected by oil spills have lost their oil cover and regained their flora and flora given sufficient time. Similarly, the Persian Gulf and Kuwait desert have recovered following the deliberate destruction of oil wells during the Gulf War of 1990–91. This is discussed in Chapter 9.

## 7.6 The rehabilitation of mine-damaged land

As the examples given above attest, mineral extraction has a substantial and direct impact on the landscape in the form of excavations, tailings ponds and spoil heaps. However, it is possible to reclaim land so disfigured and, in some cases, the former land-cover characteristics may be restored. This is relatively rare but nevertheless scars on the landscape can be beautified and an economic use for land can be restored. Reclamation or rehabilitation is generally a complex and often expensive process.

Each type of mining activity presents specific problems; further problems may be unique to the site of extraction. The many types of impact caused by mineral extraction, some of which are discussed in Sections 7.3, 7.4 and 7.5, require equally varied rehabilitation strategies. There are, however, several features common to almost all types of mine-damaged land. One of the most important relates to soil. In most cases, topsoil is removed prior to the commencement of mining and its replacement is paramount in reclamation schemes. A second common feature of many reclamation schemes is aftercare, which usually involves active management, in particular the continued addition of nitrates through either the use of artificial fertilizers or the encouragement of leguminous plants which have root nodules that support populations of nitrogen-fixing bacteria. In general, these procedures are components of reclamation programmes that attempt to accelerate the natural process of plant succession. As Bradshaw (2000) observes, succession occurs even in inhospitable environments and on harsh substrates, and reclamation schemes must thus focus on providing a 'helping hand'. The case studies that follow provide examples of a range of reclamation schemes to illustrate some of the success stories.

The widespread occurrence of coal mining has given rise to many reclamation schemes in many different environments. In the UK, for example, numerous reclamation schemes have been established on land so damaged, and many enjoy a clear afteruse, often with reasonable economic returns. Amongst the problems to be overcome are unstable substrates, poor nutrient availability, especially nitrogen, and the lack of a source of appropriate plant species (see Bradshaw, 2000). The possibilities for coalfield reclamation are illustrated by those of the Black Country Development Corporation (BCDC) in the UK, which is responsible for schemes in the south Staffordshire coalfield. According to Elliott (2000), this region has experienced some 200 years of mining,

including that of coal, limestone, ironstone and clay, as well as marl, and sand and gravel extraction. These activities have resulted in a wide range of problems, including the presence of mineshafts which cause instability, infill containing industrial waste, mine drainage channels, buried canal channels, infilled marl pits, and sand and gravel pits. Despite these problems, reclamation schemes have been established by the BCDC. Three of these have been described in detail by Elliott and the essential characteristics are given in Table 7.7. In these cases the afteruse of damaged land includes the provision of land for the construction of industrial/business premises and improvements in transport systems. Some abandoned coal mines have also been used to create wetlands, one example of which is Druridge Bay in northeast England. Here, a complex of wetland sites has been established since the 1970s, and these are now important over-wintering sites for wildfowl. According to Peberdy (1998), the Druridge Bay complex represents a small proportion of the c. 15 × $10^3$ ha of wetlands that have been created in the UK as a result of mineral extraction, notably sand and gravel, and other forms of dereliction.

Reclamation schemes have also been widely undertaken in Germany's coal-mining areas, as has been documented by F.G. Bell et al. (2000), and Schulz and Wiegleb (2000). The former have described reclamation schemes in the Ruhr, near Essen, at the Graf Moltke and Mont Cenis mines. As in the case of south Staffordshire, the techniques involved the containment of contaminated sediment, compaction of safe infill material and the creation of land for industrial parks. In contrast, Schulz and Wiegleb have described reclamation schemes in Lower Lusatia (Brandenburg), where c. 85 per cent of the region is being reclaimed for agriculture and forestry, with the remaining 15 per cent being given over to 'nature development'. This region has a long history of opencast lignite mining; beginning in 1789, the extraction has affected an area of 800 km² directly and up to 2500 km² due to lowering of the groundwater level. However, by 1989, most of the 17 mining sites had ceased to be operational and reclamation proceeded.

The re-use of land for agriculture and forestry in this region is especially important because these activities provide alternative employment to the now defunct mining industry.

One of the major problems that has to be overcome is the presence of minespoils, which are acidic and sulphurous. Drainage from these spoils contaminates wide areas, resulting not only in high acidity but rapid leaching of nutrients such as calcium, magnesium and

---

**A   Leabrook Road, West Midlands**

**Size:**   29 ha of urban land

**Use:**   Coal mining and later site of industry

**Problems:**   Numerous known and unknown mineshafts; 10–12 m of infill over glacial clay over Coal Measures, industrial waste from iron works, chemical works paint manufacture, a pottery, a galvanizing plant and several waste tips

**Techniques used in reclamation:**
The Black Country Spine Road (BCSR) was stabilized by drilling and pressure grouting; some mineshafts were treated. Waste material, because of its variability, was collected from an opencast operation, sorted and kept on-site in a specially constructed repository. One-third was treated elsewhere. *In situ* waste was layered and compacted according to tight specifications. Water was pumped from the site and a vent trench constructed to facilitate the collection of methane.

**After use:**   New business premises for the Parkway Development.

**B   Great Bridge marl pit plus Toll End sewage works and Tame Valley Canal**

**Size:**   2.5 ha of pit, 30 m deep

**Use:**   Marl extraction, sewage treatment

**Problems:**   The pit from which marl excavation had occurred was infilled with a variety of wastes, some of which are hazardous; BCSR construction and associated development could cause groundwater contaminated with chemicals such as phenols to mobilize, and thus contaminate other drainage water.

**Techniques used in reclamation:**
Pumping and treatment of waste water to eliminate hazardous substances was a major stage in reclamation. This required a temporary river diversion and water-monitoring system. Holding tanks were used to store the waste water prior to its disposal into the trunk sewer. Several mineshafts were treated to increase stability and the bridge foundations were strengthened.

**After use/ improvements:**   Construction of BCSR and improved flood defences facilitating development nearby.

**C   Patent shaft and Moorcroft**

**Size:**   35 ha Patent shaft, 14 ha Moorcroft site

**Use:**   Opencast shaft for coal extraction, steelworks, chemical works, gravel extraction

**Problems:**   Mine remnants with chemical contamination and abandoned canal channels and railway lines. Chemical wastes included sulphides; uncharted mineshafts.

**Techniques:**   Mineshaft discovery and treatment with heavy-duty caps for shafts likely to occur under subsequent buildings. Canal channels etc. were excavated, refilled and recompacted; some material was removed for disposal elsewhere whilst other *in situ* material was used for infill. Road foundations were also laid. Canal upgrading linking the two sites was undertaken.

**After use:**   An automotive components park with 18 land parcels.

---

**TABLE 7.7**  The characteristics of three reclamation schemes in south Staffordshire, UK. (Adapted from Elliott, 2000)

other cations. One approach used to counteract this process is to add substances rich in calcium and magnesium, such as limestone and pulverized fuel ash from power plants. The addition of nitrogen, phosphate and potassium fertilizer is also important to provide essential plant nutrients. These practices have facilitated afforestation with oak on *c.* 60 per cent of the acid-mine spoils, though how permanent this will be in the long term remains to be seen. The monitoring of drainage water characteristics has led Knoche *et al.* (2000) to conclude that, 'The ecosystems gradually shift from a state of geochemical-dominated processes to a mode of increasing biological control.'

Other management techniques used in the reclamation of land damaged by coal mining include the provision of a topsoil (see Bradshaw, 2000), though this is expensive, and the use of herbicides. Torbert *et al.* (2000) have shown that on reclaimed surfaces in the coal-mining areas of the Appalachian coalfields of Virginia, USA, reforestation can be sufficiently successful to provide the basis of a forestry industry, thus providing landowners with an income once mining has ceased, and improving landscape aesthetics. Using a number of sites reclaimed in different ways, three pine species (*Pinus taeda*, *P. virginiana* and *P. strobus*) were planted and subjected to identical treatments with artificial fertilizers and weed control with herbicides. The results of Torbert *et al.* show that treatments with a herbicide over an initial three-year period were particularly effective for

encouraging tree growth through their suppression of 'weed' competitors for light, water and nutrients. Moreover, Akala and Lal (2000), using examples of reclamation on coal-mined land, have drawn attention to the potential of schemes involving pasture and forestry for increasing carbon storage in soils. Their results are summarized in Table 7.8, which shows that considerable increases in soil carbon storage occur over a 25-year period for both types of land use. The soil carbon is only one element of carbon storage in an ecosystem (see Section 1.5) as storage within vegetation itself is significant. Thus it would appear that the growing of forest on reclaimed land is especially beneficial because it encourages high carbon storage in both the soil and above-ground vegetation. This potential is significant in the context of the Kyoto agreement to allocate carbon dioxide quotas; the production of carbon dioxide can be offset by programmes to sequester carbon. Where better to effect this sequestration than in mine-damaged land, especially in developed nations whose consumption of fossil fuels is prodigious? This combination would help to mitigate the deep Ecological Footprint (see Section 3.6) of those nations.

Another example of reclamation is that of the Eden Project in Cornwall, UK (Eden Project, 2001). This county is famous for its china clay pits, which have been the focus of many reclamation schemes. At Bodelva, near St Austell, in a former pit of 15 ha, there now lies a series of

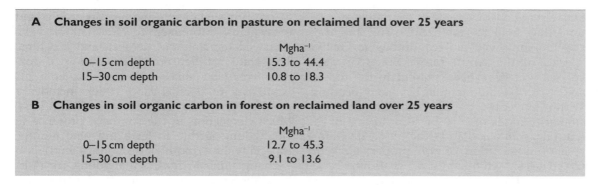

**A   Changes in soil organic carbon in pasture on reclaimed land over 25 years**

|  | Mgha$^{-1}$ |
| --- | --- |
| 0–15 cm depth | 15.3 to 44.4 |
| 15–30 cm depth | 10.8 to 18.3 |

**B   Changes in soil organic carbon in forest on reclaimed land over 25 years**

|  | Mgha$^{-1}$ |
| --- | --- |
| 0–15 cm depth | 12.7 to 45.3 |
| 15–30 cm depth | 9.1 to 13.6 |

**TABLE 7.8** Carbon storage in reclaimed soils on coal-mined land in Ohio, USA. (Adapted from Akala and Lal, 2000)

reconstructed biomes, several in huge conservatories, which is referred to as 'a gateway into the world of plants and people'. Since its opening as a visitor centre in March 2001 it has proved to be a huge success; however, many problems had to be overcome in its construction. The site had an inverted-cone shape, there were many different levels within it, with unstable slopes, and it was liable to flooding. It was necessary to construct a drainage system and create a soil. The former involved the placement of a layer of matting under the landfill to direct water into the drains and to filter out sediment. A level base was created by a 'soil' created from a mixture of china clay waste and $85 \times 10^3 \, m^3$ of compost. The two covered biomes house plants from the humid tropics and warm-temperature regions (e.g. Mediterranean regions), while the roofless biome, comprising 12 ha, houses plants from cool-temperature regions. Land-cover characteristics have thus been changed dramatically at Bodelva.

Considerable success has also been achieved in areas of bauxite extraction in Western Australia (see Section 7.3), where $c$. 450 ha of land is reclaimed annually. The major mining company, Alcoa, is required to store topsoil from areas selected for bauxite extraction so that it can be used to re-cover the area when mining ceases. On the return of the soil, it is seeded with a mixture of plant species, including native legumes. This ensures an increase in soil nitrogen, organic matter and cohesiveness. Initially, annual and biennial species dominate the vegetation and, although they are not characteristic of the native vegetation, they help to stabilize the soil and thus provide improved soil conditions for native species of the jarrah forest. Seeds of these species will have been present in the topsoil and may be augmented by hand seeding. Within ten years, the plant community has a similarity index of $c$. 50 per cent to that of the natural jarrah woodland (Koch *et al.*, 1996). In a recent investigation of soil properties, Ward (2000) has shown that by $c$. 8.5 years after reclamation, soils on reclaimed areas had developed similar chemical properties to those soils beneath unmined jarrah forest, especially in

relation to nitrogen and pH. These results once again highlight the importance of soil in reclamation projects, and, in the specific case of Western Australia's jarrah forest, the soil seed-bank is also vital (see review in Smith *et al.*, 2000).

Examples of the reclamation of mine-damaged land in the developing world are considerably less easy to find than those in the developed world. This is because regulations are often less restrictive and because the enforcement of regulations is not always a priority. Nevertheless, mining results in damaged land (see examples of impact in watersheds in Section 7.4) and, in some countries, efforts at reclamation are under way. Sharma *et al.* (2000), for example, state that $c$. $184 \times 10^3$ ha of land in India's arid zone has been abandoned since 1940 and that reclamation is made especially difficult because of lack of water. Thus the 20 ha experimental site near the village of Barna (26.2°N, 73.5°E) is particularly important as a flagship project for the region. Figure 7.8 is a schematic representation of the most significant feature of this programme. As in the case of the Eden Project (see above), the creation of a 'topsoil' and water management were essential. Sharma *et al.* report that $c$. seven years after the establishment of the scheme there is cause for optimism because the plant cover is increasing, plants are regenerating and woody species are increasing.

Given how precious land is in China, many provinces are initiating reclamation schemes. Those of Shanxi province are discussed by Miao and Marrs (2000), who state that the objective is to 'develop sustainable and healthy arable-land ecosystems'. Moreover, Ye *et al.* (2000a) have reviewed the status of derelict land in China. They state that there are 238,000 mining companies overall in China, which have created $c$. $200 \times 10^3$ km² of derelict land which includes a good proportion of valuable agricultural land. Much dereliction in the southern province in Guangdong is due to lead and zinc mining, especially the disposal of the tailings. In trials at Lechang mine in north Guangdong, Ye *et al.* have shown that reclamation can be effected if particular attention is paid to soil treatment and choice of species. Their results on five soil

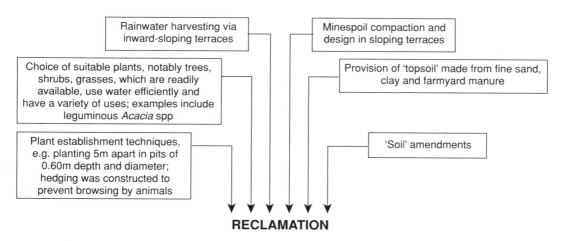

**FIGURE 7.8** The main features of land reclamation in a region of limestone extraction in northeast India. (Adapted from Sharma *et al.*, 2000)

ameliorants showed that burnt coal residue or fly ash provided the most effective barriers between tailings and topsoil in terms of encouraging plant growth. Pig manure and mushroom compost also produced good results but only when used in conjunction with a metal-tolerant species such as Bermuda grass (*Cynodon dactylon*). This species, of the three grasses and clover tested, gave the best results overall. Further work by Ye *et al.* (2000b) has shown that high acidity can be counteracted by the addition of lime and this also resulted in improved plant growth. Further examples and a discussion of the reclamation of mine-damaged land in China are given in Wang *et al.* (2001).

## 7.7 Conclusion

Land damaged through mineral extraction is widespread globally and has increased substantially in parallel with the diversification of humanity's material culture and increase in population. Although the extent of land so damaged is relatively small in comparison with land-cover alteration through agriculture and forestry, the impact is severe both within and beyond the area of actual extraction. Indeed, entire drainage basins can be contaminated and the aquatic biota diminished substantially. The most widespread cause of land degradation is coal mining; and since the 1950s the extraction of oil and natural gas has contributed to this land-cover alteration. The appropriation of carbon (Section 1.5) is thus a major cause of land-cover alteration, a penalty wrought on the environment in return for wealth generation. Both the exploitation and use of these fossil fuels deepens the Ecological Footprint of any specific nation. The extraction of metals, aggregates and precious stones are also important causes of land-cover and drainage-basin alterations.

However, not all impacts of mineral use are the products of extraction. Transport, ore processing and industrial use – which also create waste products, and contamination through accidents and the provision of plant – all cause land-cover alteration. Such impacts are likely to be greater in extent than the actual extractive activity. Moreover, all of these facets of mineral use affect society in a host of different ways, as many of the examples given above attest. The deforestation of extensive areas around sites of mineral extraction, the intensification of agriculture, game exploitation and settlement expansion may all result. These developments, in turn, affect land cover and land use over wide areas.

Mine-damaged lands can be rehabilitated. Section 7.6 contains examples of what can be achieved; even if the resulting land cover/land use is very different from that which existed prior to any human intervention, it is preferable to dereliction and hazardous conditions. Unsurprisingly, the basic tenets of reclamation focus on the fundamentals of ecology, notably the processes of succession, with all the attendant issues of soil composition and vegetation adjustment. Many of the examples given above show that there are many possibilities provided there is research and financial investment.

## Further reading

**British Geological Survey** 2001: *World Mineral Statistics 1995–1999*. British Geological Survey, Nottingham.

**Cutter, S.L. and Renwick, W.H.** 1999: *Exploitation, Conservation, Preservation. A Geographic Perspective on Natural Resource Use*, 3rd edition. John Wiley and Sons, New York.

**Fox, H.R., Moore, H.M., McIntosh, A.D.** (eds) 1998: *Land Reclamation. Achieving Sustainable Benefits*. Balkema, Rotterdam.

**Haigh, M.J.** (ed.) 2000: *Reclaimed Land: Erosion Control, Soils and Ecology*. A.A. Balkema, Rotterdam.

**Hinrichs, R. and Kleinbach, M.** 2002: *Energy: Its Use and the Environment*, 3rd Edition. Thomson, London.

**Urbanska, K.M., Webb, N.R. and Edwards, P.J.** (eds) 1997: *Restoration Ecology and Sustainable Development*. Cambridge University Press, Cambridge.

# Land use related to urbanization

## 8.1 Introduction

The preceding chapters on land-cover/land-use change as a result of agriculture and forestry (Chapters 4, 5 and 6) reflect a major human impact on the Earth's surface. The resulting land cover does, however, resemble the former land cover because it comprises primary producers, i.e. vegetation communities. Chapter 7 represents a different tableau; mineral extraction and processing occur in restricted areas but with an intensity that often results in the obliteration of the natural land cover. The world's cities have a similar desolatory impact, but one that is much more extensive and increasing daily.

The earliest nucleated settlements came into existence *c.* 10,000 years ago. These were not cities in the modern sense of residential areas and industrial facilities, but small agglomerations of permanent dwellings and communal buildings. There is much debate as to why such settlements developed. Had hunter-gatherer communities reached a level of co-operation that required communal space and constructions or was the initiation of permanent agriculture the stimulus (see Section 2.6)? Whatever the reason, and the two are probably related, this initiation of permanent settlement repre-

sents a new phase in both cultural development and land-cover/land-use change. Thereafter the major civilizations of the Old World centred on metropolises whose legacies, often as archaeological monuments, remain in evidence today.

In Asia, Africa and Europe urban landscapes of wood and brick have waxed and waned throughout history. Many of the world's great cities bear witness to the passage of time. Athens, Rome, London, and many more, retain evidence of past urban landscapes and the evolution that has occurred therein. The urban landscape is as restless as the rural; the older the city the more transformations it has experienced and the more valuable it is as a cultural archive.

Other world cities are not so old. The Americas and Australia were transformed from the 1700s as a result of European annexation. Aboriginal people had established early settlements and habitation sites, but New World landscapes were altered dramatically by European settlers. New York, Buenos Aires, Lima, Mexico City and San Franciso are young cities in comparison with many of those in the Old World. They were planned from the start, often on the basis of the grid system and often incorporating features such as cathedrals and town halls that were typical of the European homelands. Many of these cities, especially in the developing countries of Latin America, and many others in Africa and Asia, have experienced rapid population growth in the last few decades. Uncontrolled settlement has occurred; shanty towns have ensued with few or no services, and poor-quality, often hazardous, buildings. At the same time, many of these cities have begun to industrialize, providing employment opportunities, a major reason why they draw people from the countryside.

The cityscape can be diverse; land use is generally varied rather than monotonous. As Figure 1.4 shows, the domestic and industrial land uses can be subdivided into numerous categories. Then there are residential areas, administrative buildings, transport facilities (e.g. train stations), sewage works, sports facilities (e.g. stadia, playing fields) and rubbish-disposal sites. To this must be added the metalled road networks, railway tracks and car parks. However, the cityscape is not all tarmac, concrete, glass and brick; there are green spaces and water features. Canals, lakes, gardens, cemeteries, parks and forests provide fluid but stark contrasts that mitigate the static grey, red-brick or concrete buildings.

All cities are dynamic. Cityscapes alter in tune with society's needs; they grow, shrink and change character as social and economic factors influence community characteristics, wealth generation and political institutions. Cityscapes also react to technological change, especially innovations in transport, industrial development and communications. Most importantly, they are consumers of resources and generators of waste. They are unsustainable in terms of the substantial Ecological Footprint they generate (see Sections 1.4 and 3.6), yet they are the foci of wealth generation and political power.

## 8.2 The archaeological context: early beginnings

The hunter-gatherer societies that existed at the end of the last ice age c. 12,000 years ago, and even earlier, undoubtedly lived in groups and created habitation sites in strategic places and in caves. The evidence for pre-agricultural habitation is widespread, especially for the period c. 20,000 years ago to 10,000 years ago (see Gamble, 1999; De Laet et al., 1994). From c. 10,000 years BP, however, permanent settlements became increasingly abundant. Examples of these early settlements can be found in the Middle East (as shown in Figure 8.1), Egypt, the Indus Valley and China (see details in Maisels, 1999). Many of these came into existence c. 10,000 years ago. Early settlement sites have also been discovered in Latin America, the most well known of which are associated with the Olmec and Mayan cultures, beginning approximately 3500 years ago. However, recent research has uncovered an even more ancient urban complex at Caral, estimated to be c. 5000 years old, in the Supe Valley on the north coast of Peru (see below). The archaeological remains of other civilizations also attest to the power of humanity to alter land cover and create centres of political influence; examples include Machu Picchu in Peru, an Inca construction, and the Angkor complex in Cambodia, though both are considerably later in construction – notably c. AD 1500 for the former and c. AD 1000 for the latter – than the sites referred to above.

At many archaeological sites there is evidence for land-use change within the settlement during its lifetime. In some cases substantial changes occurred, reflecting a range of socio-economic factors such as population growth, political power and wealth generation through resource exploitation. Several examples occur in the Middle East; they are especially interesting because they record conditions before and after the advent of agriculture. Netiv Hagdud and Jericho in present-day Israel and Abu Hureyra in present-day Syria (see Figure 8.1 for locations) all provide information on this transition, which had such a profound affect on both environment and society (Section 2.6). At Netiv Hagdud there were large oval and small circular houses constructed of mud and brick with foundations of limestone slabs and storage pits for grain from pre-domesticated barley (Bar-Yosef et al., 1991, quoted in Maisels, 1999). Plant and animal remains attest to the exploitation of a wide range of resources. The wealth of the

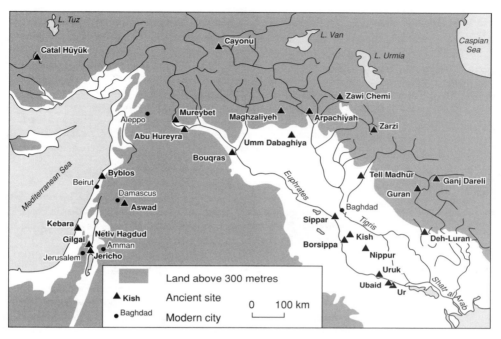

**FIGURE 8.1**  Many of the Ancient sites of the Middle East

remains found at this small settlement has led Maisels to write:

> Truly the inhabitants of this Garden of Eden wanted for nothing. A mix of aquatic, terrestrial and avian species were available throughout the year, while barley became available in spring and early autumn, the fruits, nuts and acorns from September until December.

The inhabitants of this village were primarily hunter-gatherers who were adopting an agricultural mode of existence by storing wild grains and then cultivating them prior to true domestication (see comments in Maisels, 1999).

At Abu Hureyra, a larger settlement than Netiv Hagdud, several time-transgressive phases of occupation have been recorded (Moore, 1991, quoted in Maisels, 1999). The characteristics are summarized in Table 8.1, which shows that the initial three phases reflect considerable changes in the settlement as it expanded. Following abandonment, a new village (Abu Hureyra 2) was established and, within 500 years, domesticated cereals and

pulses were being grown, and domesticated animals such as sheep and goats were being husbanded. Abu Hureyra 2 is one of many villages engaged in early agriculture in this region. Many of these have been described by Mellaart (1994), who suggests that within just a few millennia a settlement hierarchy with major centres, peripheral satellite towns and villages, and possibly hunting camps, had developed. One of the major centres was Jericho, occupying some 4 ha. It developed from a pre-agricultural camp or settlement to a substantial defended town. Similar houses to those of Abu Hureyra were built inside stone walls with towers, outside which was a ditch 3.2 m wide cut into rock. The presence of a containing wall and ditch suggests a need for defence, but this has not been verified. It is most likely that Jericho was a centre for trade in agricultural produce, and abundant plant remains of domesticated crops have been identified from the site (see Mellaart, 1994).

Recent excavations *c*. 200 km north of Lima, Peru, have led to the discovery of what has been referred to as the first urban centre in the

## THE EPIPALAEOUTHIC

**Abu Hureyra 1**

**PHASE 1   11,500–11,000 years BP**

Characteristics:     Interlinked pit dwellings *c*. 2 to 2.5 m in diameter and up to 0.70 m deep. Postholes are present around the edge and in the centre; these probably supported timber poles that held a thatched roof of reeds. The area of the village was *c*. 49 m$^2$ and it was constructed by hunter-gatherers who exploited a wide range of resources from the nearby riverine, forest and plain environments. Querns, stone tools, animal bones and carbonized seeds were present, plus storage pits of 1 m diameter.

**PHASE 2   11,000–10,400 BP**

Characteristics:     Houses are tightly grouped but are above rather than below ground as in Phase 1. Hearths were present in the centre of the site and artefacts discovered include a range of stone and other tools.

**PHASE 3   10,400–10,000 BP**

Characteristics:     The village opened up as suggested by more widely spaced floors and numerous hearths. Huts were similar to those of Phase 1, i.e. wooden posts with cladding in between. Charcoal present shows that willow, poplar, maple and tamarisk were being exploited. Year-round occupation is probable and a wide variety of resources were exploited, including gazelles and many different plant types.

**HIATUS *c*. 10,000–9500   Village abandoned**

**Abu Hureyra 2**

A new village came into existence. It was larger at 12 ha than Abu Hureyra 1 and located on a low mound. Initially, gazelle hunting was the primary activity but by 9000 BP there was a shift to herding domesticated animals, such as sheep and goats, and cultivating a range of domesticated plants, i.e. einkorn wheat, rye, pulses and barley. The buildings were mid-brick houses located along narrow lanes and courtyards, and many houses had polished floors. Occupation of the site continued until *c*. 7000 BP.

**TABLE 8.1** The development of Abu Hureyra. (Adapted from Moore, 1991; quoted in Maisels, 1999)

Americas. Like Abu Hureyra, the settlement pre-dates the invention of pottery; both are thus described as pre-ceramic. Caral is, however, later than Abu Hureyra; dated at *c*. 4090–3640 years BP (*c*. 2627 to 1977 calibrated years BC) Caral lies 23 km from the coast on a river terrace above the floodplain of the Supe River, as shown in Figure 8.2, and is one of the largest of 18 pre-ceramic sites in the area (Solis *et al.*, 2001). The central zone of this urban complex occupies *c*. 65 ha, with stone buildings constructed of river cobbles and rubble; the outer retaining walls were faced and covered

with several layers of plaster. According to Solis *et al.*, the buildings are of three types: monumental, notably several large platform mounds, residential and non-residential. Each mound is associated with residential structures, as indicated by domestic debris, with evidence for alteration over time. Other structural features, including sunken circular plazas, terraces and enclosures are present, and although their function is unknown it would appear that they were non-residential. The abundant remains of plants and animals in Caral have led Solis *et al.* to suggest a reliance on irrigated agriculture in

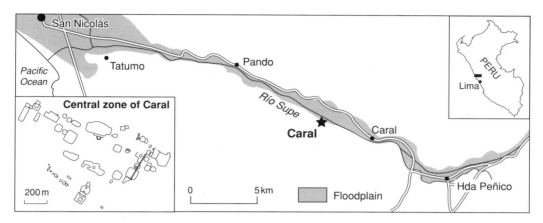

**Figure 8.2** Location and detail of Caral, an early urban settlement in the Supe Valley of Peru. (Adapted from Solis *et al.*, 2001)

the Supe Valley, based on squash, beans and cotton, and on marine resources brought in from the coast. On the basis of radiocarbon age estimations from Caral and their relationship with radiocarbon assay from coastal sites in Peru, Solis *et al.* believe that Caral represents the earliest large settlement to rely heavily on agriculture and not entirely on marine resources (i.e. a shift from marine foraging, which is equivalent to terrestrial hunting-gathering, to permanent agriculture). Indeed, this may be reinforced when excavations and age estimation are conducted at the numerous other sites in the Supe Valley. Solis *et al.* state that, 'The Supe Valley was thus the locus of some of the earliest population concentrations and corporate architecture in South America.'

Many great urban complexes were constructed much later in the northern Andean region of South America. The Incas were highly organized politically and militarily. They came to power in 1476 and established the city of Cuzco in Peru's southern Andes as their capital (for location see Figure 8.3). Today, the region abounds with Inca fortresses and settlements, the construction of which bear witness to Inca engineering skills and their foresight in constructing buildings that could withstand earthquakes. One of the most awe-inspiring settlements created by the Incas was Machu Picchu. Today, this is a World Heritage Site to which tourists flock. It is perched on a flat-

topped peak and has been reconstructed to provide a first-hand experience of an Inca settlement. Its location and further details are given in Figure 8.3, which shows that buildings of a religious nature are predominant. Presumably, the inhabitants of Machu Picchu relied on terraced agriculture in the Urubamba valley and provided a religious centre and safe haven for those who lived in the surrounding countryside.

Whilst it is impossible here to pay tribute to the numerous great cities constructed, and often abandoned, by past civilizations, mention must be made of at least one of the urban complexes of Southeast Asia. Of special note, because of its extent, splendour and, until recently, relative isolation, is the Angkor complex. As part of the Khmer Empire, the Angkor complex was constructed at the height of its power and is considered to have been in use between AD 802 and 1432 (Rooney, 1994). The name 'Angkor' means city or capital and the complex is situated near the modern town of Siem Reap, as shown in Figure 8.4. Angkor Wat and Angkor Thom dominate the complex; the former is considered to be a funerary temple complex for King Suryavarman II, the builder of Angkor Wat, whilst the latter is a fortified city with buildings for priests, military and government officials as well as administrative buildings and a royal residence. The setting, architecture and decoration of all the Angkor buildings attest to a society with an

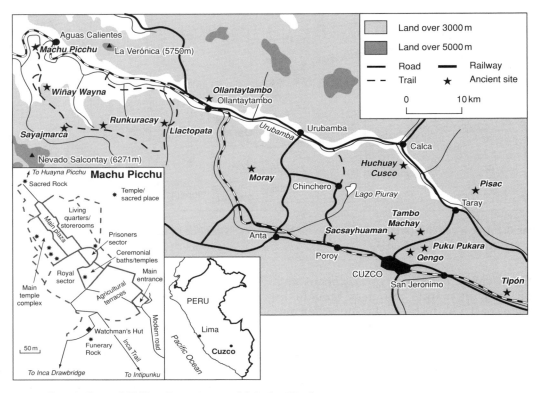

**FIGURE 8.3** Peru's Sacred Valley, Inca sites and Machu Picchu

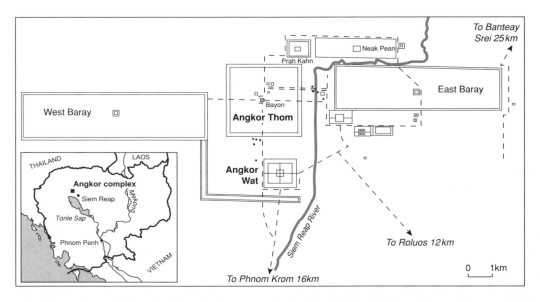

**FIGURE 8.4** Location and detail of the Angkor complex in Cambodia. (Adapted from Rooney, 1994)

urban elite, advanced skills in building construction, craftspeople skilled in creating bas-reliefs and an agriculturally productive hinterland. That considerable looting of the complex continues to occur, despite Cambodia's prospect for a peaceful future, is if nothing else testament to the splendour of its ancient heritage.

## 8.3 Historic world cities

According to *The Times History of the World* (1999), urban societies developed separately in four different regions of Eurasia. These are Mesopotamia *c*. 3500 BC, Egypt *c*. 3100 BC, the Indus Valley *c*. 2500 BC and China *c*. 1800 BC. In general, urban societies arose in fertile river valleys and were associated with productive agriculture which generated surplus food and facilitated the division of labour and trade. Amongst the earliest cities in these regions were Memphis in Egypt, Uruk in Mesopotamia,

Harappa in the Indus Valley and Cheng-Chou in the Yellow River Valley of China. However, these ancient cities exist today mainly as ruins or archaeological sites, and it is generally considered that the 'modern' city began with the emergence of Athens as a city-state *c*. 500 BC. Whilst the Greek world was already extensive by this time, having been established by the Minoan and Mycenaean civilizations between 3000 and 950 BC, Athens in 478 BC united many city-states to provide security against the Persians (Morris, 1994). It developed a naval fleet to provide protection and became the centre of the so-called Athenian Empire to which poets, philosophers and politicians flocked. It remains a historic world city today, as discussed below.

Rome is another historic world city. It originated in 753 BC, but was not in control of the Italian peninsula until *c*. 264 BC. At its apogee, the Roman Empire extended throughout southern and central Europe to the British Isles in the west and to Turkey in the east, as shown in Figure 8.5. Many of the established

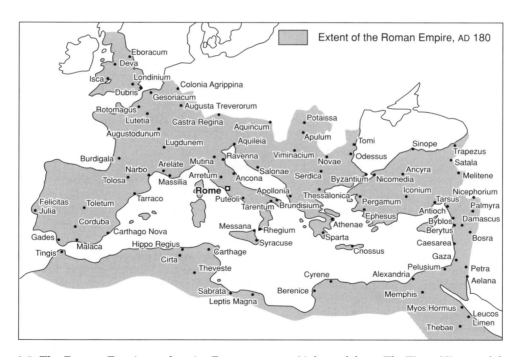

**FIGURE 8.5** The Roman Empire and major Roman towns. (Adapted from *The Times History of the World*, 1999)

settlements of this vast area were founded or influenced by Roman architects and engineers (some of these towns and cities are also shown in Figure 8.5). Rome itself is a remarkable city; at its heart lie the vestiges of ancient Rome, which attracts millions of tourists and thus continue to generate wealth for Rome's citizens as the historical monuments of Athens do for its burghers. Rome is also a modern, sophisticated city, and the old and new are juxtaposed in a surprising congruence.

London must also be included as a great world city. It too has a long history; it existed prior to the Roman annexation of Britain and continued as a major port after colonization. There is a wealth of archaeological remains from London's Roman era, including visible remains of the Roman wall. Like Athens and Rome, the last 2000 years have left their mark and in all cases land use has altered substantially, not least to accommodate burgeoning populations and technological change. It is difficult to envisage the Mediterranean-type shrubs and trees that characterized land now occupied by the urban sprawls of Athens or Rome, or the lime-rich deciduous woodland of that part of the Thames Valley that is now London.

Athens is one of the world's oldest cities, yet it is essentially a modern city. This apparent paradox stems from the fact that it was founded *c.* 7000 years ago, enjoyed a heyday *c.* 2300 years ago, but has experienced its greatest expansion since the 1850s following its appointment as capital of Greece in 1833. As Figure 8.6 shows, Athens is situated in the central-eastern part of Greece, in the coastal plain of Attica. It enjoys an aesthetically pleasing setting; not quite on the coast but overlooking the Saronic Gulf, it lies in a basin fringed by limestone mountains. At its height of importance in the fifth century BC, Athens had a population of *c.* 275,000; it was thus a megalopolis at a time when most towns in the Mediterranean world had no more than *c.* 150,000 people (see Hall, 1999, for details). Ancient Athens was part of a *polis*, an economic and political unit, of 2756 km$^2$, which united all of Attica and conjoined town and country (see Section 8.5 for modern counterparts). Hall

points out that both Plato and Aristotle envisaged a *polis* as being self-sufficient, with perhaps trade for vital non-local resources (i.e., in modern terms, *a polis* should have an Ecological Footprint that more or less balances its biocapacity, see Section 3.6). For some *polis* this might have been the case but not so for Athens; imports of grain, wine and other products were essential for its maintenance. Its political organization reflected a sophistication and acknowledged synergy between people; Hall (1999) opines that the people, the Athenians, were the city's greatest asset and that its educational and cultural reputation ensured a steady influx of philosophers, artists etc. The city thus became a hotbed of learning.

Other Greek cities of this period were planned cities. Miletus and Priene are quoted by Morris (1994) as the two most important planned cities based on the plans and principles of the Milesan architect, Hippodamus. This involved the use of the gridiron plan with an *agora* (market-place) in a central location and other public buildings, and various distinct residential districts, all within the city walls. Athens, however, was not subject to such wholesale planning; even after phases of destruction, buildings such as major temples were replaced as replicas of those destroyed. Moreover, the city grew organically, especially in its fifth-century BC heyday, as new residential districts splayed out haphazardly from the centre into the Attica plain (see Morris, 1994, for details). The decline of Athens, in terms of political superiority, occurred because of its defeat by Philip II of Macedon in 338 BC. Athens was influenced later by Rome, and many Roman buildings were constructed. In AD 267 the Goths destroyed much of the city, which was rebuilt, although the new city was smaller than the old. According to Morris, it occupied *c.* 16 ha within a new wall. Athens also experienced an Islamic period between 1461 and 1830, when it was a Turkish city. By the time the city reverted to Greek rule there were only *c.* 4000 inhabitants. Its establishment as the capital of Greece in 1833 opened a new era of growth as reflected in the data on population change given in Table 8.2. The city has experienced rapid growth in the post-war

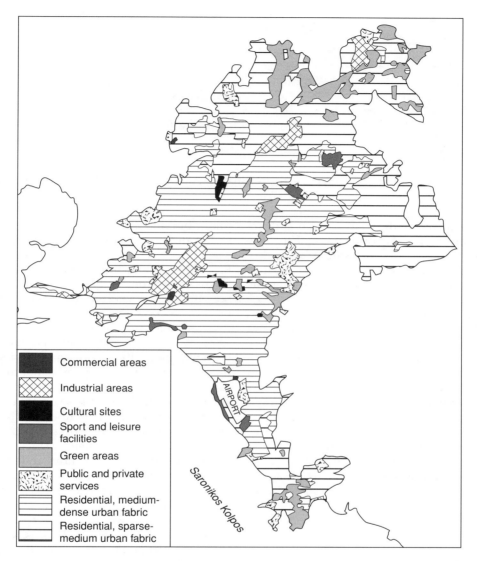

**FIGURE 8.6** Simplified land-use map of the Attica basin, Greece, 1999
The map was realised by GEOAPIKONISIS Ltd under a study contract (13350-97-11 F1 ED ISP ER) with the Space Applications Institute Center of Earth Observation Unit JCR – Ispra (Varese) Italy. It was produced through appropriate refinement (computer assisted photointerpretation) of the output of an automatic classification based on the IRS-1C (April 1977) multispectral data enhanced to the resolution of the panchromatic mode of the same system.

period; illegal settlements in its hinterland were absorbed and industrialization has occurred.

The characteristics of present-day Athens are given in Figure 8.6. It is recognized that recent growth has caused problems, especially the lack of open areas, congested roads and related pollution issues. To address these problems, the implementation of a regional plan began in 1985. A major goal of this plan is decentralization, whereby a number of 'centres' are established at community and neighbourhood levels (Envibase, 2001a), with linking road networks and open spaces. Other goals include the improvement of air quality, modernization of

| Date | Population | | |
|---|---|---|---|
| 1830 | c. 4000 | | |
| 1862 | c. 42,000 | | |
| 1971 | c. 68,000 | | |
| 1922 | c. 108,800* | | |
| **Present** | | **Area km²** | **% Green area** |
| Greater Athens | 3,098,775 | 427 | |
| Municipality of Athens | 784,110 | 37,732 | 10.3 |

* Caused by the expulsion of Greeks from Asia Minor

**TABLE 8.2** Data on population change in Athens, Greece. (Adapted from the Envibase Project, 2001a)

the business and industrial sectors and re-invigoration of the historic areas. Thus the land-use characteristics of Athens will alter substantially in the early twenty-first century as Athens repeats its role as host of the Olympic Games in 2004.

Rome has many characteristics in common with Athens. First, it has ancient origins, and ancient buildings and archaeological sites are abundant in its centre. Second, Rome experienced considerable growth after it was designated as Italy's capital in 1870. However, it has not been influenced by Islamic architecture; it experienced a second Golden Age, associated with the Renaissance, and has an abundance of green/open spaces. Although, Rome achieved considerable importance as the seat of Kings (753–510 BC) and as a Republic (509–27 BC) its first Golden Age was as the imperial capital of an empire (27 BC to AD 330). Its unique place in history is summed up by Hall (1999) who states that:

> Rome was such an urban anachronism: it served as a kind of rehearsal or trailer for all the cities that would come much later, and all the problems they in turn would face. For it was, simply, the first giant city in world history.

Compared with Athens' population of 40,000 in the fifth century BC (see above), Rome had a population of c. 1,000,000 by AD 100 and was thus particularly deserving of the term megalopolis. It benefited from Greek influence in its architecture, but also displayed a grandeur that was uniquely Roman, reflecting a strong record in

public building that witnessed Rome's achievements and those of its pre-eminent citizens.

As Figure 8.7 shows, Rome is situated on the River Tiber, which links it with Astia, its port, c. 30 km distant on the shores of the Tyrrhenian Sea. Its ancient wall encloses seven hills and even in its days as a seat of kings there was a division into rich and poor quarters. Early Rome was not a planned city and, like Athens, it grew organically and provided living conditions of a cramped and poor standard. Further rapid growth occurred after the republic was created, by which time the seven villages of the seven hills had provided Rome with a centre and the focus for public works. Figure 8.7 gives a generalized plan of Rome c. AD 280, though as Morris (1994) points out, numerous phases of demolition and re-organization of public buildings and people had occurred by this time. Ancient Rome was as dynamic as its modern counterpart and was certainly a 'world city' with its fora, mausolea, libraries, baths, circuses, temples and fortified walls (for a detailed description see Morris, 1994), the remains of many of which are still in evidence today. Nevertheless, the splendour was juxtaposed with squalor; life for poor Romans was overcrowded, unhealthy, hazardous and subsistent.

By AD 330 the capital of an ailing empire was in decline as Constantinople usurped Rome's position. According to Hall (1999), quoting Paoli (1990), Rome's population had fallen to 17,000 by 846, reflecting a rapid fall from grace. Some 700 years later, Rome experienced a renaissance as an urban centre. Population increased once again, but never to the level of

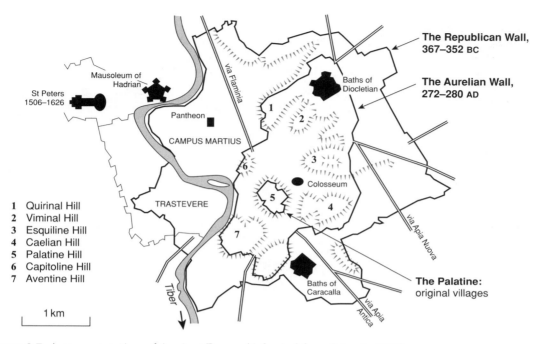

Mausoleum of Hadrian

St Peters
1506–1626

Pantheon

CAMPUS MARTIUS

**The Republican Wall,
367–352 BC**

Baths of
Diocletian

**The Aurelian Wall,
272–280 AD**

1

2

TRASTEVERE

3

Colosseum

6

5

4

7

via Apia Nuova

1   Quirinal Hill
2   Viminal Hill
3   Esquiline Hill
4   Caelian Hill
5   Palatine Hill
6   Capitoline Hill
7   Aventine Hill

1 km

Tiber

Baths of
Caracalla

via Apia
Antica

via Flaminia

**The Palatine:**
original villages

**FIGURE 8.7** A representation of Ancient Rome. (Adapted from Morris, 1994)

Rome's earlier Golden Age. According to Morris (1994), Rome's population reached 124,000 by 1650. The Renaissance in Europe generally prompted a significant shift in approaches to urban form and architecture, notably the use of the gridiron system, enclosed spaces and the primary straight street (see for example Hall's discourse on Florence). Whilst Rome did not quite match the achievements of Florence, it nevertheless acquired a renewed vigour that is reflected in its modern urban fabric. Like Paris, Rome developed a primary street system with many piazzas; numerous churches and cathedrals were constructed, and great architects and artists were employed as the city was reconstructed under the auspices of various papacies (for details see Morris, 1994). Further change and expansion has occurred since 1870 when Rome was established as the capital of the Italian state. Data on modern Rome are given in Table 8.3, which shows that it is now nearly three times the size of its ancient counterpart in terms of population and area occupied, and that the city has a wealth of green, open spaces. Moreover, the

abundant use of trees to line major streets during the past 100 years has been noted by Attorre *et al.* (2000), who have shown that various periods of development have favoured particular tree species. Their findings are summarized in Table 8.4. Why such associations should occur is open to question; it may simply be that each regime wished to be different from its predecessor but Attorre *et al.* suggest that the results of their survey should encourage the distinction between Rome's urban areas to be continued, not only on the basis of architecture etc., but also on the basis of their tree species. They recommend heterogeneity rather than homogeneity in the green elements of urban landscapes (see also Section 8.6).

Another historic world city that warrants mention here is London, though to do justice, in a few hundred words, to such a complex subject, is impossible. Like Athens and Rome, London was established as a settlement in prehistoric times. Sites where Palaeolithic tools and camps were established are known from a number of locations, e.g. Piccadilly,

| A | General Characteristics | | |
|---|---|---|---|
| | Area | 1290 km² | Rome and port of Ostia |
| | Built-up area | 338 km² | i.e. 27% of total area |
| | Green/open | 947 km | i.e. 73% of total area |
| | | | |
| B | Specific land-use characteristics % | | |
| | Built area | 29 | |
| | Highways and railways | 1 | |
| | Green urban areas: sport/leisure | 3 | |
| | Agricultural areas | 49 | |
| | Forest, scrub | 15 | |
| | Beaches, dunes | 2 | |
| | Water bodies | 1 | |
| | | | |
| C | Other data | | |
| | Population (1995) | 2,806,466 | |
| | Cars | 1,908,000 | |

**TABLE 8.3** Data on modern Rome. (Adapted from Envibase, 2001b)

| | | Approximate |
|---|---|---|
| **Period** | **Approximate date** | **Tree species** |
| Fascist | 1930–50 | Pine (*Pinus pinea*) |
| De Vico | 1920–30 | Holm-oak (*Quercus ilex*) |
| Umbertine | 1890–1920 | Plane (*Platanus acerifolia*) |
| Papal | 1871–90 | Elm (*Ulmus minor*) |

**TABLE 8.4** The tree species in Rome's tree-lined roads in relation to recent cultural/political periods. (Adapted from Attorre *et al.*, 2000)

Wandsworth and Stoke Newington; similarly, evidence of Mesolithic, Neolithic, Bronze Age and Iron Age activity in the Thames basin is abundant. This points to occupation from *c.* 8000 years BP. However, the Romans were responsible for establishing a substantial settlement that was called Londinium. According to *The Times History of London* (1999), the earliest occupation of London probably comprised a number of temporary defended camps for soldiers patrolling a Thames crossing. A bridge was probably in place by *c.* AD 50 and in AD 60 Queen Boudicca razed the first Londinium. By the end of the century the second Londinium had become the provincial capital and within another 100 years the settlement had been fortified. The city became a major port and commercial centre, and housed a cosmopolitan population in the area of modern London known as the City (see Figure 8.8). John Morris (1982, quoted in Inwood, 1998) believes that the population in Londinium's heyday was *c.* 100,000, though a decline ensued as the Roman Empire fragmented (the rule of Rome ended in 410). Today, the population of Greater London is *c.* 6.5 × 10⁶. It is characterized by tremendous contrasts in architecture, community characteristics in terms of ethnicity, wealth, the availability of green spaces, shopping facilities and a congested transport network. These characteristics have evolved and developed during more than 2000 years of history.

Hall (1999) highlights three specific periods

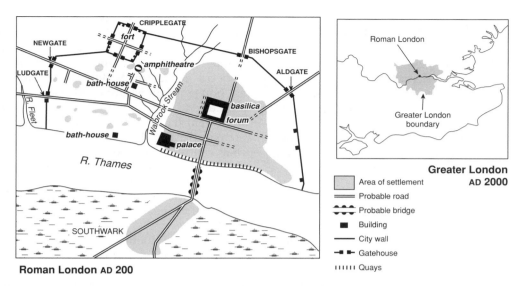

**FIGURE 8.8**  London in Roman times and in 2000. (Adapted from *The Times History of London*, 1999)

of that history, notably London as a centre of innovation in the theatre between 1570 and 1620, London as a utilitarian city between 1825 and 1900, and London as the capital of capitalism between 1979 and 1993. The first of these established the city as a cultural mecca, a claim that remains true today. As Hall points out, sixteenth-century London was a hotbed of poets, playwrights and impresarios; an amateur activity came of age by becoming professional. By 1600 there were *c.* 200,000 people, and by *c.* 1660 this had risen to *c.* 500,000. Elizabethan London was also a city of contrasts: the rich and poor were juxtaposed as were the literate and illiterate. New buildings proliferated, including the construction of numerous theatres. Hall makes the point that capitalism was initiated in London at this time, a regime that re-asserted its importance in the 1980s. London experienced further massive growth in the 1800s due to industrialization and a shift of people from the countryside into the city. Expansion was rapid and gave rise to a range of social issues such as poor housing quality, health problems and rising crime. During this period the suburbs expanded, e.g. Belsize Park, and villages hitherto outside the

city became amalgamated, e.g. Finchley (see Inwood, 1998).

Two world wars have also affected London's land use, just as the Great Fire did in 1666. Parts of the city were rebuilt and even more heterogeneity of land use resulted. This has since been fuelled further as the nature of employment has altered. In the 1970s and 1980s, for example, the shift from manufacturing to service industries began – a process that continues today throughout Britain. In addition, London's Docklands witnessed decline and depression, and then an attempt at resurgence through the renovation of dockland warehouses etc. for high-priced housing, and the provision of leisure facilities and specialist, expensive shopping. In brief, London today is one of the largest cities in the world; its land use reflects its history of growth and development based on intellectual, economic, royal and government interests. It has residential areas of all descriptions, premises for industries and services, government buildings of all types and public buildings of worldwide importance. To this must be added the green spaces: parks woodlands, gardens and cemeteries, as discussed in Section 8.6.

## 8.4 Recent world cities

Many of the world's cities do not have as long a history as those described in Section 8.2. European annexation of lands in the sixteenth and seventeenth centuries in the New World especially, and later in Australia and New Zealand, led to the foundation of many new settlements that subsequently developed into cities. From their beginnings many of these cities were planned cities, though they have often been influenced by conditions in the home country. Examples of such cities include Havana (1514) in Cuba, Guatemala City (1776), Concepcion (1765) in Chile, and Mendoza (1561) and Buenos Aires (1580) in Argentina (see detail in Morris, 1994), which were established by the Spanish conquistadors. These, and many others, were designed according to a royal ordinance of Phillip II of Spain, which established guidelines for the structure of 'towns' throughout Spain's empire. Today, cities and towns alike throughout Latin America have a plaza (a so-called square, which is oblong and occupied by grass/vegetation) or a serious of plazas, depending on the size of the settlement. Around the plaza, churches or cathedrals were constructed, and streets were organized on the basis of the gridiron plan (see Section 8.2).

Many cities of North America were also influenced by the Spanish invaders in the mid-sixteenth century, and others by the French and British. The gridiron plan prevailed in both North and South America as the basis for city design and planning. As the name suggests, the gridiron plan comprises a rectilinear network of roads at right angles to each other. In comparison with the design or structure of the cities described in Section 8.2, especially the older districts, the gridiron system confers a high degree of regularity rather than *ad hoc* arrangement. Cities with a gridiron basis in North America include Toronto, Chicago and Los Angeles, whilst cities such as Boston and New York have had a gridiron system superimposed on an irregular settlement pattern, which reflects their status as the earliest settlements founded in North America. It is also important

to note that many modern, i.e. eighteenth to twentieth century, cities have been constructed on the basis of a predetermined plan. The most obvious examples are Brasilia in Brazil, and Canberra in Australia. Both are capital cities and house administrative buildings for their respective nations. On a smaller scale, the so-called 'garden cities' of the UK, such as Letchworth and Welwyn, are also planned urban centres though often incorporating pre-existing villages. Whatever the planning system used, concrete, tarmac and numerous other artificial surfaces have replaced the former land cover of vegetation and/or agricultural land.

In the case of Buenos Aires, in Argentina, the city has replaced the grassland of the pampas along the Rio de la Plata. First founded in 1536, the settlement of Nuestra Senora Santa Maria del Buen Aire did not prosper and gave away to the better-located Asuncion, now the capital of Paraguay and also on the River Plate (see Figure 8.9). The second attempt to re-establish Buenos Aires took place in 1580 on a site just a few miles north of the original settlement. On appointment of, first, a royal governorship from Spain in 1618 and then as Spain's fourth vice-regal capital in 1776 (Morris, 1994), the city expanded rapidly. By 1900 emigration from Spain resulted in an increase of population to more than a million to make Buenos Aires Latin America's largest city. Today, it has a population of *c*. $3 \times 10^6$ and Greater Buenos Aires has a population of *c*. $11 \times 10^6$. The latter accounts for *c*. 33 per cent of Argentina's total population and the area of land occupied is 7,729 km² of what was rolling pampa (*Pampa Ondulada*).

As Figure 8.9 shows, the heart of Buenos Aires is a compact arrangement of avenues dividing the area into squares. The Avenida de Mayo is the main road; at its eastern end are the Casa Rosada (presidential palace), the cathedral and the Cabildo around the Plaza de Mayo; at the western end is the congress building. The city has undergone massive expansion in the last 50 years. Morello *et al.* (2000) have examined the process of urbanization that has occurred since 1869 and the impact this has had on the *Pampa Ondulata*, which is one of Argentina's richest agricultural areas.

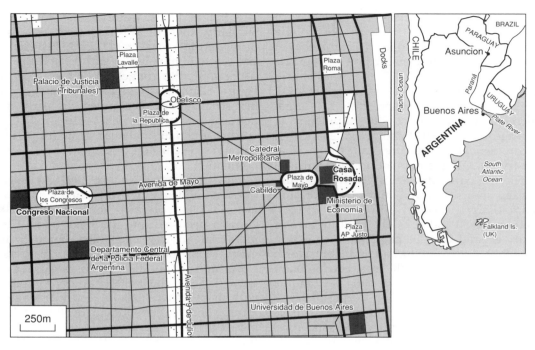

**FIGURE 8.9** Buenos Aires. (Adapted from Lonely Planet City Map Series, 2001)

They refer to this impact as the 'landscape footprint', i.e. the productive land, water, and natural and semi-natural landscapes consumed by the city. (This is quite different to the Ecological Footprint described in Section 3.6.) The landscape footprint is especially significant in the so-called peri-urban area, the equivalent of an ecotone between ecosystems, where the urban and rural coalesce, and neither is at its best; the urban sector lacks services such as water and sanitation provision, while the rural sector has poor ecosystem services such as reduced carbon storage and photosynthesis as land is exploited for other resources such as illegal rubbish dumping, soil and clay removal for urban gardens and brick manufacture, and land parcelling in anticipation of price rises as the city expands. As Morello *et al.* point out, the amount of land parcelling is itself a measure of urban expansion; in 1869, for example, Buenos Aires was characterized by 11 such plots in three main areas: the centre along the edge of the Rio de la Plata, which was the port area, and

corridors following the Sarmiento railroad to the west and the Roca railroad to the south. Subsequent expansion occurred along railroad routes and, later, road networks all radiating out from the city. Land parcelling accelerated between 1947 and 1970 and involved 61 km² per year between 1970 and 1991; it declined to 16 km² per year. Much of this expansion has taken place on land with good soil for agriculture, so urban expansion has had preference over the production of agricultural commodities. This, Morello *et al.* suggest, indicates a short-sighted approach to urban growth, and inadequate planning strategies.

Toronto, Canada, is an example of a city planned on the basis of the gridiron system. According to *Encarta* (2001) it owes its origins to the construction in the 1750s of a fort by the French in the so-called Toronto Passage between lakes Ontario and Huron. This was destroyed in 1759 by the British who, almost 30 years later, purchased land from the indigenous people on which to found the city of York. This became the capital of what is now the state of Ontario. Toronto's location was particularly advantageous; it benefited from its proximity to the Great Lakes, the construction of the Erie Canal, which opened in 1825, and the establishment of a rail network in the 1850s. By 1834, York had become a component of a larger settlement named Toronto. Its population at this stage was *c*. 80,000 and activities focused on commercial agriculture and the lumber industry. By 1912 the population had increased to *c*. 500,000, partly due to the growth of manufacturing industries, and mining in northern Ontario (*Encarta*, 2001).

Between 1890 and 1923, the built-up area of the city expanded by 50 per cent, as shown in Figure 8.10. According to Gad and Holdsworth (1990), industrial growth was predominant within the city centre, especially within the vicinity of the railway, and on the periphery of the city (Figure 8.10). As industries, such as printing, metals/machinery, textiles and food processing, became established in various parts of the city (see detail in Gad and Holdsworth, 1990), the increasing importance of the railways was reflected in the conversion of public land

into freight yards, factories and warehouses. Similarly, the city's King Street was transformed from a major retail centre to a commercial centre dominated by offices, and the retail sector shifted to Youge and Queen Streets (see Figure 8.10). Undoubtedly, the railways and industrialization transformed the early twentieth-century land use of Toronto.

After a period of slowed growth *c*. 1914–40, Toronto's expansion resumed; by 1953 the population of the metropolitan area had reached $1.2 \times 10^6$ and in 1967 the existing thirteen municipalities were combined into six, including the original York (*Encarta*, 2001). Growth persisted into the 1960s and 1970s as industrial activity intensified; mining, automobile construction for both Canada and the USA, and financial businesses fuelled this growth. Since the 1980s Toronto, in common with many other cities in North America and Europe, has experienced a decline in heavy industry, and an increase in service and financial industries. In the post-Fordist era (Fordism, *c*. 1945–75, was an era of manufacturing especially associated with the mass production of automobiles and other consumer

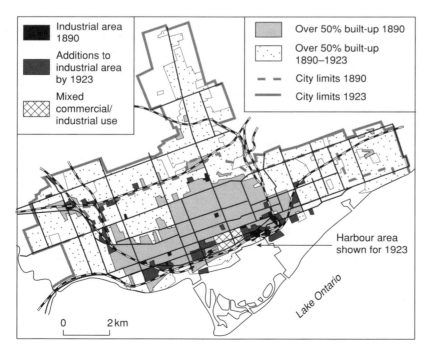

**FIGURE 8.10** Land use in Toronto 1890—1923. (Adapted from Gad and Holdsworth, 1990)

goods using dedicated equipment and under management–worker relationships that focused on links between wage increases and productivity) further land-use changes have occurred along with changes in social structure. Walks (2001), for example, believes that the Fordist distinction between urban and suburban districts, based on occupation and income levels, is no longer tenable; inner-city gentrification (revival of rundown areas through the provision of expensive residential opportunities) has led to a move back into the city by high earners, often associated with financial occupations. Toronto remains a leading city of Canada and a world city.

More recently, several new cities have been created in areas with no previous settlement, and on the basis of approved plans and designs. The major examples are Brasilia in Brazil, Islamabad in Pakistan and Canberra in Australia. All are national capitals; Canberra is the oldest. According to the National Capital Authority (2001), 911 square miles of land were set aside by the state of New South Wales in 1911; construction began in 1913 on the basis of a plan by Walter Burley Griffin, an American landscape architect, who won an international competition aimed at finding an appropriate design for Australia's proposed capital. The world wars disrupted progress and it was not until 1958 that the city began to take shape. Figure 8.11 shows that the city has Lake Burley Griffin at its centre and several foci, e.g. the City and Capital Hill, around which roads have been constructed in concentric circles. Today, Canberra has a population of *c*. 300,000 most of whom live in satellite towns each with its own town centre. In terms of land use, there are many open spaces and parks etc. (see Figure 8.11), whilst the buildings are mainly concerned with public administration/government, education and service industries. There are plans for further development involving the creation of distinct townships in landscaped valleys (see details in National Capital Authority, 2001).

Unlike Canberra, Brasilia retains elements of a gridiron system, as shown in Figure 8.12. The city was constructed between 1956 and 1960 on the basis of an award-winning plan by Lucio Costa. Like Canberra, the city is divided in two: Asa Norte and Asa Sul. It has numerous green spaces, is dominated by public administra-

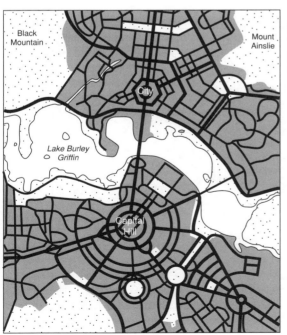

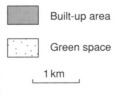

Built-up area

Green space

1 km

**FIGURE 8.11** The basic layout of Canberra, Australia

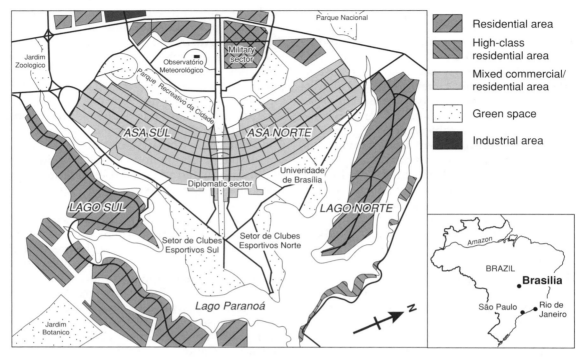

**FIGURE 8.12** Brasilia and its land use. (Adapted from Areal, 1996)

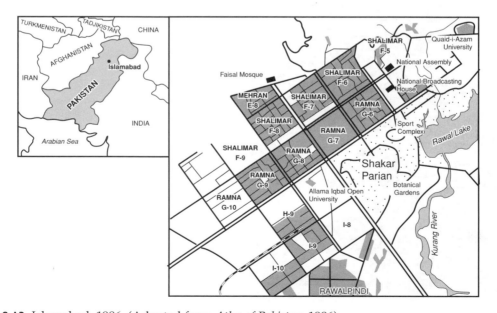

**FIGURE 8.13** Islamabad, 1986. (Adapted from *Atlas of Pakistan*, 1986)

tion/government buildings, commercial streets that are dominated by bars and restaurants, and four major shopping malls. According to Areal (1996), the city was built for *c*. 500,000 people, but rapid population growth has encouraged the development of several satellite towns; today, the total population has grown to more than $2 \times 10^6$.

Islamabad is the most recent of the three capitals. It was constructed in the early 1960s and became Pakistan's capital in 1967. The city lies in the northern part of the Pothowar Plateau and is overshadowed by the Margalla Hills. This region has a long history of human occupation, including the settlement of Taxila, which reached its apogee *c*. 600 BC. Islamabad was designed by Doxiadis Associates, a firm of Greek architects, whose plan comprised an overall triangular shape within which there was a gridiron arrangement of wide streets. The urban area of Islamabad, as shown in Figure 8.13, occupies some $905.15 \text{ km}^2$ and has a population of *c*. 900,000 (Geocities, 2001). The city is divided into eight zones, each reflecting a specific land use, e.g. the residential sector and the educational sector. Each has its own shopping area and public park. Its abundance of green spaces and wide tree-lined roads is in stark contrast to its twin city of Rawalpindi with its age-old streets, bustling bazaars and historic buildings.

## 8.5 Recent growth in cities of the developing world

In Europe during the Industrial Revolution, *c*. 1750 to 1900, there was a massive movement of people out of the countryside into the towns and cities. This rural depopulation and resulting urbanization was caused by population growth and the increased employment opportunities afforded by newly established industries. Although many of these cities have experienced a recent trend of population shift known as counterurbanization, involving not ruralization of people but a shift of population to peripheral and smaller urban centres than the metropolises, city growth has continued. In

North American cities, suburbanization has occurred. Moreover, in the developing world a wave of urbanization has characterized the past 40 years or so, and many cities in Africa, Asia and Latin America are amongst the fastest growing cities in the world, and thus reflect rapid land-cover change. Population growth in the cities themselves and rural–urban migration for employment purposes are the major causes. Moreover, many so-called rural areas in Africa and Asia have population concentrations in excess of those in the urban areas of the developed world. These are termed 'ruralopolises' by Qadeer (2000) and occur in India, Pakistan and Bangladesh.

The issue of rural–urban migration is illustrated by the case of Lima in Peru (for location see Figure 8.2), where informal settlements in the city's periphery provide a stark contrast with the surrounding desert. As in the case of many of Peru's cities, Lima is situated near the coast and in a river valley that provides a water source in an otherwise arid area. In Lima's case the river is the Rio Pizarro and its centre still boasts several architecturally interesting colonial buildings around the Plaza de Armas. Like its counterparts elsewhere in Latin America (see Section 8.4). the city was constructed on the basis of gridiron blocks; there were 117 such blocks with the central plaza close to the river. According to Morris (1994) only 67 of these blocks were developed before the overall plan gave way to less controlled development. As Table 8.5 shows, population growth has been rapid and this continues today. It is manifest as 'informal' or shanty-town constructions along the city's expanding margins; 'temporary dwellings' are constructed in the coastal desert and clothe the foothills of the Andes to the east of the city. Entire villages from the Andean mountains have migrated to Lima to create so-called *pueblos jovenes* (young towns) comprising individual houses of reed mats. These settlements are equivalent to the 'ecotonal' areas of Buenos Aires referred to in Section 8.4. They have no services such as rubbish or sewage disposal, though some services, e.g. water provision, are ferried out from Lima's centre in tankers and distributed to those who can pay for them. Most people have come in search of

| Date | Population |
|------|-----------|
| 1561 | 2500 |
| 1599 | 14,262 |
| 1614 | 25,434 |
| 1700 | 37,234 |
| 1755 | 54,000 |
| 1812 | 64,000 |
| 1940 | 540,000 |
| 1957 | 1,135,000 |
| 1961 | 1,846,000 |
| 1972 | 3,303,000 |
| 1981 | 4,601,000 |
| 1996 | 6,800,000 |

**TABLE 8.5** Population growth in Lima, Peru. (Adapted from Morris, 1994)

employment and, as time progresses and even with low incomes, the reed mats are replaced by breeze blocks and bricks to give an atmosphere of permanency when compared with the newly erected reed mats of the next wave of impromptu settlers. A side-effect of this city growth is the alteration of beach quality as domestic waste and sewage accumulate in the white sand; health problems also ensue from this problem.

The landscape footprint (Morello *et al.*, 2000) is evident in the peripheral zones of most medium-sized and large cities in the developing world. For example, Morelia in the state of Michoacán, Mexico, had a population of *c.* 550,000 (in January 2000), having grown rapidly in the preceding few decades; it occupied *c.* 709 ha in 1960 and by 1990 had expanded to 3368 ha, mostly through the coalescence of smaller settlements and expansion into agricultural land (Lopez *et al.*, 2001). The landscape footprint is also evident in many smaller towns as exemplified by those in the Niger Delta, Nigeria, where Shell's oil-extractive activities provide work (as discussed in Section 7.5). The rapidity with which this footprint expands and its characteristics of both urban and rural environments have led to the labelling of these peri-urban spaces as *desakota* regions (McGee, 1989), a term derived from the Indonesian words for country (*desa*) and town (*kota*). The coalescence of the two is blurring the

distinction between rural and urban, both in terms of land use and employment. As Rigg (1998) points out, individuals and entire households shuffle between agricultural and industrial activities; individuals may live in a rural environment but work in industry, and households may obtain their income from dual sources.

Notwithstanding this debate, it is evident that many of China's towns and cities have experienced substantial growth since the end of the Maoist era and the advent of a more liberal economy in the late 1970s. Indeed, China has been experiencing a rapid rate of developing and economic growth, especially in its coastal provinces and cities, which have also benefited from new relationships and links within the Asia-Pacific rim. Using remote sensing techniques (Landsat Thematic Mapper), Ji *et al.* (2001) have identified those areas of China that are experiencing rapid urbanization, much of which is taking place at the expense of much-needed arable land (see also Section 5.6). Their results are given in Figure 8.14, which highlights the situation in the coastal provinces of the southeast. Ji *et al.* point out that some 14 cities in this region were declared as special economic zones in 1980 by the Chinese Government and, as such, have attracted foreign investment. Here the rate of urbanization has been more than 20 per cent of the extant urban area in 1989/92, for which period the base-line data are available. In particular, in the province of Guangdong it is estimated that the average rate of urban growth has been 27 per cent with urbanization occurring in the countryside as well as in the rural–urban fringe. Urban growth is also high in Fujian province, especially around Fuqing city. The development of the latter between 1991 and 1996 has been monitored by Xu *et al.* (2000) using Landsat Thematic Mapper images. Some of their data are summarized in Table 8.6, which shows that the amount of land used for urban purposes has doubled in five years. This has occurred at the expense of arable land, although some arable land has been converted to orchards to supply fresh fruit to the growing city of Fuqing.

The city of Shenzhen is one of the designated special economic zones in Guangdong province

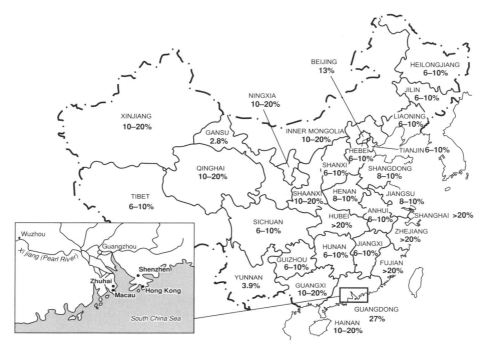

**FIGURE 8.14** The rates of urban expansion in China. (Adapted from Ji *et al.*, 2001)

**A   Development of Shenzhen**

| | 1980 | 1994 |
|---|---|---|
| Population | $0.333 \times 10^6$ | $3.35 \times 10^6$ |

Current rate of population growth: $0.35 \times 10^6$ per year

| | 1989 | 1996 |
|---|---|---|
| Urban area | $149\,km^2$ | $363\,km^2$ |

A further $118\,km^2$ of land has been cleared for construction

**B   Land-cover change in Fuqing County, Fujian Province, China**

| | 1991 ha | 1996 ha |
|---|---|---|
| Arable land | 10,097.7 | 6386.0 |
| Orchard | 1021.1 | 3569.1 |
| Forest | 6280.2 | 5054.1 |
| Urban | 1897.6 | 4100.1 |
| Water body | 1872.1 | 2005.1 |
| Barren | 817.8 | 872.1 |
| Total | 21986.5 | 21986.5 |

**C   Fuqing City Area (km²)**

| | |
|---|---|
| 4.495 | 7.864 |

**TABLE 8.6** Data on the development of Shenzhen (based on Ji *et al.*, 2001) and Fuqing County and City. (Adapted from Xu *et al.*, 2000, China)

and has experienced phenomenal growth in the past 20 years. Ji *et al.* (2001), and Sui and Zeng (2001) have examined this development, a variety of data on which are given in Table 8.6. These data reflect a 10-fold increase in population and 2.5-fold increase in the built-up area. The expansion is due to the construction of residential property and factories. In their detailed study, Sui and Zeng (2001) examined the urbanization around the city of Longhua in the Shenzhen economic zone, which they have identified as a major *desakota* region. Their study shows that growth takes two forms: expansion from established population centres; and the development of industrial and commercial centres along new roads. Such expansion also results in the fragmentation of forests and remaining natural ecosystems, which are further reduced in size when recreational facilities such as golf courses are created. Sui and Zeng also suggest that Longhua's development typifies *desakota* development in Asia generally.

Certainly, it is typical of China's southeastern cities. As Figure 8.14 shows, the Pearl River delta region, is the broader region of Guangdong province in which Shenzhen and Zhuhai, another special economic zone, as well as Hong Kong and Macau, are situated. Its present phase of development began in 1979 as China's reforms focused on decentralization to allow the country's regions to exercise a large degree of autonomy. As part of this freedom there began a shift away from agriculture and an initial trend towards diversification. Subsequent development in the Pearl River delta region has been examined by Johnson and Woon (1997), and Lin (2001). All show that development has been rapid but diverse. Johnson and Woon state that, 'The Pearl River delta is richly textured and there are distinctive responses to generalized pressures and substantial local variation which can, perhaps, be seen as a manifestation of development with [south] Chinese characteristics.' A related sentiment is expressed by Lin who states that, 'The intrusion of global forces has not homogenized local particularities. Global capitalism has to seek shelter from locally specific conditions to take root in socialist soil.' It is interesting to

| A Land cover change in Indonesia 1991–93 | |
|---|---|
| Agricultural land | −106,000 ha |
| Residential areas | + 58,000 ha |
| Industrial land | + 16,452 ha |
| Offices | + 5,210 ha |
| Other urban land uses | + 26,774 ha |

| B Gonomic growth rate in Jakarta 1987–97 | |
|---|---|
| | 6.0 to 8.3% |

**TABLE 8.7** Data on land-cover change in Indonesia. (Adapted from Firman, 2000)

note the maintenance of diversity that is the product of history and geography, yet the loss of agricultural land and forests is of cause for concern, especially in relation to China's Ecological Footprint, which is lengthening and deepening more rapidly than that of most other nations.

Unlike other nations, China has not experienced any major recessions in its economic progress since liberalization, so urbanization has proceeded unabated. However, as Firman (2000) reports, the recent recession in Southeast Asia has affected the rate of land conversion from rural to urban substantially in Indonesia. The data in Table 8.7 show that urban expansion was at a peak in the mid-1990s. As in China, the area of land used for residential and industrial purposes increased at the expense of agricultural land. Since 1997, however, an economic recession has curtailed the rate of urbanization and industrialization. Firman also reports that peri-urban areas have returned to agricultural use, encouraged by the city government. This 'urban agriculture' is focusing on maize, soya beans and vegetables, but unemployment continues to rise. In terms of land cover, the rate of conversion of land from agricultural/rural uses to urban/industrial uses has declined considerably in the last few years.

African cities have also expanded in the 1990s, but for different reasons and in different ways than those of Asia. In the latter, for example, the forces of globalization have been especially important, not least because of foreign direct investment. Whilst this situation is not so widespread in Africa, urban growth and development have been fuelled by other factors. According to Briggs and Mwamfupe (2000), structural adjustments financed by such international agencies as the International Monetary Fund (IMF) have made a major contribution to land-cover and land-use change in many of Africa's cities. They cite the example of Dar es Salaam in Tanzania, whose periods of growth since 1945 are illustrated in Figure 8.15. Whilst earlier stimuli for growth concerned mainly internal factors within Dar es Salaam itself, funds for structural readjustment programmes have been the major stimuli in the 1990s. As Figure 8.15 shows, the city has almost doubled in size, in terms of the land area occupied, during this decade. Structural adjustment programmes promote capitalism and reduce state control; they promote goods produced locally and that have a market niche; they encourage investment and generally stimulate economic growth. Important elements are thus privatization, trade liberalization, and free

currency exchange. Such a programme, known as the Economic Recovery Programme, was inaugurated in 1986; it was a programme designed by the Tanzanian Government and the IMF.

As Figure 8.15 shows, urban development prior to 1992 had occurred in a ribbon-like fashion along the main roads. This reflects the importance of road communications allowing access to the city centre. However, since 1992 and despite continued population growth, the pattern of urban expansion has altered, with infill rather than ribbon developments predominating. In addition, housing density has increased, especially along the all-important main roads. Only to the south of the city has there been little expansion due, according to Briggs and Mwamfupe, to dominant land ownership by the Zaramo people, which apparently deters potential settlers who are not Zaramo despite the fact that land prices are low. This latter reflects a cultural deterrent to land-use change, despite the structural adjustment programme. However, the infills etc. can be

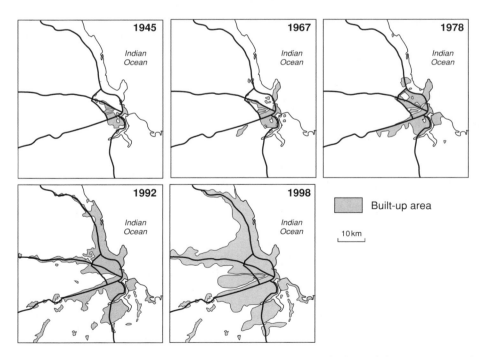

**Figure 8.15** The growth of Dar es Salaam between 1945 and 1998. (Adapted from Briggs and Mwamfupe, 2000)

judged as a direct response to this programme, especially the improvements in transport it has generated.

In particular, the state-owned transport operator, Usafiri Dar es Salaam (UDA), was privatized in 1992 and public transport was deregulated. The result was a massive expansion in the type and quantity of privately run transport systems, which extended beyond the arterial roads into the areas between. This meant that people no longer found it essential to live adjacent to the main roads and thus ribbon expansion of the urban area declined in favour of infill between main roads and increased density in existing residential areas. In addition there has been a rise in the ownership of private vehicles due to increased wealth for some groups and the introduction of a car allowance for public-sector workers as a means of enhancing salaries. This group also benefits from a house-ownership allowance, which has also encouraged urban expansion. According to Briggs and Mwamfupe, wealth accumulation has resulted from the structural adjustment programme, mainly due to trade liberalization and a resulting increase in consumerism. A favoured form of investment by these entrepreneurs is property purchase, and thus urban expansion is encouraged. Similarly, the wealthier citizens of Dar es Salaam tend to spread their investments across a range of enterprises rather than risk all on one large-scale enterprise. This has encouraged the development of numerous small businesses, another stimulus to urban development.

Urbanization is also occurring rapidly in many island nations, as Cocklin and Keen (2001) have reported for the south Pacific islands where the urban population growth rate tends to exceed the general population growth rate. In Vanuatu, for example, the latter is 2.5 per cent annually whilst the urban population growth rate is 3.5 per cent; in Papua New Guinea the values are 2.2 per cent and 3.9 per cent respectively, and in the Marshall Islands overall growth is 3.5 per cent whilst urban population growth is a massive 4.2 per cent. In many islands this urbanization has been accompanied by increasing poverty, poor infrastructure, unemployment, declining health

standards and squatter settlements. Equally important are problems associated with water supply, and sewage and waste disposal. The latter is particularly difficult because of the unsuitability of the islands' geology for landfill sites, so often swamps or areas of mangroves are used instead. Waste disposal also results in the contamination of land with oil, heavy metals and pesticides. All of these factors bring about land-cover/land-use change. Overall, the Ecological Footprint (see Section 3.6) of many Pacific island cities exceeds the biocapacity of individual islands.

## 8.6 Urban parks, forests and lakes

As discussed in the previous sections, most urban areas are characterized by green spaces and many contain water bodies that are mostly artificial. Both green spaces and water bodies, provided they are managed, are assets. At the very least they are aesthetically pleasing; they provide a break within the relative monotony of the bricks, concrete, tarmac and glass that comprise the urban environment, and thus they enhance urban landscapes. Green spaces and water bodies in urban environments may also generate income through the provision of recreation facilities and activities e.g. mazes, wildlife refuges, nature trails, angling, rowing etc. In many cases the green spaces and water bodies are artificial; many have little in common with the natural or semi-natural environments replaced by, or which surround, urban areas, whilst others are tiny replicas of indigenous vegetation communities. As well as parks there are public and private gardens, allotments, cemeteries and tree-lined streets, which contribute to the varied land use that characterizes urban areas.

In Greater London, for example, there is a wide variety of green spaces and numerous lakes and ponds. The green spaces comprise parks, which number 1,700 (174 km$^2$), including royal parks (e.g. Hyde Park and Regent's Park), heath and common land (e.g. Hampstead Heath and Wimbledon Common), the grounds of stately homes (e.g. Marble Hill Park and

| Park | Date | Size (ha) | Comments |
|------|------|-----------|----------|
| Bushy Park and Longford River | 1500–37 1639 | 445 | Enclosed from ploughed farmland by Cardinal Wolsey and Henry VIII; river created by Charles I |
| Greenwich Park | 1433 | 74.9 | The first royal park to be enclosed |
| Hyde Park | 1536 | 255 | The land was acquired for hunting by Henry VIII |
| Kensington Park | 1689 | 115.7 | Formed from land taken from Hyde Park |
| Regent's Park and Primrose Hill | 1811 | c. 162 | Arose from plans of Crown Architect, John Nash |
| Richmond Park | 1637 | 955 | Enclosed as a hunting park by Charles I; notified as a Site of Special Scientific Interest (SSSI) in 1992 |
| St James's Park and Green Park | early 1500s | 23.3 19 | Acquired by Henry VIII; laid out by John Nash in 1827 |

**TABLE 8.8** The royal parks of London. (Adapted from Royal Parks, 2001)

Holland Park), the famous botanical gardens at Kew, as well as sports pitches and allotments. Many of the royal parks also have lakes or ponds, e.g. the Serpentine, an 11.34 ha lake in Hyde Park, while nineteenth-century parks, such as the Lea Valley Park, contain reservoirs. These are detailed in *The Times History of London* (1999; see also Inwood, 1998). The royal parks are listed in Table 8.8. The first to be established was Greenwich Park; it was enclosed by Henry VI and was primarily used for hunting deer. Later, Henry VIII was responsible for establishing Bushy Park, Hyde Park, St James's Park and Green Park; again, hunting was the primary objective. Today, these parks are a welcome respite for both tourists and locals from London's hurly-burly.

Parks and tree-lined streets play a significant role in the urban environments of the USA, as Nowak *et al.* (2001) have documented. They state that urban areas occupy *c.* 3.5 per cent of the conterminous USA, i.e. 281,000 km²; they contain *c.* 75 per cent of the US population and some $3.8 \times 10^9$ trees, with an average tree canopy cover of 27 per cent. These data include trees along streets as well as those in parks and gardens. Their data, based on remotely sensed data and United States forest-cover maps of the Department of Agriculture (USDA), shows that the average tree cover in US urban areas is 504 trees per hectare, with the greatest density of 751 trees per hectare in Atlanta and the lowest density of 312 trees per hectare in New York. Nowak *et al.* have also determined the characteristics of urban tree distribution by state, data on which are given in Table 8.9. This shows that urban trees are particularly abundant in Georgia and Alabama, which also have a large proportion of their overall tree cover in urban environments. Many central and southern states have a relatively low urban tree cover, e.g. New Mexico, while several smaller states in the northeast, e.g. Connecticut and New Jersey, have a high proportion of their total tree cover in their urban areas, which are especially extensive. Nowak *et al.* suggest that the type of natural vegetation, population density and land use help to explain these patterns. Of particular importance is the presence of natural forest; cities in forested regions such as those of the northeast are particularly well endowed with urban trees. In addition, there is an inverse correlation between population density and tree density: where there are more people, there

| State | No. urban trees | % urban tree cover | % state tree cover | % urban area of state |
|---|---|---|---|---|
| Georgia | 232,906 | 55.3 | 4.7 | 5.4 |
| Alabama | 205,847 | 48.2 | 4.7 | 6.3 |
| New York | 132,466 | 26.3 | 4.5 | 7.2 |
| Nevada | 15,834 | 9.9 | 0.8 | 1.1 |
| Wyoming | 1392 | 3.6 | 0.1 | 0.3 |
| New Mexico | 5682 | 4.8 | 0.3 | 0.7 |
| Connecticut | 44,800 | 21.8 | 14.0 | 28.5 |
| New Jersey | 143,869 | 41.4 | 223 | 30.6 |

**TABLE 8.9** Data on the urban tree characteristics of selected states in the USA. (Adapted from Nowak *et al.*, 2001)

are fewer trees. City size also has an effect; the larger the city area in forested or grassland areas the greater the tree cover, whilst there is a decrease with city size in desert areas. This probably relates to unused areas in cities, which increase in number with city size, and which will be colonized naturally by trees in those ecological zones that will support tree growth, i.e. forested and grassland regions. The statistics alone attest to the contribution of trees to urban land use in the USA.

Urban forests that have close relationships with the natural vegetation of the surrounding regions are important in Sweden and Japan, though in both cases urban forests are broadly interpreted as not only forests/woodlands in urban areas but also those in the peripheral regions of cities. In both cases, ground vegetation is uncultivated and thus it is a feature distinguishing urban forests from parks. According to Rydberg and Falck (2000), Sweden's urban forests vary in age, with older forests characterizing the central parts of older urban areas and younger ones at the urban fringe. The latter occupy *c.* 3000,000 ha or approximately 1 per cent of Sweden's forested area; *c.* 180,000 ha of this forest is owned by municipal authorities, and *c.* 35 per cent is used for outdoor recreation activities and nature conservation. These and central urban forests have their origins in Sweden's natural coniferous forests, the silvicultural practices in which have influenced their management and characteristics. Rydberg and Falck distinguish between estates constructed before the 1960s

and those constructed after the 1970s. The former often comprised older trees, which have aged further, whilst the latter incorporated clear-felled areas or abandoned farmland, which have been replanted, often with Norway spruce. The chief factor that distinguishes these forests from their natural counterparts is the high degree of human use and involvement in their management. As Table 8.10 shows, Sweden's urban forests can be classified according to location and use, which are factors that also require different approaches to management. It is clear from the analysis of Rydberg and Falck that forests are very important in the lives of most Swedes.

In Japan, the traditional countryside landscape persists at the edges of large towns and cities, e.g. rice paddies, forests and irrigation canals. These, as Nakamura and Short (2001) have discussed in relation to Chiba City on the northeastern shore of Tokyo Bay, contain a wealth of biodiversity, which is threatened by urban sprawl. They advocate the establishment of core zones for conservation, wildlife corridors to connect the core zones, and urban nature reserves. Moreover, several important tourist attractions comprise forests in or close to urban areas. The attraction involves trees, especially cherry when it blossoms, and often Buddhist shrines. One such example is that of Arashiyama National Forest in the northwest of Kyoto. According to Fukamachi *et al.* (2000), this forest receives *c.* $10 \times 10^6$ visitors annually. As Figure 8.16 shows, the Arashiyama National Forest is extensive; it occupies 59.03 ha and

| | CATEGORY | MANAGEMENT |
|---|---|---|
| A | Trees close to houses, i.e. groups of trees close to buildings | Low-density coppicing, shelter-belt structure; selective felling |
| B | Neighbourhood forests, i.e. small forests within residential areas | Stratifying the forest edge: increasing diversity, thinning where necessary, reduction in fragmentation to increase biodiversity |
| C | District forests, i.e. medium-sized forests between two or more sectors of a town | Increase in species composition, including hardwoods; creation of diversity in species and density to accommodate multi-purpose uses |
| D | Recreational forests, i.e. large forests usually on the urban fringe which are used for recreation | As much use is by senior citizens, provision should include good paths and a variety of woodland environments, e.g. glades and multi-layered woodland points |
| E | Production forests, i.e. forests on the urban fringe in which traditional forestry is practised | Commercial management with some attention to multiple use |

**TABLE 8.10** Urban forests and their management in Sweden. (Adapted from Rydberg and Falck, 2000)

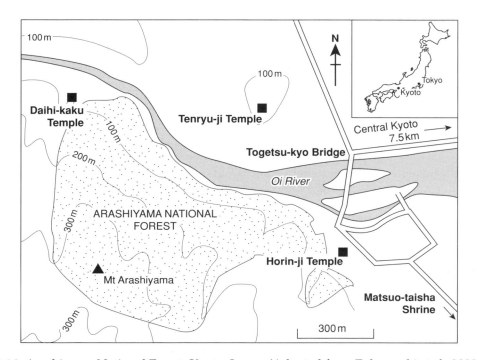

**FIGURE 8.16** Arashiyama National Forest, Kyoto, Japan. (Adapted from Fukamachi *et al.*, 2000)

| | Gencljk Park | Altin Park | Segmenler Park |
|---|---|---|---|
| Year established | 1940 | 1993 | 1983 |
| Size (ha) | 25.3 | 64 | 6.72 |
| Proximity to city centre (km) | 2 | 8 | 3 |
| Characteristics | Green areas | Green areas | Green areas |
| | Water bodies | Water bodies | Water bodies |
| | Mature plant cover | Immature plant cover | Mature plant cover |
| Facilities | Restaurants etc. | Exhibition centre | Amphitheatre |
| | Funfair | Amphitheatre | Playground |
| | Playground | Restaurants etc. | |
| | Wedding salon | Swimming pool | |
| | | Playground | |
| User characteristics | | | |
| (i) | Educated to primary-school level | Educated to high-school level | Educated to university level |
| (ii) | Lower income | Middle income | Middle income |
| (iii) | Preferred by married people | Third choice of married people | Second choice of married people |
| (iv) | 25–44 age group | 12–18 age group | 56+ age group |
| | | 19–24 age group | |

**TABLE 8.11** The features and user characteristics of three parks in Ankara, Turkey. (Adapted from Oguz, 2000)

comprises some 200 tree species, including cherry, red pine and a variety of broadleaf species. It also contains numerous bridges and temples, including the Tenryu-ji Temple built by Muso Kokushi in 1346, which has a Japanese garden within its confines. Since this time the forest has experienced many changes depending on specific management objectives. For example, many dead and dying red pine trees have been removed since the 1960s and replaced with mixed broadleaved species and Japanese cedar and cypress. Management has also attempted to address the increasing multipurpose function of the forest. As well as recreation, religious purposes and aesthetic values, Arashiyama National Park is managed for water and erosion control; dams to prevent mudslides and spur roads have been constructed, as is the case for many of Japan's forests (see Section 6.4).

How and why urban residents use parks is especially important for management and the provision of adequate facilities. This is true for urban parks worldwide, but has been specifi-cally documented for Ankara's parks by Oguz (2000). In Turkey, parks are of recent origin, which contrasts with the examples given above for Europe, North America and Japan. According to Oguz, most parks in Turkey date from the republican period, i.e. since 1923, and his survey of three parks, all with water bodies, which were constructed after 1940, shows that each is used for specific purposes. The characteristics and user characteristics of the three parks are given in Table 8.11, which shows a relationship between the age and educational level of park users and the park visited. As in other cities, parks play a significant role in the lives of Ankara's residents. The role of parks as centres of recreation and celebration is particularly apparent in Iran whose cities have parks and gardens in abundance. Residents throng these green spaces at weekends and on holidays; family outings and picnics are important aspects of family life. The famous Imam Khomeni Square in Ishfahan combines history, family traditions and green spaces; it is one of the largest squares in the world and was constructed, along with its mosques and bazaars, in 1612. This is referred to in Faramarzi (1997), who also draws attention to the gardens of Shiraz, many of which have pavilions, museums and elegant old houses.

## 8.7 Conclusion

Urban land cover/land use is especially varied and rarely bears any resemblance to the natural environment from which it was carved, except in the exceptional cases where natural vegetation has been preserved in urban forests and parks. Many cities, especially those in Europe and parts of Asia, have ancient origins. Their archaeological remains bear witness to past empires and civilizations, and their social and cultural achievements. In some cases, including Athens, Rome and London, the ancient and modern are juxtaposed; more than 2000 years of urban living has occurred in these and many other cities. Today, the vestiges of their past continue to generate wealth as tourist attractions, whilst they remain the centre of political and economic power in their respective countries.

The annexation of the Americas in the sixteenth century, and of Australia in the eighteenth, spread the tradition of European city construction into the New World and the Antipodes. In the Americas especially, the grid-iron planning system predominated; streets and avenues intersecting at right angles defined blocks of residential and/or commercial buildings as well as parks. Around the main squares the buildings of government or administration, and often cathedrals, were constructed. The layout is the same today as it was when many of the cities were founded; though, in many cases, especially those that experienced industrialization in the nineteenth and twentieth centuries, the land use has altered substantially, as is illustrated by the case of Toronto, Canada.

In the past 40 years the world as a whole has witnessed a massive shift of people from rural areas into the cities. This is especially apposite in the developing world. Natural, often high, population growth and the lure of employment are the major factors underpinning this trend. The growth is particularly marked in China – notably in its east-coast cities – which is industrializing rapidly having liberalized its economy in the late 1970s. Foreign investment and economic restructuring have also been instrumental in exacerbating urban growth throughout the developing world.

Urban environments are not all bricks and mortar; most towns and cities have green spaces and water bodies, which are important both to visitors and residents. Most are artificially created and all are intensively managed. Apart from providing recreational and often educational facilities, green spaces and water bodies are the major carbon sinks in an environment that is almost entirely a producer of carbon. They are also havens for wildlife.

Even the most planned city has a varied land use, reflecting urban multifunctioning. These uses change spatially and temporally to produce heterogeneous urban environments that provide a stark contrast to their rural hinterlands.

## Further reading

**Hall, P.** 1999: *Cities in Civilization*. Phoenix Giant, London.

**Macionis, J.J. and Parrillo, V.N.** 2001: *Cities and Urban Life*. Addison Wesley Longman, Harlow.

**Morris, A.E.J.** 1994: *History of Urban Form Before the Industrial Revolutions*, 3rd edition. Longman, Harlow.

**Pacione, M.** 2001: *Urban Geography: A Global Perspective*. Routledge, New York.

**Tate, A.** 2001: *Great City Parks*. E. and F.N. Spon, New York.

**United Nations Centre for Human Settlement (Habitat)** 2001: *Cities in a Globalizing World. Global Report on Human Settlements 2001*. Earthscan, London.

# Land-cover and land-use change caused by other human activities

## 9.1 Introduction

The preceding chapters have examined the major types of land cover and land use, along with the factors that have influenced and continue to influence land-cover and land-use characteristics. Agriculture, forestry, mineral extraction and urban growth are undoubtedly the most important human activities that contribute to the dynamism of land-cover/land-use change, but it must be remembered that there are many other factors that are also influential. Of particular importance are invasions of exotic plants and animals, the increasing problems of domestic and industrial waste disposal, tourism, the provision of facilities for sport, and the often devastating impact of war and terrorism. A discussion of any of these factors could not be accommodated in earlier chapters, yet there are few parts of the world where land cover/land use has not been affected by some or all of them. Moreover, the occurrence and intensity of these factors are likely to intensify substantially in the coming decades.

Invasions of exotic plants and animals have occurred through both deliberate and inadvertent introductions. The spread of agriculture from centres of origin in prehistory, for example,

resulted in the spread of crops and weed complexes (see Section 2.6). Palaeoecological investigations reveal the immediate general impact of the initiation of agriculture but do not, unfortunately, provide information on inadvertent introductions of so-called weeds. Throughout prehistory and history, there have been introductions of plants and animals to new lands. Some of these have been documented, e.g. the spread of vines and various nut and fruit trees by the Romans into the UK. The expansion of empires undoubtedly promoted such exchanges, but the most important and far-reaching exchanges can be attributed to European colonization of the Americas, and later Australasia. Maize, potatoes and tobacco are the most significant crops introduced from the Americas into Europe and thence into Africa, Asia and Australasia. Wheat, barley, sheep, cattle, pigs and horses went in the opposite direction. Overall, agricultural systems were altered significantly. The introduction of plants as garden ornamentals has also been widespread; many species have become naturalized, often causing the demise of native species. Many plant species accidentally introduced have had a similar affect, as has the introduction of grazing animals in some areas.

Waste disposal is an ever-increasing problem because of growing populations and the increasing use of packaging. Landfill disposal is common, but it requires land, and thus land cover and land use are altered while the tip is active and when reclamation is undertaken. The disposal of industrial waste also affects land cover and land use; it may cause dereliction, and subsequent reclamation will give rise to yet another type of land cover/land use. Similarly, war gives rise to a type of dereliction. This can take many forms, including the direct impact of armaments, e.g. bomb craters, defoliation caused by herbicide use, and the strategic destruction of resources such as oil. In addition, wars give rise to war graves, which are a permanent testimony to armed conflict. Graves of soldiers from the First and Second World Wars, for example, are widespread throughout the world and, with graves resulting from other wars, for example, are a distinctive type of land use.

In a world where leisure time is increasing, there is a growing demand for leisure facilities, especially for sporting activities, and tourist facilities at home and abroad. Increasing participation in golf, for example, has led to the creation of golf courses worldwide, including the arid Middle East. Racecourses, soccer and rugby pitches, as well as grounds for athletics all constitute a land cover/land use that is quite different from the pre-existing natural land cover. Tourism is one of the world's fastest growing industries; in some island nations such as the Seychelles and the Maldives, it is the major source of income. A suitable climate, coupled with natural landscapes and cultural features, attract visitors who, consequently, require facilities such as accommodation, entertainment etc. Sport and tourism are thus major agents of land-cover and land-use change.

## 9.2 The impact of exotic invasions on land cover

Plants and animals transported deliberately or accidentally from their natural environments to other parts of the world can often have a significant impact on land cover. Generally, the process of invasion is insidious rather than blatant, and often involves the diminution of populations of native species or even extinction. The problem is worldwide. According to Vitousek *et al.* (1997), the invasion of alien organisms is a major cause of global environmental change; it is also a significant cause of biodiversity loss. Not all introduced species have significant impacts; many may die out, but those for which there are no predators or that are better suited than indigenous species to their new environments have little competition, so they flourish. Once introduced, species are usually difficult to control and even more difficult to eradicate. Moreover, it is likely that rates of introduction will increase; in some respects they are a product of globalization, notably increased trade and travel.

Introduced animal species are many and varied (e.g. the grey squirrel in the UK; rabbits, water buffalo and cane toads in Australia; and rats worldwide, but especially to oceanic islands). In general, animal introductions do not have a major impact on land cover in so far as they do not radically alter vegetation communities. Introduced plants, in contrast, can have a substantial impact on land cover as they become naturalized (see Richardson *et al.*, 2000, and Sax and Brown, 2000, for a discussion of this term), though as Prieur-Richard and Lavorel (2000) point out, the mechanisms whereby this process occurs are poorly understood. This in itself makes control strategies difficult to formulate. It is also impossible to estimate the global scale of plant invasions. However, limited data exist for specific regions. Sax (2001) quotes a rate in excess of 1000 vascular plants for each of California, Hawaii and New Zealand. Other areas for which published data are available, and which suggest that exotic invasions are having a significant impact on land cover, include the UK, the Mediterranean and South Africa. Some examples are discussed below.

In South Africa the naturalization of exotic species has had a particularly marked impact on endemic fynbos vegetation communities. The fynbos (fine bush) comprises a flora unique to south and southwest South Africa, where a

Mediterranean-type climate prevails; *c.* 20 per cent of the *c.* 8500 species are endemic, many of which are proteas (Calow, 1998). This Cape Floristic region is one of world's so-called 'hot spots' of vegetation diversity, but there are numerous threats to its survival. These include agriculture, urbanization, the harvesting of wild flowers, and poor burning regimes. Amongst the invading plants are exotic tree species used in forestry especially *Pinus* spp. (pines), *Hakea* spp. and *Acacia* spp. For example, *Pinus halepensis* has been invading the fynbos since 1855 (Richardson, 1998). This, and other invasive species such as *Pinus radiata* have a significant impact on land cover when they occur in dense stands. On the one hand, the hydrological function of dense stands of trees is quite different to that of the pre-existing shrub-dominated fynbos and so there are implications for water management at the catchment scale. On the other hand, the development of a tree canopy inhibits the growth of native species and so reduces biodiversity. This latter has been recorded by Holmes *et al.* (2000) who have monitored sites in the Western Cape province where restoration programmes have been initiated; in crude terms the density of native plants at control sites was *c.* 33 per cent more than at invaded sites. This in itself represents a substantial alteration of land cover. Further alterations occurred, with restoration programmes that involved three types of treatment: burn standing; fell with remove and burn; and fell and burn. The first of these caused least damage and was most conducive to fynbos recovery. However, it is not ideal as it also helps to disseminate the propagules of invasive species and the standing dead trees inhibit the weeding of alien seedlings. Holmes *et al.* add that fynbos recovery could be improved if indigenous seed from local soil seedbanks were used to augment the seedbank of fired areas. The dynamics of invasion, burning and restoration all thus involve alterations in land-cover characteristics.

Non-native plant invasions have been widely recorded in Australia and New Zealand. As Williams and West (2000) have discussed, the Australian flora now includes more than 10 per cent non-native species and the New Zealand flora has almost as many exotics as it has indigenous species. Many have become established in such significant proportions that they are now considered to be environmental weeds (see also Fowler *et al.*, 2000). Some of these are listed in Table 9.1. However, it must be noted that environmental weeds, so called because they invade natural ecosystems and have a detrimental impact on biodiversity and ecosystem function, also include indigenous species introduced to regions beyond their natural range or into which disturbance has encouraged their spread. The problem is especially acute in Australia and New Zealand because their floras (and faunas) have evolved in relative isolation and so there are large

|   | | **Australia** | **New Zealand** |
|---|---|---|---|
| **A** | **Basic data** | | |
|   | No. of indigenous species | 25,000 | 2050 |
|   | No. of naturalized exotics | 2200 | 2071 |
|   | No. of environmental weeds | 1060 | 240 |
|   | | | |
| **B** | **Examples of environmental weeds** | Blackberry | Heather |
|   | | Gorse | Marram grass |
|   | | Lantana | Prickly pear |
|   | | Mesquite | Ragwort |
|   | | Mimosa | Scotch broom |
|   | | Athel pine | Water hyacinth |

**TABLE 9.1** Data on invasive and indigenous plant species in Australia and New Zealand. (Adapted from Williams and West, 2000)

numbers of endemics (see Table 9.1). Moreover, both countries were settled by Europeans, and the transfer of crops and animals involved the transfer of other plants and animals as well, including ornamentals; both have a wide range of vegetation types in which exotics can find niches and both have encouraged the diversification of trade, which has increased exchanges between the Americas and between African countries, notably South Africa. The development of trade links with Asian countries may result in further invaders.

Radiata pine (*Pinus radiata*) and blackberry (*Rubus fruticosus*) are amongst the non-native plant invaders altering land cover in Australia, especially in the southeast where plantations of radiata pine have been established. Seeds escape from the plantations and compete with native eucalypts while blackberry, introduced from Europe more than a century ago, has become a significant component in native eucalypt forests. Both species are a threat to remnant eucalypt forests, as Lindenmayer and McCarthy (2001) have highlighted. Their study in the Tumut region *c.* 700 km, west of Canberra, focused on the Buccleuch State Forest, which was established between the early 1930s and 1985. It comprises *c.* 45,000 ha, itself representing a major land-cover change from the original eucalypt forest, some 192 fragments of which remain within the forest boundary. It is these fragments that are particularly susceptible to invasion by radiata pine and blackberry. The results of this study are summarized in Table 9.2, which shows that the fragmented euca-

lyptus forests are especially susceptible to invasion when compared with continuous areas of native forest. Moreover, since plots adjacent to old sections of the radiata plantations were most heavily infested, the problem of radiata spread is likely to increase as plantation sections age. Lindenmayer and McCarthy advise that the establishment of new radiata plantations should be coupled with controls such as the planting of sterile pine trees in order to avoid an escalation of invasion in the future

Eucalypt woodland is also being replaced by Scotch broom (*Cytisus scoparius*) at Barrington Tops, New South Wales (Downey and Smith, 2000). It has been in Australia for approximately 200 years and now covers *c.* 10,000 ha at Barrington Tops where it competes with snow gum (*Eucalyptus pauciflora*, also known as white sallee) woodland with acacia shrubs. Attempts to control it through biological control via insect herbivores have not been particularly successful; burning can have short-term beneficial effects, but actually encourages broom survival in the long term. Downey and Smith conclude that such disturbances act as a positive stimulus unless several years of post-fire herbicide applications are used. This would prevent the seed bank of broom being replenished and thus provide opportunities for native species to regenerate.

Prickly acacia (*Acacia nilotica*), a leguminous tree, is another environmental weed in Australia. It was originally introduced from Africa to provide shade and fodder, but is now considered to be a threat to the grasslands of

| | % of plots invaded by blackberry | % plots invaded by *Pinus radiata* |
|---|---|---|
| **A** Continuous eucalypt forest | 7 | 0 |
| **B** Remnant eucalypt forest | 41 | 5 |
| **C** Plantation of *Pinus radiata* | 59 | – |

*Notes:*
Edge effects occurred in relation to both invaders
Both took advantage of relatively open canopies and both were responsible for higher infestations when invasions occurred from older rather than younger areas of plantation

**TABLE 9.2** Data on the invasion of *radiata* pine and blackberry in remnants of eucalyptus forests in southeastern Australia. (Adapted from Lindenmayer and McCarthy, 2001)

central and northwestern Queensland. Radford *et al.* (2001) have examined seed-dispersal mechanisms to predict the future spread of the species. Their results show that trees adjacent to water, e.g. those in riparian environments or watering holes, produce more seed than those in drier situations, and that cattle spread the seed in their dung with peaks close to watering points. Radford *et al.* suggest that as water appears to be a key factor in the spread of prickly acacia, climatic change involving increasing wetness may hasten its spread. In the current climatic conditions, individual trees are invading the naturally treeless semi-arid *Astrebla* grasslands and in some riparian locations they even form thickets.

In New Zealand, Toft *et al.* (2001) have examined the impact of the weed tradescantia (*Tradescantia fluminensis*) on insect communities in the south of North Island. This species is one of the most significant forest weeds in New Zealand and occurs abundantly in the fragmented forests of both North and South Island. It is a herb, introduced from Brazil, that forms a continuous ground cover and thus inhibits the growth of other species of the ground flora. The sites investigated by Toft *et al.*, just north of Palmerston North, are podocarp-broadleaf forest remnants; all have been affected by other introduced animal species, namely possums, cattle and rabbits, as well as by logging. Tradescantia is abundant, with only three other species being recorded in the ground flora at all sample sites. Although the results of the study are inconclusive they support the suggestion that the diversity of beetles and fungus gnat species is related to the diversity of vascular plants. Thus depletion of ground cover diversity as tradescantia invades is likely to lead to diversity declines in insect faunas, possibly with implications for ecosystem functioning as grazing and detrital energy flows are altered.

Islands smaller than New Zealand have also experienced severe invasions of exotic species. St Helena and Ascension Island – both volcanic islands in the South Atlantic – have been severely affected by both plant and animal

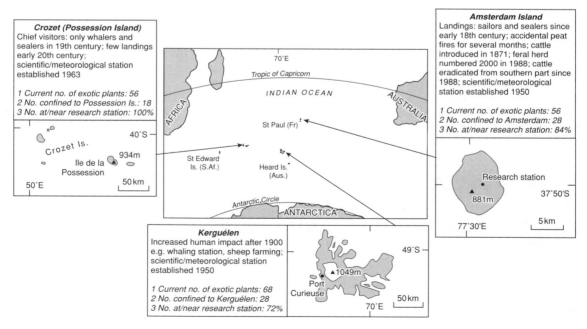

**FIGURE 9.1** Background information and data on exotic plant invasions for three French sub- Antarctic islands. (Adapted from Frenot *et al.*, 2001

invasions, as have the Galapagos Islands close to the Equator, *c.* 970 km west of Ecuador, and Madagascar in the Indian Ocean. The latter, for example, has one of the richest floras in the world with *c.* 80 per cent of its species being endemic. Cacti and eucalyptus as well as a range of grasses have become naturalized and are aggressive competition for native species. More recently, particularly remote islands have been subject to increasing contact with the outside world; these include the islands of the Sub-Antarctic region.

A recent study by Frenot *et al.* (2001) on Crozet (Possession Island), Kerguelen and Amsterdam Islands (see Figure 9.1 for location) has shown that each has been affected uniquely in time and space. Although all the exotic species recorded on all three islands are natives of north European temperate environments, less than half are common to two or three islands, and all are grasses and herbs. Almost all the exotic species have been introduced accidentally, mostly by the ships that visit the islands from France bringing people and supplies. Frenot *et al.* observe that this is probably why such a high proportion of exotic species occurs at or near the research stations. Once on the island, other vectors, such as the wind or birds, may facilitate spread. Other important factors in exotic plant presence include the forage brought with the first sheep from the Falkland Islands in 1907; imported food for hens may have had a similar impact. In relation to land-cover alteration, however, several naturalized exotic species have altered the species composition of native communities substantially. They include several grasses, e.g. *Agrostis stolonifera* in the grazed zone of Amsterdam and *Holcus lanatus* in the ungrazed zone. Clearly, the introduction of grazing animals to this island has been a significant factor in exotic plant establishment. This is also indicated by the proliferation of *Holcus* spp. in the vicinity of the abandoned sheep farm on Kerguelen.

As in several of the studies referred to in this section, the data from these Sub-Antarctic islands implicates disturbance as an important factor in the establishment of exotic species. As human populations continue to increase, the

disturbance of natural and semi-natural ecosystems will intensify, and the prospects for the extinction of species and the 'globalization' of the world's flora will intensify.

## 9.3 Land-cover change associated with waste disposal

One of the most important components of the Ecological Footprint (see Section 3.6) of nations, and especially of cities, is the large volume of waste that is produced. Waste can take many forms and each requires specific actions for its disposal; almost all disposal methods affect land cover. The most important types of waste include domestic rubbish, industrial waste, such as chemicals, and sewage. The latter is treated in specific ways in sewage plants, which are particularly abundant in the vicinity of urban areas in developed nations. The disposal of domestic rubbish is generally achieved either through landfill or incineration; both require land. The disposal or treatment of industrial waste is varied because of the diverse nature of the waste. Moreover, the disposal of industrial waste in the past is now often considered to have been inadequate, and where it hampers development projects reclamation is necessary, as in the case of the reclamation of mined land (Section 7.6). The examples given in this section illustrate the impact, of what is a widespread problem, on land cover and thus reflect specific elements of the Ecological Footprint.

Until the 1840s human excrement in the UK was disposed of in privies located outside houses. Seepage often contaminated drainage systems, including drinking water, and caused the spread of diseases such as typhoid and cholera, sometimes to epidemic proportions, especially in cities. To address the issue of public health, storm drains in cities began to be used for sewage disposal and thus the process of disposing of sewage into watercourses began. Inevitably, this practice also caused problems as watercourses became polluted, and in some parts of the world this still occurs today. Eventually, sewage-treatment plants were constructed in order to treat the effluent prior to its discharge into watercourses or the sea. The first such plants came into operation in the early 1900s and, today, most sewage in developed countries is treated to some extent before the waste water is released. In many parts of the developing world, however, raw sewage may be disposed of indiscriminately and treatment plants are uncommon. Where sewage-treatment plants are constructed they replace the previously existing land cover or land use.

The objective of sewage-treatment plants is to remove pathogens, and to decrease the solid content and the biological oxygen demand (BOD), i.e. the capacity of effluents to use up oxygen. The various processes needed to achieve these objectives require large tanks, filtration systems and oxidation ponds, as well as the sewers to channel the waste to the treatment plant. In Hong Kong, for example, $c.$ 98 per cent of the $2.2 \times 10^6 \, m^3$ of sewage produced daily is collected via 1320 km of sewers and treated in 200 plants before release into the sea (Hong Kong Government, 2001a).

Whilst sewage-treatment plants are a distinct type of land use, so are the landfills used for the disposal of rubbish. The volume of domestic rubbish is increasing daily, partly because of population increase and partly because of increased packaging, including the use of plastic, which is largely non-biodegradable. The disposal of rubbish varies enormously worldwide; the most adequate provision is in developed nations where collection services and disposal via infill or incineration are widespread. In many developing countries rubbish disposal occurs on a more *ad hoc* basis, with varying levels of collection services, sometimes little or no control over dumping, and poorly managed tips. Worldwide, however, the problem of domestic rubbish disposal is acute.

Table 9.3 gives data on the quantity and composition of domestic (municipal) solid waste in the USA. First, the data show a substantial increase in rubbish production during the last 40 years, and especially in the 1990s. Second, the single most important component of this rubbish is paper. These data generally reflect the characteristics of waste production in an affluent nation. Table 9.3 also

### A  Trends in municipal solid waste

|  | lbs/person/day | Total waste generation (tons × 10) |
|---|---|---|
| 1960 | 2.7 | 88.1 |
| 1970 | 3.3 | 121.1 |
| 1980 | 3.7 | 151.6 |
| 1990 | 4.5 | 205.2 |
| 1999 | 4.6 | 229.9 |

### B  Composition %

| | | | |
|---|---|---|---|
| Paper | 38.1 | Rubber, leather, textiles | 6.6 |
| Yard waste | 12.1 | Glass | 5.5 |
| Food waste | 10.9 | Wood | 5.3 |
| Plastics | 10.5 | Other | 3.2 |
| Metals | 7.8 | | |

### C  Method of disposal

| | Tons × $10^6$ | % |
|---|---|---|
| Landfill | 121 | 54.8 |
| Recycling | 62 | 28.2 |
| Combustion | 37 | 17.0 |
| Total | 220 | |

**TABLE 9.3** Data on municipal solid waste production in the USA. (Adapted from Environmental Protection Agency, 2001)

shows that more than half of this waste is disposed of in landfills, and that the proportion being recycled is *c.* 28 per cent and increasing. Incineration is used to destroy the remainder. Although this is less consumptive of land than landfill, it is not a popular choice because of the production of heat-trapping (greenhouse) gases and the possible production of other substances, e.g. polychlorinated dibenzo-p-dioxins (PCDDs) and polychlorinated dibenzofurans (PCDFs), which are harmful to human health (see comments in Wright and Nebel, 2002).

Landfill sites generate another set of land-use issues, which range from their siting to reclamation and afteruse. Landfill is not an ideal solution to rubbish disposal, as Allen (2001) has discussed. Most importantly, dry burial slows down degradation and so delays the onset of a stable state for the landfill which, in turn, inhibits reclamation. The entire process is not an appropriate component of sustainable development or of sustainable cities (see Wilson *et al.*, 2001). Even sealed landfills can

cause problems, including an accumulation of methane gas, which can be explosive, and the contamination of soils and drainage systems through leaching. Another important consideration is landfill siting and afteruse. As Misgav *et al.* (2001) have discussed, most landfills are sited in urban areas and so their reclamation must be compatible with urban environments, notably safety and amenity provision. They suggest that the most appropriate afteruse for such landfills is some type of open-space provision but they also point out that possible afteruse should be a component of determining where a landfill should be sited initially.

In common with many other nations, the total production of solid waste in New Zealand and the per capita production have increased markedly in the past two decades. Landfill disposal has increased as a result. Figure 9.2 shows the spatial distribution of landfills in New Zealand and the amount of waste deposited in each in 1998. Unfortunately, no data are available on the area occupied by each

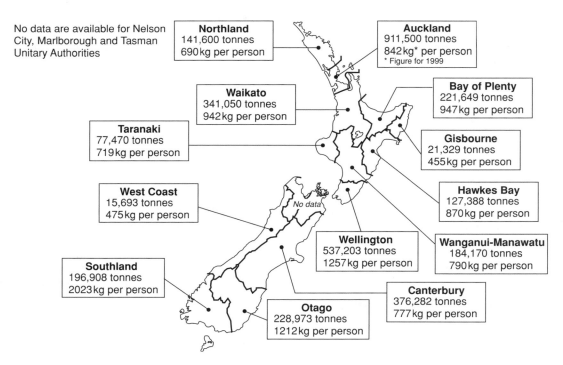

**FIGURE 9.2** Data on solid waste in landfills in New Zealand, 1998. (Adapted from New Zealand Government, 2001)

landfill, but the volume of waste deposited provides a proxy measure. Inevitably, the amount of waste deposited reflects the distribution of New Zealand's population. For example, Auckland, with a population of $1.175 \times 10^6$ is New Zealand's largest city and biggest producer of waste; c. 900,000 tonnes of waste was deposited in its vicinity in 1998. What is not so clear, however, is why the amount of rubbish generated per person is so variable. The values range from 455 kg per person per year for Gisborne to a huge 2023 kg per person per year for Southland, though the prevalence and effectiveness of recycling or minimizing programmes are likely to be significant.

Many cities in the developing world have inadequate arrangements for rubbish disposal, a situation exacerbated by rapid urban growth (see Section 8.5). In some instances the accumulation of rubbish in large, unregulated tips has given rise to recycling activities, by individuals or families who 'pick' the rubbish, and which amount to small businesses. One example of a dump site is that of Payatas in the northeast of Quezon City in Metropolitan Manila, Philippines. According to a review by the Vincentian Missionaries (1998), the Payatas is an open pit of 15 ha, the largest and oldest dump operating in Metropolitan Manila. It provides homes and an income for some 4000 scavenger families who extract recyclable materials and sell them to established waste recovery and recycling concerns, or who run small businesses such as junk shops. To avoid exploitation of these people, the Vincentian Missionaries have assisted them to set up the Payatas Environmental Development Programme, which includes a Federation of Scavengers. It has an internal loan system and, in general, has improved conditions for the Payatas people.

## 9.4 The impact of war and terrorism on land cover and land use

An often overlooked impact on land cover and land use is that of war and, increasingly,

terrorism. The impacts on land cover may be direct, e.g. the destruction of cities, production of bomb craters, defoliation due to herbicide use and the destruction of resources such as oil reserves. Indirect impacts include the creation of refugee and prisoner of war camps, as well as war graves. War and terrorism have thus exacted, and continue to exact, a significant price in terms of land-cover and land-use change, which may vary in a temporal context from being relatively ephemeral to enduring over centuries.

The direct consequences of war are many and varied. In urban environments the bombing of specific targets, such as armaments factories, or non-specific bombing, alters land use. Amongst the cities of Europe that bear witness to Second World War bombing are London, Coventry, Dresden and Berlin. In the latter, for example, 125,000 people died and half of the buildings in the city were destroyed along with about one-third of the industrial plant. During the London Blitz alone (September 1940 to March 1941), some 20,000 people died. During the whole of the Second World War, some 29,890 people died and, as Figure 9.3 shows, considerable damage occurred in the City of London itself (see also Inwood, 1998). More recently, cities such as Phnom Penh in Cambodia, Saigon (now Ho Chi Minh City) in Vietnam, Kabul in Afghanistan and Kuwait City have borne the brunt of armed conflict in the last 40 years. Terrorist attacks have also wrought considerable damage on urban landscapes. The IRA's bombing of London's Baltic Exchange, for example, in 1992 led to the construction of a new office block, and the destruction of part of Manchester's city centre in 1996 has resulted in the construction of a new shopping centre. However, these and many other incidents involving terrorist activities pale into insignificance when compared with the devastation and loss of life caused by the deliberate crashing of two aircraft into both towers of New York's World Trade Center on 11 September 2001. More than 3000 people were killed and New York lost one of its most familiar landmarks. The rubble will eventually be replaced, but life in New York, or elsewhere, will never be quite the same again.

The countrysides of nations at war have also

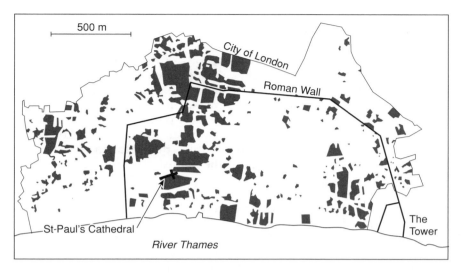

**FIGURE 9.3** The extent of bomb damage during the Second World War in the City of London. (Adapted from *The Times History of London*, 1999)

been transformed through various forms of military activity. During the Vietnam War (1964 to 1973), for example, landscapes were altered by the creation of bomb craters that became duck ponds in peacetime, and by the use of defoliants on forests and croplands to reduce the cover and food available to the Viet Cong. According to Shaw (1989), some $25 \times 10^6$ craters of between 6 and 30 m wide were created during the Vietnam War, more than one million hectares of forest were destroyed and two million hectares of agricultural land made unproductive. Many of the craters remain in evidence today, and much recovery of defoliated areas has occurred, though both US servicepeople and Vietnamese people may have suffered health problems as a result of contamination by dioxins present in the defoliant Agent Orange (the herbicides 2, 4, 5-trichlorophenoxyacetic acid and 2, 4-dichlorophenoxyacetic acid). Elsewhere, the machinery of war has been abandoned; for example, there is a 'tank graveyard' close to Asmara, the capital of Eritrea; this comprises several hectares covered with defunct tanks, armoured cars and jeeps. It is a legacy of the 30-year struggle by Eritrea to free itself from neighbouring Ethiopia, which ended in independence in 1993.

The Gulf War of 1990–91 brought about devastation of a different kind. Caused by the invasion of Iraq, an alliance led by the USA responded to defend Kuwait. During the conflict, Iraqi troops released $c.\ 11 \times 10^6$ barrels of oil into the Persian Gulf, which resulted in the contamination of $c.\ 1280$ km of the Gulf's coastline, mainly in Kuwait and Saudi Arabia; moreover, retreating Iraqi soldiers set alight more than 700 oil wells (McClain, 2001). The immediate impact of this abuse of the resources was thus considerable, not only by contaminating coastlines with a destructive film of oil but also by destroying wildlife, impairing marine ecosystems and altering climate. Indeed, the specific nature of this destruction has led to its description as 'ecoterrorism'. On the tenth anniversary of this conflict, McClain notes that much recovery had occurred; the Gulf's coral reefs, which are amongst the most fragile of the Earth's ecosystems, had recovered as had many coastal areas (see review in Price, 1998). However, inland Kuwait had been less fortunate because of the formation of oil pools and oil lakes in the desert sands. McClain states that $c.\ 300$ oil pools and lakes were created, which are now gradually sinking into the sand. Some recovery of vegetation communities is evident but desert environments are less forgiving than those of coasts because they do not have the vital component of water and its cleansing capacity. The contamination of soils and desert sands with substances such as heavy metals will have long-term effects.

War has as much of a devastating impact on people as it has on the environment. As well as loss of life *in situ*, war often causes the large-scale displacement of people who thus become refugees. There are numerous areas of the world today in which displaced people congregate in camps. For example, the civil war in Afghanistan has caused millions of people to flee into Pakistan and Iran, both of which are now hosts to between one and two million Afghans. Similarly, recent conflicts in the Balkans have generated millions of refugees in many of the former Yugoslav republics. The creation of refugee camps itself represents substantial land-cover/land-use change; many are extensive and many become fixtures. This factor has been examined by Kibreab (1997) in relation to the Sudan. He points out that environmental degradation occurs for a variety of reasons, not least of which is the insecurity associated with conflict and land ownership or lack of it. Other factors, notably poverty, are more the outcome of degradation than its cause. Kibreab also points out that land-cover/land-use change may occur in areas not directly affected by war. He states 'not only do insecurity and wars destabilize socio-economic environments, but they also undermine natural resource conservation activities because of loss of family labour'. He cites the situation in the Eritrean and Ethiopian plateau, where productive agricultural systems have been maintained by limiting soil erosion through conservation, e.g. terracing, terrace maintenance and water management. However, during periods of war, able-bodied males are conscripted or join armies voluntarily, and so reduce the availability of labour for countryside maintenance. Soil erosion and degradation ensue and result in a substantial decline in productivity.

Other environmental impacts of refugees include the reduction of wildlife populations, as bushmeat becomes an increasingly important component of the diet. This has been evident in the case of civil strife in the Democratic Republic of the Congo in 1997 and 1998. Van Krunkelsven *et al.* (2000) quote reports of the large-scale slaughter of wildlife, including elephants and gorillas, by groups of renegade fighters and refugees hiding in the Maiko and Kahuzi-Biega National Parks. Both of these parks were also affected by an influx of Rwandan refugees following the ethnic conflicts that began in 1994. A wildlife census of the latter was undertaken in 1996 and compared with an earlier one in 1990 by Inogwabini *et al.* (2000) who showed that no major decline in gorilla numbers had occurred. This may reflect adequate protection for gorilla tourism, though it is highly likely that a subsequent uprising in the Democratic Republic of Congo itself has had an adverse affect on annual populations through poaching.

The direct impact and immediate aftermath of war are indeed stark, but there are also longer-lasting reminders in the shape of war graves. These are testament to past conflicts and provide a measure of the human cost. War graves occur throughout the world; those that relate to conflicts with the UK are maintained by the Commonwealth War Graves Commission. As Figure 9.4 shows, these graves and memorials occur in 150 countries. Many relate to casualties during the Second World War, such as those in the Channel ports, which contain the largest number of burials. These casualties were mainly sustained during the 1939–40 campaign. The Dunkirk Memorial, for example, commemorates the soldiers of this British Expeditionary Force who have no grave, notably some 4516 soldiers who included five soldiers from the Royal Indian Army Service Corps with the remainder belonging to UK land forces. The memorial is located at the entrance to the British war graves section of Dunkirk Town Cemetery (see Figure 9.4). Here lie 800 soldiers from various parts of the Commonwealth, the UK itself, as well as from Czechoslovakia, Norway and Poland. Moreover, along the coast and along the border with Belgium are cemeteries containing and commemorating the dead of the First World War (1914–18).

Testaments to war are widespread. Through-

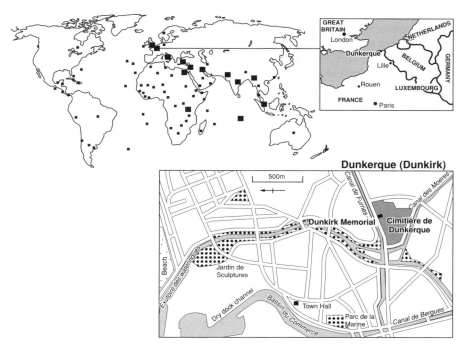

**Figure 9.4** The location of Commonwealth war graves and specific detail of Dunkirk. (Adapted from Commonwealth War Graves Commission, 2001)

out Russia, for example, there are imposing war memorials in most major cities and towns. That of St Petersburg, the Defenders of Leningrad Movement, is one of the most well known. Located 8 km south of the city centre, it commemorates those who lost their lives during the siege of St Petersburg by the Germans during the Second World War. More than 500,000 people died of shelling, starvation or disease, and many are buried in the mass graves of the Piskaryouka Cemetery, northeast of the city. In Iran there are numerous cemeteries that testify to the Iran/Iraq conflict of 1980–88. Examples include the Golestan-é Shohada, the Rose Garden of Martyrs, in Ishfahan and the cemetery of the boy soldiers in Hamadan. South Korea, placed as it is with China to the west and Japan to the east, has a long history of conflict (including its involvement in the Vietnam War) and has established a huge memorial in Seoul to record its bellicose past. The memorial is a museum located centrally in substantial grounds. The exhibits include weaponry and armoury dating to the fourth century AD, various early wooden war ships, which reflect Korea's reliance on and vulnerability to sea-borne exploits, and numerous Second World War aircraft parked in the memorial's grounds.

## 9.5 The impact of sport on land cover and land use

Sport is one of the most important components of recreation, time for which is increasing significantly as working hours are reduced and as disposable incomes rise. These are also the reasons why tourism is increasing, as discussed in Section 9.6. Indeed the two may be linked, as sporting activities such as hunting and skiing are often components of tourism. Sport can take a variety of forms, many of which are outdoor and require specialized facilities. Those that have the greatest impact on land cover/land use are field sports, and can exert a significant affect on either urban or rural landscapes. Even activities such as rambling, trekking and climbing can influence land cover, notably through the trampling of vegetation. This aspect of land-cover alteration has been reviewed elsewhere (e.g. in Mannion, 1997). More obvious impacts of sport include the construction of pitches for soccer, rugby, cricket, hockey, polo, horse racing etc. Moreover, golf course creation has expanded markedly in the past 30 years or so, as has motor racing and skiing. Inevitably, a comprehensive review of the impact of sport on land cover and land use is impossible here, although the examples given below illustrate a range of impacts that sporting activities can effect.

Skiing is a widespread countryside-based sport. It is prevalent in Europe and North America where upland areas receive a suitable snow cover that combines with suitable slopes. Traditionally, skiing has been the preserve of the European Alps where major resorts have been significant for a century; downhill and slalom (involving the use of zig-zag markers) courses or pistes are particularly common, while cross-country skiing has its origins in Scandinavia. Skiing is now well established in Canada, the USA, Russia, Australia and Iran. In terms of the impact skiing, and ski resorts, have on land cover, most research has been undertaken in European Alpine environments. In Switzerland, which reflects similar problems to other European Alpine environments, landcover change is caused by the construction of downhill ski runs. First, the use of machines for grading the slopes disrupts the vegetation cover and, second, it disturbs the structure of topsoil and its seedbank. The latter reduces rates at which natural revegetation occurs. According to Fattorini (2001), the transplanting of grasses, legumes and herbs is more effective in revegetation schemes above the tree line than reseeding, and native species have the highest rates of survival. The restoration of a complete vegetation cover is, however, likely to take at least two decades. Other environmental changes associated with ski-piste construction may also contribute to land cover change. For example, Ries (1996) has reported that soil

erosion and solifluction have been exacerbated by the construction of ski runs and ski lifts at Schauinsland in the Black Forest.

Hunting for sport is another activity that influences land-cover characteristics. Whatever the morality of the activity, it provides an important source of income. In many parts of the world it is more lucrative than farming. How it affects land-cover characteristics is illustrated by sport hunting in Scotland, where it is important in upland moorland environments. The objective of the hunt determines the management required for vegetation communities. If deer are being raised, heather moorland is burnt on a large scale to produce hundreds of hectares of even-aged heather; if grouse are favoured, then patch burning is desirable to produce juxtaposed immature and mature stands favoured by the birds for foraging and nest building respectively. Landscapes so managed are thus allied to the dynamics of firing, as is the species composition of the vegetation communities.

Many parts of southern Africa are managed for sport hunting. In South Africa, Zimbabwe and Namibia this type of land use, especially in semi-arid environments, is often more lucrative than agriculture. In Namibia, for example, commercial herding is a major employer and income generator; beef cattle and sheep rearing dominate the industry, with large farms or ranches to provide sufficient forage. Stocking rates are low with one animal per hectare, or fewer, being common in a country characterized by water shortages. In the past two decades many stock-rearing farmers have turned to game ranching involving gemsboks, zebras and springboks. These animals provide meat and hides, which are sold commercially. However, their greatest value is as tourist attractions, and sport hunting on so-called guest farms also generates income through the provision of other services such as accommodation and meals. In terms of land-cover characteristics, reversion of farms/ranches to natural vegetation, i.e. savanna grasslands predominantly, occurs as agriculture is abandoned.

Field-based sports include horse racing and golf, which require a considerable amount of land. Occasionally found within urban areas, they are more usually located beyond city

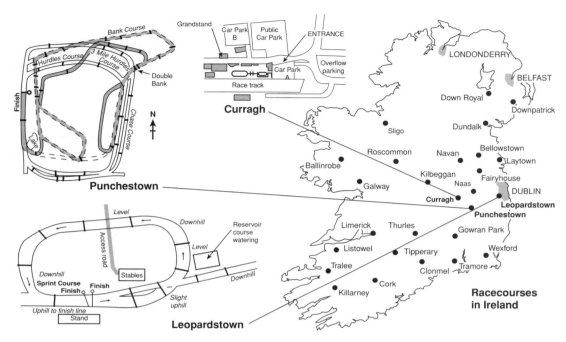

**FIGURE 9.5** The distribution of racecourses in Ireland. (Adapted from Irish Racing, 2001; Curragh, 2001; Leopardstown, 2001; Punchestown, 2001)

boundaries but within easy reach. Horse racing is a popular sport worldwide, but is particularly associated with the United Kingdom and Ireland. Figure 9.5 shows the distribution of courses in the latter. Each occupies at least 10 ha and comprises a course that is turfed, as well as a viewers' stand, hospitality buildings, stables and car parks (see detailed maps in Figure 9.4). The Curragh, in County Kildare, is the headquarters of Irish Flat Racing; it is owned by the Irish Turf Club established in the 1760s (Curragh, 2001), though it is considered to have a long history going back to the third century AD when chariot races were held in the area. In contrast, the courses at Punchestown and Leopardstown offer 'over the sticks' racing, which involves the inclusion of fences, some with ditches, along the course.

Golf courses occupy a similar area, if they offer the standard 18 holes; 9 or 27 holes may also be offered. Like racecourses, golf courses are mainly grassed areas with managed fairways and greens (see Figure 9.5); there are also 'rough' grass areas and bunkers of sand as well

as occasional water features and woodland copses. Golf is enjoyed worldwide and golfing holidays have ensured that the sport is combined with tourism. For example, it is a major attraction in Scotland where there are more than 400 golf courses. There are 40 in Fife alone, of which those of St Andrews are especially well known. Figure 9.6 gives the location of Fife's courses and details of St Andrews' golf courses, including the famous Royal and Ancient Golf Club. This was founded in 1754, though golf has been played in the area since the 1400s. Strictly speaking, the golf courses here are golf links since they are located on the coast overlooking the North Sea. St Andrews has Europe's largest golf complex, which occupies *c.* 283 ha (700 acres). There are five courses, including the infamous Old Course, which is the venue of the British Open Championship. There are also two turf nurseries, a driving range and practice area, two greenkeeping centres, two clubhouses and three shops (Scotland's Golf Courses, 2001).

Land-cover characteristics in urban areas are

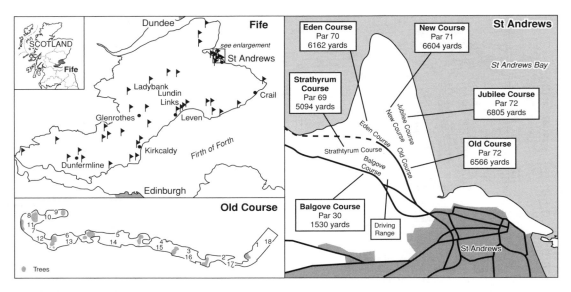

**FIGURE 9.6** The golf courses of St Andrews. (Adapted from Fifeguide, 2001; St Andrews, 2001)

also heavily influenced by land use associated with sport. *The Times History of London* (1999), for example, shows the importance of stadia, cricket pitches etc. in London's landscape. The same is true for most large cities. Indeed facilities for sport are varied and numerous, as is illustrated by the case of Hong Kong island. This has an area of only 75 km², yet hosts a wide variety of sports facilities, as shown in Table 9.4. Some of these are indoor while others are outdoor. Note the large number of soccer pitches for such a small island, which reflects the popularity of the sport in Hong Kong. Similarly, the importance of cricket in South Africa, with a population of $40 \times 10^6$, is reflected in the many pitches listed by the United Cricket Board of South Africa in some 50 cities, many of which have more than one pitch; in Harare, with a population of *c.* 900,000, there are ten pitches registered in the Zimbabwe Cricket Union (Republic of South Africa Cricket Organisation, 2001).

It is also important to recognize that the character and distribution of urban sports facilities are just as dynamic as their commercial and residential neighbours. This is high-lighted by a study of major league sports facilities in cities of the USA by Newsome and Comer (2000). They note that, since the late 1980s, there has been an upsurge in the construction of sports grounds associated with the major league sports of baseball, hockey, basketball and football. A major driving force behind this activity is the need to increase revenues through the provision of improved facilities whilst the investment reflects the increasing importance of sport in peoples' lives. However, the analysis of construction trends by Newsome and Comer shows that not only are new facilities being generated, but also that their location has shifted from the essentially suburban locations favoured in the 1945–80 period to downtown locations. Their analysis of locations of major league stadia shows that in 1997 some 51 per cent of stadia were in downtown locations compared with only 42 per cent in 1965. Amongst the reasons given for this relocation trend are corporate sponsorship and the revitalization strategies of many US cities. As relocation and revitalization occur, urban landscapes and land cover change accordingly.

| Recreation and sports facilities | |
|---|---|
| **Active facilities (as at 31 March)** | **2001** |
| Amphitheatres | 24 |
| Archery ranges | 6 |
| Courts: | |
| badminton | 506 |
| basketball | |
| handball | 28 |
| squash | 324 |
| tennis | 263 |
| volleyball | 232 |
| Stadia (outdoor) | 2 |
| Turf pitches (Natural and artificial) | 69 |
| Hard-surfaced soccer pitches | 213 |
| Hockey pitches | 2 |
| Rugby pitches | 2 |
| Indoor games halls/indoor recreation centres/leisure centres | 82 |
| Roller-skating rinks | 35 |
| Sportsgrounds | 24 |
| Swimming pool complexes | 35 |
| Water sports centres | 4 |
| Golf driving ranges | 5 |
| Obstacle golf course | 1 |
| Bowling greens | 9 |

**TABLE 9.4** Sports facilities in Hong Kong. (Adapted from Hong Kong Government, 2001b)

## 9.6 The impact of tourism on land cover and land use

Tourism has been one of the world's major growth industries since the Second World War, and especially since the 1960s. As discussed above in relation to recreation activities, increasing leisure time and increasing wealth, along with improved transport such as air travel, have been the major stimuli underpinning the growth in tourism. Data from the World Tourism Organization (see Table 9.5) show that the major generating and receiving areas are developed nations, notably those of Europe and North America, but there has also been a marked increase of tourism to and from developing countries in the 1990s. Today, even the most remote parts of the world are opening up to tourism; the wealth so generated is becoming a mainstay of many local, regional and even national economies. Antarctica, Siberia, Mali, the Falkland Islands and Mongolia are amongst tourist destinations advertised today, whereas a decade ago such destinations would have been considered more or less inaccessible. The impacts that tourism has on the environment are many and varied; to a large extent they depend on the type of tourism, which is itself diverse though all forms of tourism affect land-cover/land-use characteristics directly or indirectly, because the environment is a fundamental component of the tourism experience (see comments in Butler, 2000). Tourism can be subdivided into two categories: beach or sun tourism; and specialist tourism, of which heritage tourism and ecotourism are components. Each reflects a specific and self-explanatory focus, though ecotourism in its pure form involves not only the appreciation of landscapes and indigenous culture, but also the minimal Ecological Footprint by visitors. Tourism affects land use/land cover because tourists require transport and accommodation, however basic the latter might be, and they generate various forms of waste. Hotels and campsites transform landscapes, as do associated sewage and waste-disposal facilities (see Section 9.3).

Resort development is a major impact of tourism, whether it involves lodges in South African game parks, the wholesale urbanization of coastal areas, as has occurred in Mediterranean countries such as Spain and Turkey, or the creation of cities like Las Vegas in the USA or Sun City in South Africa as venues for gambling. Urbanization, as discussed in Chapter 8, involves a deepening and broadening of the Ecological Footprint of a region or nation, and contributes to a high carbon index (Section 3.6). There is thus an environmental price to pay, at least in terms of waste disposal and carbon production, for tourist-resort development, which replaces pre-existing land cover or land use, usually either natural vegetation communities or some form of agriculture. The development of Spanish resorts since 1960 illustrates the land-cover/land-use change that has occurred in this part of the Mediterranean basin, which was amongst the first areas to be developed for mass tourism (see Pearce, 1995). Figure 9.7 shows the location of Spain's major resort areas, the most important of which are the Costa Brava, Costa del Sol and Costa Blanca. Although there are resort foci, such as Torremolinos and Alicante, planned and unplanned built-up areas have spread ribbon-like along these three coastal regions. High-rise and low-rise buildings providing accommodation, restaurants and entertainment predominate. The former fishing villages have been engulfed as tourism has overtaken the traditional activities of fishing and agriculture.

| A | 1999 tourism numbers/receipts | Tourist arrivals $\times 10^6$ | Market share % | Growth | Tourism receipts US\$ $\times 10^9$ | Market shares % |
|---|---|---|---|---|---|---|
| | Europe | 394.1 | 59.3 | 2.7 | 232.5 | 57.2 |
| | Americas | 122.9 | 18.5 | 2.4 | 118.0 | 26.8 |
| | East Asia/Pacific | 97.2 | 14.6 | 11.1 | 67.8 | 15.4 |
| | Africa | 26.9 | 4.0 | 7.8 | 9.8 | 2.2 |
| | Middle East | 17.8 | 2.7 | 16 | 8.6 | 1.9 |
| | South Asia | 5.7 | 0.9 | 8.3 | 4.3 | 1.0 |
| | World | c. 665 | 100 | | 441 | 0 |

| B | Top tourism destinations: | France, Spain, USA, Italy, China, UK |
|---|---|---|
| C | Top tourism earners: | USA, Spain, France, Italy, UK, Germany |
| D | Top tourism spenders: | USA, Germany, Japan, UK, France, Italy |

| E | Tourism originators: | Numbers of tourists $\times 10^6$ | Market share |
|---|---|---|---|
| | Europe | 368.9 | 58.0 |
| | Americas | 124.6 | 19.6 |
| | East Asia/Pacific | 92.9 | 14.6 |
| | Africa | 16.2 | 2.5 |
| | Middle East | 10.1 | 1.6 |
| | South Asia | 5.5 | 0.9 |
| | Other | 18.3 | 2.9 |

**F   Trends 1950–99**

| | |
|---|---|
| Tourist arrivals 1950: | $25 \times 10^6$ |
| Tourist arrivals 1999: | $664 \times 10^6$ |
| Average annual growth rate: | 7% |

**G   Forecasts For 2002**

| | Tourist arrivals | Average annual growth rate | Market share % |
|---|---|---|---|
| Europe | 717.0 | 3.0 | 45.9 |
| Americas | 282.3 | 3.9 | 18.1 |
| East Asia/Pacific | 397.2 | 6.5 | 25.4 |
| Africa | 77.3 | 5.5 | 5.0 |
| Middle East | 68.5 | 7.1 | 4.4 |
| South Asia | 18.8 | 6.2 | 1.2 |

World tourist arrivals: $1560 \times 10^6$

*Note:* data for A–D are for 1998 and/or 1999

**TABLE 9.5**  Data on world tourism (from World Tourism Organization, 2001)

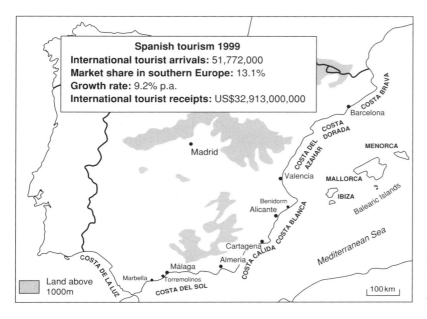

**FIGURE 9.7** Tourism in Spain and the location of the Costa resort areas. (Data from World Tourism Organization, 2001)

While the Spanish resorts developed within and around pre-existing towns and villages there are other resorts that have been purpose-built. Two examples are Sun City in South Africa and Cancún in Mexico. Sun City was created as a gambling resort in Bophuthatswana, an independent black homeland prior to 1994 when it became part of South Africa. Its independence from South Africa, a status not, however, recognized internationally, freed it from the latter's former restrictive gambling laws. This, plus its location near the Johannesburg–Pretoria region, which is heavily populated, the abundant cheap labour and the semi-arid climate, provided excellent conditions for a gambling resort. As Figure 9.8 shows, the facilities of Sun City include golf courses and swimming pools, as well as casinos, entertainment complexes and four large hotels. One of these hotels is the Place of the Lost City, which occupies *c*. 25 ha. As well as accommodation, this hotel complex comprises grounds containing created tropical forest, waterfalls, lakes and streams, which lead to the Valley of the Waves where there is a large-scale wave machine. This represents a dramatic change from the pre-existing natural vegetation of savanna.

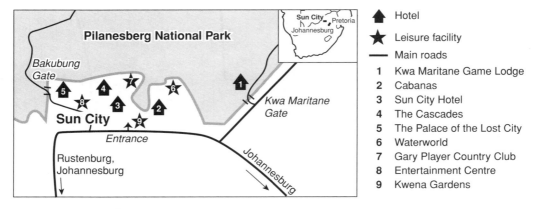

**FIGURE 9.8** Sketch map of Sun City, South Africa

Las Vegas, USA, is another city heavily reliant on gambling and related activities. Unlike Sun City, it has a history prior to its development as a world-famous gambling centre (see 1vol, 1995–97). It was a stopping-point for watering horses in the early 1800s and the first buildings were constructed by Mormons in the 1850s; the town grew rapidly with the arrival of the railway in the early 1900s. The origins of modern Las Vegas really began in 1931 when the state of Nevada legalized gambling and introduced 'quickie' divorce laws; the first casino was built in 1941. The success of El Rancho Vegas encouraged further casino/hotel building and since the 1950s many themed resort hotels have been constructed, demolished and reconstructed or replaced. Today resort hotels, each accommodating more than a thousand visitors, are thriving, having diversified to include family entertainment facilities as well as gambling (see 1vol, 1995–97, for details).

Cancún on Mexico's Yucatán Peninsula was developed in the mid-1970s on the basis of a computer-based search of Mexican coastal areas for a suitable resort site. The barrier-island site is now a premier resort in Mexico and receives some $2 \times 10^6$ visitors per year, mainly from the USA and Europe. As well as its advantages due to its location on the Caribbean coast, Cancún enjoys a strategic position in relation to the region's numerous archaeological sites, as shown in Figure 9.9. However, it is a dichotomous city. There are two distinct parts: the reasonably well-planned hotel zone or Isla Cancún, which has an upmarket ambience, and Cancún City or Centro, which is residential and a 'service' provider, especially in terms of labour, and which is much less elegant than the hotel area. Both urban areas have replaced tropical coastal forest as did the Mayan settlements elsewhere in the Yucatán (Figure 9.9). These now also contribute to wealth generation in the area. In addition, the wetlands, particularly mangrove communities, in the lagoon were cleared as the Isla Cancún was developed. The Yucatán coast has also

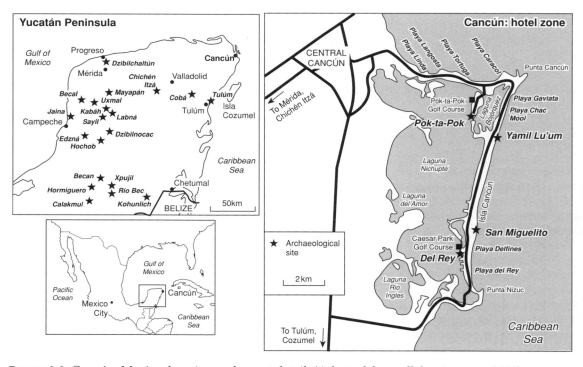

**FIGURE 9.9** Cancún, Mexico; location and resort detail. (Adapted from allaboutcancun, 2001)

been subject to other alterations linked with tourism. Meyer-Arendt (2001) has given details of changes along the north coast associated with development for domestic tourists since 1945. This has involved the construction of summer homes along a 20 km coastal strip around the port of Progreso (see Figure 9.9 for location). Environmentally, the most significant impact has been shoreline modification associated with house construction and groyne provision. The latter is designed to reduce

beach-sand movement and comprises substantial rock and timber structures know as *espolones*. These have become so numerous that the beach areas are no longer attractive; more than 100 have been constructed since 1990 alone, yet have failed to prevent beach erosion.

Unfortunately, coastal environments are particularly vulnerable to change when tourism development occurs, as the above examples illustrate. Moreover, islands can be subject to such extensive changes that the ensuing

| | | |
|---|---|---|
| **A** | *Dredging* | |
| | **Purposes** | The creation of ship and boat channels |
| | | To obtain sand for beach nourishment |
| | **Impact:** | Sediment removed by dredging is deposited in wetlands as has occurred south of St John's. Here, a salt pond of c. 0.8 km$^2$ has been infilled and an additional area of mangrove wetlands of 0.65 km$^2$ has been buried. Some deposition encapsulates garbage from cruise ships. Reef areas dredged for sand, which is used in the construction of resorts, have much-reduced growth. |
| **B** | *Sewage dumping* | |
| | **Purpose:** | Essential disposal of waste |
| | **Impact:** | Some sewage is treated but raw sewage is also dumped in wetlands, e.g. in McKinnons Salt Pond, from where mangroves have disappeared. The problem has been exacerbated by the failure of sewage-treatment plants. Flooding with seawater has caused sewage overflow and the contamination of beaches as well as fish kills. |
| **C** | *Resort creation* | |
| | **Purpose:** | To provide accommodation and bathing breaches etc. |
| | **Impact:** | Infilling of salt ponds reducing areas of mangroves, using sand and coral for building and creating turbidity. The latter reduces coral growth, e.g. in Mosquito Bay, and the range of marine life forms. In some cases, e.g. Emerald Bay, stabilized mangrove communities are being converted to unstable breaches by suction dredging. |
| **D** | *Sand mining* | |
| | **Purpose:** | For the production of cement for use in the construction industry |
| | **Impact:** | Sand mining is occurring on beaches that may already have a diminishing natural supply. Large pits are created and sediment budgets are destabilized through cutting off the tide's access to salt ponds by track construction for trucks. Ponds thus undergo eutrophication (nutrient enrichment), e.g. at Ffryes Bay and Club Antigua. |
| **E** | *Boating and diving* | |
| | **Purpose:** | Tourist activities, e.g. yachting and diving |
| | **Impact:** | Anchoring on reefs to facilitate passenger diving, which is targeted at the 'best' reefs. Divers may also damage reefs by standing or walking on them and by collecting coral 'souvenirs'. |

**TABLE 9.6** Tourism-linked activities and their impact in Antigua. (Adapted from Baldwin, 2000)

degradation of environmental systems may threaten the viability of tourism. This is the unsustainable face of tourism and is exemplified by the problems recently experienced in Antigua and Barbuda in the West Indies. According to Baldwin (2000), tourism generates some 80 per cent of the islands' gross domestic product and employs c. 50 per cent of the labour force. The tourism industry is entirely focused on beach resorts and activities, and consequently resorts have been constructed close to the shoreline; each hotel requires a beach, yet beach erosion is a serious threat to this industry. It has been accelerated by human disturbance of the dynamics of the interrelated coastal wetlands of many fringing coral reefs and intermediate beaches. Such disturbance occurs because of dredging, the disposal of sewage, resort creation, sand mining, boating and diving, the details of which are summarized in Table 9.6. Thus, resort construction and operation, as well as tourist activities, are conspiring to destroy the resource base on which the tourism industry is based. The detail provided by Baldwin indicates poor tourism management, and lack of overall planning and control, despite the fact that failure of the tourism industry will cripple the islands' economy.

In National Parks, tourism has direct and indirect impacts as Lilieholm and Romney (2000) have discussed. This is especially evident in relation to wildlife as animal behaviour may be altered, disease may be introduced, fire may become more frequent as may vehicular use, and feeding patterns may be disturbed if tourists feed animals or if waste disposal is inadequate and waste is left available for animal foraging. However, where animal populations are allowed to extend beyond the carrying capacity of a park, to enhance the tourist experience through increased sightings, the vegetation, and hence land cover, is likely to be altered. This has occurred in some of Africa's National Parks as elephant populations have been allowed to increase. Not only do they begin to pose threats to local farmers but they also uproot trees and overgraze. Lilieholm and Romney also report that expanding elk and bison herds in Yellowstone National Park may be preventing the regeneration of willow and aspen. In turn, this decline has caused a drop in beaver populations. In addition, path and track creation cause depletion of vegetation in localized areas. Hall and Farrell (2001) also draw attention to the depletion of woody material in the vicinity of campsites in the Cascades Mountains of Oregon, USA. This they suggest is a widespread activity that alters land cover in localized spots, yet it has received little attention when compared with trampling or rubbish disposal.

## 9.7 Conclusion

Although the disparate issues discussed in this chapter appear to be unrelated, all are nevertheless responsible for land-cover/land-use change all over the world. Some of these factors, notably the invasion of alien or exotic species, have an impact that is generally gradual in relation to human life span and that operates at timescales of half-centuries or centuries. Nevertheless, such invasions not only alter land-cover characteristics through altering the species composition, usually by reducing biodiversity and sometimes by causing species extinction, but also cause alterations in animal and/or insect communities, and thus in energy flows and the carbon cycle (see Section 1.5). Moreover, the dynamics of why and how such invasions occur are not well understood and, consequently, their control is elusive. In contrast, the direct impact of war and terrorism on land cover/land use is immediate; urban and rural landscapes are transformed with an alarming immediacy. There may be some recovery, but testaments to war are widespread globally, not least of which are the war grave cemeteries that are poignant reminders of the loss of human life that conflict inevitably involves.

The other factors involved in land-cover/land-use change examined in this chapter operate at temporal scales intermediate to those of exotic invasions and war/terrorism. Waste disposal, sport and tourism are undoubtedly significant causes of land-cover and land-use change. All of these activities deepen and broaden Ecological Footprints and increase carbon indices (Section 3.6). All reflect increasing leisure time and wealth in the developed world and indicate the power of human communities not only to generate wealth but also to dictate land use. These three factors are often interlinked; sport and tourism attract people and people produce waste of various types. Although much tourism is characteristic of non-urban areas, e.g. game, national and wildlife parks, and even the remote areas of the Arctic and Antarctic, it is also heavily concentrated in cities and purpose-built resorts such as those of coastal and skiing areas, and even wildlife parks etc. are required to find lodges and/or camping grounds. Urbanization and loss of natural land cover are thus encouraged. At the same time the disposal of sewage and garbage so generated may cause another form of land-cover/land-use change, sometimes to the detriment of the resource attracting the tourists.

The impact of all of these factors is set to increase well into the twenty-first century. Development, especially an increase in education standards, increasing wealth and increasing leisure time, as well as improved travel possibilities and media promotion, will all contribute to globalization of which the factors discussed herein are a substantial component.

## Further reading

**Holden, A.** 2000: *Environment and Tourism.* Routledge, London.

**McDougall, F.R., White, P.R., Franke, M. and Hindle, P**. 2001: *Integrated Solid Waste Management*, 3rd edition. Blackwell, Oxford.

**Mooney, H.A. and Hobbs, R.J.** (eds) 2000: *Invasive Species in a Changing World.* Island Press, Covelo, California.

**Van Driesche, J. and Van Driesche, R.** 2000: *Nature out of Place: Biological Invasions in the Global Age.* Island Press, Washington DC.

**Wittenberg, R. and Cock, M.J.W.** (eds) 2001: *Invasive Alien Species: A Toolkit of Best Prevention and Management Practices.* CABI, Wallingford.

**Wright, R.T. and Nebel, B.J.** 2002: *Environmental Science. Toward a Sustainable Future*, 3rd edition. Pearson Education, New Jersey.

# Conclusions and prospect

## 10.1 Introduction

The heterogeneity that characterizes the Earth's surface can be described in terms of land cover and land use. Natural factors such as climate, soils and geology conspire to produce distinctive types of land cover manifest mostly as natural communities of plants and animals or ecosystems. These, in turn, are dynamic in time and space, and exert a reciprocal influence on climate, soils and geology. That life and the environment enjoy such an intimate relationship is the essence of Gaia (Lovelock, 1995). Moreover, all of these factors have been linked within the global biogeochemical cycle of carbon throughout geological history, a relationship that is at the heart of global management in the twenty-first century.

The advent of hominids some $5 \times 10^6$ years ago and the later emergence of modern humans, *Homo sapiens sapiens*, as controllers of ecosystems are major thresholds in environmental and human history. Humans emerged as a significant agent of environmental change mainly because of their appropriation of the carbon cycle through technological innovation. This began when hunter-gatherers first harnessed fire to herd animals; fire was also a powerful tool that enabled the early agriculturalists to clear natural vegetation for crop growing. Massive population growth through

the centuries, coupled with scientific and technological developments, have allowed humans to alter large parts of the Earth's surface. Natural land cover has been transformed into land use over large areas, and one type of land use has been transformed into another type of land use. The human imprint is far-reaching and adds another form of dynamism to the natural processes operating on the Earth's surface. Human activity either alters the rate of natural processes, such as soil erosion, or creates new processes such as the invasion of exotic plant and animal species. This occurs within a global context affected by natural change such as climatic change.

There are many ways in which humans influence land cover and land use, as documented in three classic publications: Marsh's (1864) *Man and Nature; Or, the Earth as Modified by Human Action*; Thomas's (ed., 1956) *Man's Role in Changing the Face of the Earth*; and Turner et al.'s (eds, 1990) *The Earth as Transformed by Human Action*. The approach adopted in this book focuses on agriculture in the developed and developing world, forest exploitation and afforestation, mineral extraction and urban development as the most important causes of land-cover and land-use change. As these activities have become widespread, through the process of development and in response to a host of socio-economic factors, many additional types of impact have arisen. These include the invasion of exotic plant and animal species, waste disposal, tourism and sport; and just in case the human imprint is already not sufficiently deep, war and terrorism cause unnecessary and entirely destructive land-cover alteration.

The reasons why land-cover and land-use change occurs are often more varied than the changes themselves. Different stimuli may give rise to similar changes in a given area; conversely, similar stimuli may generate heterogeneous changes in a given area. The stimuli may be natural or socio-economic, or a combination of the two. Climatic change is a particularly important natural stimulus, though climatic change is itself likely to be heavily influenced by human perturbation of the global biogeochemical cycle of carbon, now and in the coming decades. Other natural processes that cause or contribute to land-cover/land-use change include soil erosion, desertification, acidification and eutrophication but, as in the case of climatic change, all of these processes are deeply influenced by human activity. Indeed, distinguishing alterations due to natural causes from those due to human causes is difficult, but important for the purposes of management.

Socio-economic stimuli are many and varied. They relate to subsistence, and the generation of power and wealth. Subsistence is existential but how it is achieved varies throughout the world depending on climate and natural vegetation, soils etc. Once surplus food is generated the food-procurement type is no longer subsistent but becomes part of a wealth- and power-generation system. Wealth generation is intimately linked with development, which in turn involves increasing standards of living; nutritional levels, educational standards and release from poverty all improve with development. Equally important is the rate of population growth; slow rates of growth tend to be associated with little conversion of natural land cover to land use, and more change of one land use to another. The converse is also true. Industrialization, science and technology are also linked to wealth generation and development; they all generate land-cover and land-use change either directly or indirectly.

These factors are also linked with the way people organize themselves, i.e. whether the population is dispersed, as in agricultural or rural contexts, or is concentrated, as in urban environments, and how government policies express people–environment relationships. The trend towards urbanization began with the Industrial Revolution of the mid-1700s, but the shift of people to cities has achieved a new vigour. Today, urbanization is increasingly apparent worldwide and especially in the developing world. Other socio-economic influences on land cover and land use include religion, tourism, all types of waste disposal, and the provision of sports facilities. All are linked to levels of wealth and to means of wealth generation. Political alliances that control trade or agricultural subsidies etc. are also important. The land-cover consequences of war and terrorism are far-reaching but serve only to acknowledge the consequences of loss of life and property. The cause of conflict is quite another matter; religion, political ideology etc. seem somewhat abstract when the resulting destruction depletes both human and natural resources of both victor and vanquished. Besides, there are rarely true winners, and the world in general is impoverished.

The factor uniting natural and socio-economic causes of land-cover and land-use change is the global biogeochemical cycle of carbon. The relationship between people and environment can be expressed in terms of society's ability to manipulate this, and other, biogeochemical cycles. There have been important thresholds in this relationship in the past but the land-cover and land-use changes of the future are all-important because of the role of the carbon cycle in climatic change. In terms of future prospects for land-cover and land-use change, all of the factors referred to above will continue to be significant. World population growth will be highly significant, though regional variations will generate an uneven effect, with polarization between the developed and developing worlds. In addition, scientific and technological advances will have a worldwide impact, notably biotechnology in agriculture, and computer-generated information and transmission in general. Future patterns of energy production and consumption, improvements in standards of living and continued conflict involving war and terrorism will all contribute to future land-cover and land-use characteristics. All will have an impact on the global carbon cycle, which is increasingly being considered as a political issue. International, national and local measures to manage Earth's resources are becoming as much about politics as they are about environment and science. The carbon cycle is at the heart of this management and its politicization, and will be a significant component of land-cover and land-use alteration in the future.

## 10.2 Conclusion

### 10.2.1 Summary of land-cover and land-use change

Land-cover characteristics have varied enormously throughout the Earth's *c.* $5 \times 10^9$-year history. Geological strata and their contained fossils attest to this state of overall flux, which has been punctuated by periods of relative constancy. Geologically, the last $3 \times 10^6$ years have witnessed particularly momentous environmental change; the regular increases and decreases in global temperatures in tandem with the advance and retreat of ice sheets, and the rise and fall of tree lines generated continuous change in land-cover characteristics (see Mannion, 1999c). These naturally driven transformations are intimately linked with the global biogeochemical cycle of carbon. It is interesting, and possibly no coincidence, that this particularly dynamic geological period also witnessed the advent of hominids, primates with human-like characteristics, and eventually *Homo* spp., culminating in modern humans (*Homo sapiens sapiens*) *c.* $300 \times 10^3$ years ago (see Lewin, 1999, for a review of human evolution).

For a large part of their relatively short history humans have been integral components of the world's ecosystems; their food-procurement strategies developed from scavenging to active hunting and included gathering. Eventually, sophisticated hunter-gatherer societies emerged, with skills that facilitated the manipulation of biomass, often through the use of fire. This involves direct manipulation of the carbon cycle and land-cover alteration. As humans developed increasingly sophisticated food-procurement strategies, through plant and animal domestication and permanent agricultural systems *c.* 10,000 years ago, modification of the global carbon cycle intensified, and land-cover alteration intensified and spread. The many different types of agricultural systems that exist today reflect the varied ways in which humans have manipulated the Earth's surface to obtain food, fibre and beverages etc. Agriculture remains the most significant human activity for the transformation of natural land cover into land use.

Manipulation and clearance of the world's forests are linked to the spread of agricultural systems and substantial human intervention in the global carbon cycle. Forest removal, either partial or complete, and forest creation have occurred and continue to occur. Forest removal or manipulation is caused primarily by the spread of agriculture, but other activities also contribute. Forests support large and small-scale industries linked to pulp and paper production, and wood production for structural and domestic purposes. In tropical regions, even when only selective or partial logging is practised, the opening up of pristine forests encourages other forms of exploitation, especially shifting agriculture. Mineral extraction can have similar consequences. In contrast, afforestation programmes create or replenish forest cover, though this rarely, if ever, has the same characteristics as natural forest. Indeed, afforestation tends towards monoculture, whereas natural forests have high biodiversity. In relation to the global carbon cycle, forest loss causes depletion in the terrestrial store and an increase in the atmospheric store; afforestation has the reverse effect.

Extractive industries also have an influence on land-cover characteristics. In terms of spatial extent, this impact is minor in relation to agriculture or forest manipulation, or indeed to urban spread. Nevertheless, mineral extraction has both direct and indirect impacts on land cover. All extractive industries involve land-cover change as vegetation is removed to facilitate mining. The degree of alteration depends on the type of extraction, notably whether it is deep- or shallow-surface mining and whether any form of processing is undertaken on-site. Land-cover change at a distance from the site of extraction is also commonplace. Drainage systems are particularly vulnerable, and the characteristics of their flora and fauna can be altered significantly because of changes in water chemistry. Moreover, the transport of substances such as oil and gas via pipelines causes changes in land cover, as does the construction of plant, such as oil refineries and gas storage facilities. This usually occurs at considerable distance from the site of mining, as is also the case with the processing of metalliferous and non-metalliferous ores.

Energy availability is a key factor in ore processing and results in alteration of the global carbon cycle. The more energy a nation or region consumes, the greater is its Ecological Footprint (Section 3.6). This can be mitigated to some extent through the reclamation of land rendered derelict by mineral extraction, though this is costly and time consuming; each type of extraction creates a unique set of problems. Nevertheless reclamation is worthwhile, and viable afteruses with economic returns can contribute to wealth generation. In addition, extractive industries are often responsible for land-cover alteration over a much wider area than that occupied by the mine. Apart from the construction of plant, deforestation is likely to occur as agriculture is established to provide mining camps with food. This pattern is common in developing countries where mineral extraction occurs in remote areas.

There are many other human activities that contribute to land-cover change, not least of which is urbanization. The developed world is already largely urbanized and the developing world is rapidly urbanizing. The former process accelerated with the onset of the Industrial Revolution in the 1750s; the latter process is a phenomenon of the past 50 years but now many cities in the developing world are growing rapidly and overtaking some of the megalopolises of the developed world. When urbanization, defined here simply to involve an agglomeration of family and communal buildings, can be identified in the record of human history is open to debate. There is little in the archaeological and/or palaeoenvironmental record to suggest any significant urbanization before 10,000 years ago. There is a link between urbanization and the development of permanent agriculture, possibly because the latter is particularly efficient for food production. There is both a proximal supply of labour and a market. How the two relate is probably not through a cause-and-effect mechanism but through mutual evolution that favoured survival, influence and wealth generation. Many great civilizations have left legacies of their former glory: the Egyptians, Mayans, Khmer, Persians, Incas, Aztecs, Greeks and Romans, to name but a few. Their architecture, funerary monuments and middens attest to their activities, and continue to generate wealth for their modern descendants through heritage tourism.

Modern cities are a product of the past and present. Many have long histories: Athens, Rome and London, for example, have histories extending back more than 2000 years. Their structures and land-use characteristics, which include vestiges of their ancient persona, reflect not only their own histories but also the history of Europe and that of the world. Urban land use is diverse; it reflects temporal and spatial development as well as power relations and political alliances. War and pestilence juxtaposed with economic boom and bust have shaped cityscapes worldwide. In their fabric and their museums, art galleries and theatres, cities are ever-growing cultural archives of a singular nature.

This is equally true of the cities of the New World and Australasia, which were mostly planned from the start and whose structure and function reflect the influence of colonial powers. From the late sixteenth century many cities were established, in the Americas and from *c*. 1750 in Australia. Following instructions from Spain, many cities in the Americas were founded on the basis of the gridiron plan, involving a central plaza around which government buildings and a cathedral or church were constructed. The use of the gridiron pattern was widespread in North as well as South America. Initial functions of these settlements reflected coastal or lake locations, often at the mouths of large rivers, which provided access to resource-rich hinterlands. Forests, furs, mineral extraction and agriculture provided the early wealth of many of these cities.

Developments in agriculture, the process of industrialization and the establishment of communication networks, initially railways and later roads, contributed substantially to the dynamism of cities worldwide, but especially in Europe and North America, and prompted land-use change. Population increase, fuelled by *in situ* increase and rural depopulation, coupled with the development of urban transport systems such as buses and underground railways, also contributed to urban land-use change. The advent of the car and its increasing use since *c*. 1950 altered land-use patterns again and encouraged the development of suburbia. Cities, especially those in the developed world, became the seats of wealth generation, which had hitherto been based mainly on agriculture. This wealth came from a myriad of different industries and was based on natural resources, such as minerals, coupled with energy use. Carbon consumption in cities increased substantially as fossil fuels, beginning with coal and turning later to oil, were exploited to provide energy for both industrial and domestic use. At this stage, the Ecological Footprint deepened and broadened as the generation of waste products, notably carbon dioxide and sewage, as well as domestic rubbish and industrial waste, increased.

*All* of these factors generate land-use change. Cities are in large part consumers of land, water and energy, and generators of waste; they have a major impact on the global carbon cycle. However, cities have parks, gardens, lakes and woodlands as components of urban land use. They also contain primary producers, which draw down carbon dioxide from the atmosphere – in contrast with most other urban activities, which generate carbon dioxide. Although there is no question that there is a balance between carbon dioxide consumption and production, as the latter is overwhelmingly dominant, urban parks, gardens etc. are important components of cities. They provide an aesthetically pleasing digression from roads

and buildings; they provide recreation sports and entertainment facilities, and sometimes educational opportunities. Many are also linked with a city's history and are thus part of its cultural heritage. They can be colourfully but accurately described as green oases in a grey urban sea.

Whilst agriculture, forestry, mineral extraction and urbanization are the major causes of land-cover and land-use change, there are many other stimuli that have a significant effect. Some factors exert a catastrophic influence, but others operate subtly. The invasion and subsequent proliferation of alien plant species occurs subtly; often the invasion is hardly recorded until it reaches substantial proportions. However, this is a worldwide problem caused by deliberate and inadvertent plant introductions. Land-cover characteristics are altered as introduced species compete successfully with native species, which may become extinct; insect and bird populations may alter as a consequence and biogeochemical cycles, along with hydrological processes, may be affected detrimentally. At the other end of the impact spectrum, war and terrorism have immediate and catastrophic effects on land cover and land use. There are few countries that remain unaffected by war and there are an increasing number being affected by terrorism. Direct impacts are varied, with particularly significant changes occurring in urban environments that have been bombed or have suffered explosions. Damage to buildings can be widespread and subsequent regeneration of affected areas usually results in a new urban landscape. In the countryside, vegetation destruction may occur through defoliation, the machinery of war may be abandoned and bomb impacts will leave craters of varying sizes. The human dimension of war is also recorded in so far as urban and rural bombardment claims the lives of both civilian and military personnel. War memorials and war graves attest to this loss of life worldwide. They constitute a unique type of land use.

Other causes of land-cover and land-use change include leisure activities such as sport and tourism, and waste disposal. The latter tends to cause land-cover/land-use change in the vicinity of urban areas, which require plant for sewage treatment and disposal as well as sites for either incinerators or landfill for rubbish disposal. As standards of living have improved and packaging, especially with plastic, has increased, the amount of domestic waste has grown considerably. Consequently, there has been an increase in the number of landfill operations, the favoured waste-disposal method. Once full, such sites can be reclaimed to generate yet another type of land cover and land use.

Provision of land for sport contributes significantly to the heterogeneity of urban and peri-urban land use. Facilities include stadia, pitches of various sorts and indoor facilities such as gymnasia, swimming pools and health clubs, though in terms of spatial extent, golf courses and racecourses are of particular note. Facilities for sport are, in general, more prolific in developed nations than in developing nations, or where there is a rich elite. This is because disposable wealth and leisure time are the primary factors that generate a need for such facilities. Many are associated with the provision of other services such as restaurants, accommodation, car parks etc., so there is variation of land use within the sporting facility. Moreover, the presence of a facility, such as a golf course or racecourse, may influence land-use characteristics in the wider area where wealth generation focuses on provision for visitors, e.g. hotels, shops and restaurants.

Tourism can have a similar but more pronounced effect. The resource that attracts tourists may even be overwhelmed; beaches become crowded rather than secluded, forests may be cut down to provide fuelwood for tourist hotels and lodges, and coral reefs may be degraded by pollution and overexploitation. Thus a balance needs to be achieved in order to offer a resource, the tourism opportunity, but to prevent its degradation. Planning and management are all-important. However, both the natural and cultural worlds provide many opportunities for tourism development, which is amongst the fastest-growing industries worldwide and the largest generator of wealth and employment for some countries. Transport and the provision of services such as accommodation are reflected in the land use of tourist areas, as is waste disposal. This land use may replace natural vegetation or agriculture. No tourism is without impact, but tourism with the greatest Ecological Footprint, such as resort-based beach tourism for the masses, causes the most extensive land-use change.

## 10.2.2 The driving forces of land-cover and land-use change

As the many examples given in this book show, the driving forces of land-cover and land-use change are many and varied. Natural environmental change characterizes the Earth's history as climate has altered and continents shifted. The record of the last $3 \times 10^6$ years in particular reflects the operation of a dynamic natural system. The related driving forces are climate and the global carbon cycle, which conspire to raise and lower sea levels, to alter the altitudinal and latitudinal locations of tree lines, and to cause the disaggregation and reformation of the world's ecosystems. Figure 10.1 presents this relationship schematically.

The socio-economic causes of land-cover and land-use change are also directly or indirectly related to the global carbon cycle. The appropriation of carbon, made possible through technological advances, has not only been essential for subsistence but has also underpinned wealth generation. The three major categories of stimuli referred to in Figure 1.7 are also dynamic; not all operate in a given place at a given time but some are operative everywhere all of the time. As stated in Section 1.6, all of these factors are components of that most complex of relationships, which is that between people and environment, or the noosphere, and which is ultimately linked with the carbon biogeochemical cycle. This relationship is represented schematically in Figure 10.2.

The major difference between natural change and that which is anthropologically driven is a matter of scale, especially temporal scale. The former operates at global as well as regional scales, while the latter is particularly evident at local and, to some extent, regional scales. Temporally, and despite the geologically rapid

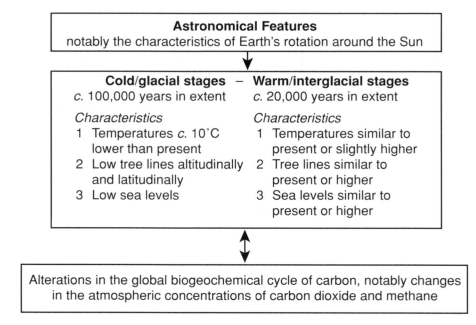

**FIGURE 10.1** Schematic representation of major natural land-cover change during the last $3 \times 10^6$ years

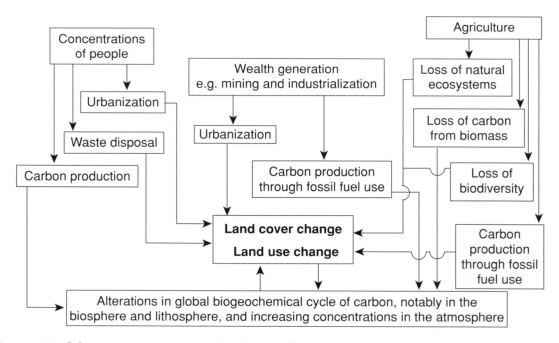

**FIGURE 10.2** Schematic representation of land-cover/land-use change caused by socio- economic factors, and their relationships with the global biogeochemical cycle of carbon

scale of change in the last $3 \times 10^6$ years, anthropological factors operate at much shorter timescales. Even within a few years, land-cover and land-use change can be dramatic, as is illustrated by the alarming data on deforestation (see Section 6.3 and Table 6.1). Rapid increases in urbanization (Chapter 8, especially Section 8.5), tourism (Section 9.6) and waste generation (Section 9.3) also illustrate this point.

This disparity between natural and anthropogenic rates of environmental change is of particular concern because the latter provides little opportunity for natural or semi-natural systems to adjust. In particular, the disruption to the global carbon cycle is now becoming manifest as global climatic change (see Section 10.3). The disruption of other biogeochemical cycles such as sulphur, nitrogen and phosphorus are causing other pollution problems, including acidification and cultural eutrophication (see Mannion, 1999b, for a review of acidification, and Bouwman and van Vuuren, 1999, for a review of cultural eutrophication). This disruption of biogeochemical cycles is paralleled by disruption of physical processes such as soil erosion (reviewed in papers in Lal, 1999) and desertification (reviewed by Food and Agriculture Organisation, 2001).

## 10.3 Prospect

Without doubt, the next few decades will witness extensive changes in the Earth's land cover and land use. The pressures for change are enormous as all aspects of the people—environment relationship alter and as the Ecological Footprint of many nations is likely to increase. Moreover, the evidence is mounting for global climatic change, which is already influencing plant and animal distributions. As well as climatic change, the impact of which is difficult to predict, the most important factors that will influence land-cover and land-use change are global population growth, related changes in agricultural systems, including their spread and the impact of biotechnology, the continued industrialization of many devel-

oping countries and associated patterns of energy use and increasing wealth, improving communications networks and the role of environmental issues at all political levels. Many other land-cover and land-use changes will be generated by the repercussions of alterations in these major factors. For example, waste disposal will increase as populations increase and the requirement for land for sport and tourism will accelerate as increasing wealth and leisure time become available.

### 10.3.1 Climatic change

Recent concerns about the possibility of global climatic change have given rise to major research initiatives that focus on collating evidence for change, likely impacts on natural and socio-economic systems, and possible means of mitigation. Much of this work has been undertaken under the auspices of the Intergovernmental Panel on Climate Change (IPCC). The results and recommendations of this international group of experts are widely accepted, and are the basis on which planning strategies at national and international scales are being formulated. The following summary is based on IPCC (2001a; 2001b; 2001c) and the Hadley Centre of the UK Meteorological Office (Hadley Centre, 2001); researchers at the latter also contribute to the IPCC.

There is evidence for an increase in global temperatures from a variety of sources, including instrumental data. Overall, the global average surface temperature has increased by $0.6 \pm 0.2°C$ during the 1900s, with the 1990s being the warmest decade and 1998 being the warmest year. Proxy data, e.g. tree rings, corals, ice cores and historical records, along with instrumental data, indicate that the 1990s were the warmest decade of the last millennium. Table 10.1 gives further details of climatic change, which imply that it has not been uniform globally. Moreover, the cause, or causes, of this change are considered to involve greenhouse gas emissions and other factors; until *c.* 1950 a combination of greenhouse gas emissions and increasing solar output were most important, whereas since 1950 greenhouse gases, sulphate aerosols and ozone are

| | |
|---|---|
| **A** | Global average temperature of the lowest 8 km of atmosphere has changed by + 0.05 ± 0.10°C per decade but the increase in global average surface temperature has been greater, i.e. by 0.15 ± 0.05°C per decade. The difference occurs mostly over tropical and subtropical regions. |
| **B** | There have been decreases of *c.* 10% in the extent of the duration of lake and ice cover in the mid- and high latitudes of the Northern Hemisphere during the 1900s. |
| **C** | Mountain glaciers in non-polar regions retreated during the 1900s. |
| **D** | Spring and summer sea-ice in the Northern Hemisphere has decreased in extent by *c.* 10–15% since the 1950s. Some evidence suggests that Arctic sea-ice thickness in late summer and early autumn has declined by *c.* 40% with a more limited decline in winter sea-ice thickness. |
| **E** | On average global sea level rose between 0.1 and 0.2 m during the 1900s. |
| **F** | Global ocean heat content has increased since the late 1950s. |
| **G** | There have been changes in precipitation: (i) between 0.5 and 1% per decade increase in mid- and high latitudes of the continents of the Northern Hemisphere in the 1900s; (ii) an increase of 0.2 to 0.3% per decade over tropical land areas in the 1900s; (iii) a decrease of *c.* 0.3% over subtropical regions in the Northern Hemisphere in the 1900s. |
| **H** | It is probable that the mid- and high latitudes of the Northern Hemisphere have experienced *c.* 2–4% increase in the frequency of heavy precipitation events. |
| **I** | There is the possibility of a 2% increase in cloud cover over mid- to high-latitude continental regions in the 1900s. |
| **J** | There appears to have been a reduction in the frequency of extreme low temperatures and a small increase in the occurrence of extreme high temperatures. |
| **K** | Warm episodes, e.g. of El Niño–Southern Oscillation (ENSO) have been more frequent and longer lasting since the mid-1970s. |
| **L** | Some increases in drought and severe wetness occurrences. |
| **M** | The frequency and intensity of droughts have increased in the last few decades. |

**TABLE 10.1** Aspects of climatic change during the 1900s. (Adapted from Intergovernmental Panel on Climate Change, 2001a)

significant. The latter three are products of human activity and their production has increased substantially, often associated with land-cover and land-use change.

On the basis of these trends, the IPCC predicts that, on average, global warming will be in the range of 1.4 to 5.8°C for 1990 to 2100. The variation occurs because predictions are based on various scenarios of greenhouse gas production. The Hadley Centre predicts global warming in the range 2.0 to 4.0°C. In both cases, the least global warming is predicted to occur if carbon emissions are reduced (in relation to the present day). The highest rates of warming are associated with increased use of fossil fuels and continued increases in greenhouse gas emissions: intermediate warming occurs with intermediate scenarios. Under any of these conditions, global warming will have a significant impact on land cover and land use. However, this will not be globally uniform;

high latitudes are likely to warm more significantly than low latitudes. The degree of impact will also depend on the adaptability of environmental and socio-economic systems. Thus some regions will experience a greater degree of change than others. The IPCC (2001b) deliberations that relate to land cover and land use are summarized in Table 10.2. No region will escape change, and land cover and land use will alter significantly from their present configurations because of global climatic change alone. However, the scenarios are generalizations; their eventual reality or otherwise will depend on a range of factors, including climatic feedback mechanisms, which could exacerbate the magnitude of climatic change, and mitigation strategies.

The latter have been summarized by IPCC (2001c). They fall into two groups: mechanisms to reduce the production of greenhouse gases, i.e. through the use of 'clean' energy; and mech-

**A  Africa**
Low adaptability generally
Decrease in grain yields; reduction in food security
Reduction in water availability; increase in desertification especially in Southern, North and West Africa
Increased droughts and floods; impact on agriculture
Increased extinctions of plants and animals

**B  Asia**
Low adaptability in some areas; average adaptability overall
Increased floods, droughts, forest fires and cyclones will all cause land-cover alteration
Decreases in agricultural productivity will compromise food security in some areas
Decreased water availability
Exacerbation of threats to biodiversity
Reduction in permafrost in the north

**C  Australia/New Zealand**
Some positive advantages in terms of crop productivity, but only in the short term
Water resources may decrease
Heavy rains, tropical cyclones will increase
Biodiversity will decline in vulnerable areas, e.g. alpine, wetland and semi-arid habitats, coral reefs

**D  Europe**
Adaptive capacity is high
Increased differences between north and south in terms of water availability
Large reduction in alpine glaciers and permafrost; increase in river flooding
Some positive effects on agriculture in the north but decreases in productivity are likely in southern and
    eastern Europe
Altitudinal and latitudinal shifts of vegetation communities are likely, though some habitats – e.g. tundra,
    wetlands – will be reduced
Altered temperature and precipitation regimes may alter tourist requirements

**E  Latin America**
Adaptive capacity is generally low
Loss and retreat of alpine glaciers will reduce water availability from glacial meltwater
Increased frequency of floods and droughts; increased frequency and intensity of tropical cyclones
Crop yields are likely to decrease; subsistence farming is particularly vulnerable
Increased biodiversity loss
Coastal flooding of major cities

**F  North America**
High adaptive capacity
Some benefits in Canada for crop production and forestry, but declines in more southerly regions due
    to drought
Long-term benefits may be poor
Loss of glaciers, ice caps and permafrost; varied impact on water resources
Coastal cities are prone to flooding; property damage due to weather could increase
Loss of habitats such as prairie wetlands, tundra

**G  Polar regions**
These regions are particularly vulnerable to change
The Arctic, Antarctic Peninsula and Southern Oceans are particularly susceptible to change
Ice-sheet thickness and extent will decline; similarly permafrost
Invasion by more warmth-demanding species
Ice melting will affect all regions

**TABLE 10.2**  Likely regional alterations caused by global climatic change in the future. (Adapted from Intergovernmental Panel on Climate Change, 2001b)

anisms to absorb greenhouse gases, notably the provision of carbon sinks such as forests and other types of land use. Obviously, the latter would alter land cover and land use. Mitigation measures will also depend on international agreements as only action at this scale will provide meaningful results. Afforestation is considered a valuable contribution to biospheric carbon sinks and is a means of offsetting carbon emissions (see Sections 6.4 and 6.5). Moreover, if there is a shift to fuels other than fossil fuels, their extraction and transport will have less impact on land cover than at present (see Section 7.5) and there will be less perturbation of the global sulphur biogeochemical cycle, resulting in reduced acidification.

Climatic change and possible mitigation strategies will both contribute to land-cover and land-use change in the next few decades.

## 10.3.2 Population change

Throughout history, changes in population have stimulated land-cover and land-use change. Obviously, population increases have underpinned much such change, but declines in population have also had a local or regional influence. For example, the decline of the Mayan civilization in Central America and the abandonment of specific areas allowed tropical forest to recolonize Mayan cities such as Tikal, Guatemala; similarly, the decline of the Khmer Empire in Southeast Asia facilitated the return of a vegetation cover to Angkor (see Section 8.2).

In the next few decades, however, global population will increase substantially. As in the case of climatic change (Section 10.3.1), this increase will not be globally uniform but will

| A   United Nations population predictions (2001) | | | |
|---|---|---|---|
| | | Values 10⁹ | % rates of annual change |
| | 2000 | 2050 | |
| Developed nations | 1.2 | 1.2$^{ab}$ (1.1–1.3) | 0 |
| Developing nations | 4.9 | 8.2$^{b}$ (6.8–9.6) | c. 1.3 |
| World | 6.1 | 9.3$^{b}$ (7.9–10.9) | c. 1.2 |

| B   Population Reference Bureau (2001) | | | |
|---|---|---|---|
| | | Values 10⁹ | |
| | 2001 | 2025 | 2050 |
| Developed nations | 1.19 | 1.25 | 1.24 |
| Developing nations | 4.94 | 6.57 | 7.79 |
| World | 6.14 | 7.8 | 9.04 |

| C   Food and Agriculture Organisation (2001) | | | |
|---|---|---|---|
| | | Values 10⁹ | |
| | 2000 | 2025 | 2050 |
| Developed nations | 1.31 | 1.36 | 1.32 |
| Developing nations | 4.74 | 6.46 | 7.59 |
| World | 6.05 | 7.82 | 8.91 |

*Notes:*
[a] This is an average value, as some countries will experience negative growth
[b] These predictions are based on medium variant conditions, i.e. medium fertility rate projection, which requires stabilization of the fertility rate at replacement level; the values in parentheses are the ranges predicted using low to high fertility rates

**TABLE 10.3** Projections of likely global population change. (Adapted from United Nations, 2001; Population Reference Bureau, 2001; Food and Agriculture Organisation, 2001b)

be concentrated in developing countries. In contrast, some countries in the developed world, notably those in Europe, will experience population declines. Various agencies, such as the Food and Agriculture Organisation (FAO), the United Nations (UN) and the Population Reference Bureau (PRB), have projected likely population change for the next few decades. These predictions are summarized in Table 10.3, which shows that, in the next 25 years, world population is set to increase by *c*. $1.7 \times 10^9$; this represents an increase of almost 28 per cent on the current population of $6 \times 10^9$. By 2050, all the agencies predict an additional increase of *c*. $1.2 \times 10^9$. The total increase of *c*. $3 \times 10^9$ by 2050 represents *c*. 50 per cent of the present value.

On a 25-year timescale or a 50-year timescale these substantial population increases will undoubtedly stimulate land-cover and land-use change. Major repercussions of either of these huge population increases will occur in agriculture (see also Section 10.3.3) as existing agricultural systems intensify production and as they are created at the expense of remaining natural ecosystems. Much more carbon will need to be appropriated as food, i.e. edible carbon, than occurs at present. Moreover, the data in Table 10.3 show that the greatest increase in population will be in the developing

nations, which will therefore have the greatest need to adjust land cover and land use as they strive to feed their people and, in many cases, to maintain the production of agricultural commodities for export.

Further data from the agencies referred to in Table 10.3 show that even in developing countries there is considerable variation in the rates of population increase. As Figure 10.3 shows, the greatest increases in population are likely in Sub-Saharan Africa and Middle Eastern countries. These nations are also particularly vulnerable to climatic change (see Section 10.3.1) and so are likely to be faced with serious problems that will inevitably become manifest in land-cover and land-use change. Particularly serious is the near-certainty that large areas of remaining natural ecosystems will be lost and that poverty will remain unalleviated as standards of living fail to rise or even decline.

Other demographic factors that are likely to influence land cover and land use are alterations in the structure of the workforce. For example, some nations are experiencing high mortality rates from AIDS/HIV. In Botswana and Zimbabwe the workforce has been considerably reduced; if this pattern continues, agricultural land use may decline or the rate of increased productivity/cultivation necessi-

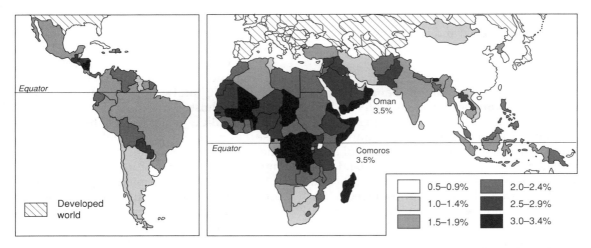

**FIGURE 10.3** Percentage rate of natural population increase in developing countries. (Data from the Population Reference Bureau, 2001)

tated by population increase may be reduced. Moreover, the pattern of urbanization that is currently evident is likely to accelerate as industrialization in the cities of developing countries proceeds and attracts labour. (This is discussed in Section 10.3.6.)

Although increases in population in developing countries will have considerable repercussions, the land cover and land use of those countries that will experience little or no population growth will not remain unchanged. For example, declining populations may give rise to a reduction in the amount of land used for agriculture, a process that has already begun in many European countries because of food surpluses (see Section 10.3.3). In addition, changing demographic characteristics are likely to alter land use, especially urban land use; for instance, an ageing population will require an increase in suitable accommodation such as retirement homes and nursing homes. Similarly, the trend towards smaller family units, such as single-parent families, will encourage building programmes within and around cities.

### 10.3.3 Agriculture

As stated above, global population increase will inevitably exert a profound influence on agricultural systems. Likely changes can be examined under two headings: first, there are possibilities for intensification and extensification; second, science and technology will affect the spatial aspects of agricultural systems and the practice of agriculture.

Intensification and extensification relate directly to global population change. They are processes that are occurring currently and that are likely to accelerate in the coming decades. In many developed nations, e.g. in Europe and North America, there has been a trend of extensification involving the removal of land from production, i.e. set-aside, through deliberate policies. Farmers have been encouraged to find alternative land uses, such as those related to recreation and tourism. This can be manipulated through the provision or retraction of subsidies, as occurs through policies such as the European Common Agricultural Policy (CAP),

or through the designation of Environmentally Sensitive Areas (ESAs) in which agriculture is limited. As populations in these countries remain stable or decline, and as science and technology provide opportunities to increase productivity (see below), the extensification process is likely to accelerate. The situation is, however, quite different in many developing countries, where populations are growing rapidly. Here, there are two possibilities: agricultural systems are likely to expand at the expense of remaining natural ecosystems; and it will be necessary, if not essential, to intensify production in existing agricultural systems. The first possibility is not particularly desirable because of the loss of biodiversity and further impairment of the Earth's biosphere to store carbon and provide other equally vital ecosystem services such as biogeochemical cycling, all of which contribute to the regulation of global climate. Intensification has drawbacks because of possible pollution through fertilizer use, the contamination of water supplies with pesticides, accelerated soil erosion and salinization caused by irrigation etc. Moreover, the widespread adoption of genetically modified crops and animals may bring problems as well as solutions to social and environmental challenges (see below). In reality, both expansion and intensification are likely to occur; there is a cogent argument to favour the latter (see, for example, Avery, 1997) in order to preserve the remnants of the world's natural vegetation and so ensure the services it provides.

Scientific and technological developments will continue to influence the extent and character of agricultural systems just as they have in the past. Whilst a complete discussion of all such possibilities is not feasible here, reference must be made to biotechnology, especially genetic modification, which is already beginning to become manifest in the field. As Table 10.4 shows, there are many applications of biotechnology in agriculture. Genetic modification is aimed at increasing productivity, through mechanisms to combat pests, e.g. insect resistance, or through the improvement of environmental tolerance, e.g. drought or high salinity. Alternatively, it is directed at

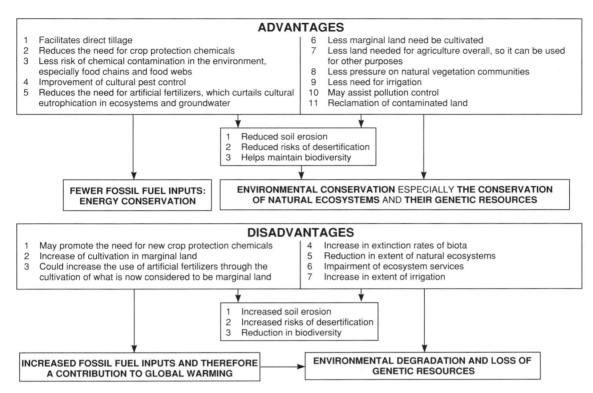

**ADVANTAGES**

1  Facilitates direct tillage
2  Reduces the need for crop protection chemicals
3  Less risk of chemical contamination in the environment, especially food chains and food webs
4  Improvement of cultural pest control
5  Reduces the need for artificial fertilizers, which curtails cultural eutrophication in ecosystems and groundwater

6  Less marginal land need be cultivated
7  Less land needed for agriculture overall, so it can be used for other purposes
8  Less pressure on natural vegetation communities
9  Less need for irrigation
10  May assist pollution control
11  Reclamation of contaminated land

1  Reduced soil erosion
2  Reduced risks of desertification
3  Helps maintain biodiversity

**FEWER FOSSIL FUEL INPUTS: ENERGY CONSERVATION**

**ENVIRONMENTAL CONSERVATION** ESPECIALLY **THE CONSERVATION OF NATURAL ECOSYSTEMS** AND **THEIR GENETIC RESOURCES**

**DISADVANTAGES**

1  May promote the need for new crop protection chemicals
2  Increase of cultivation in marginal land
3  Could increase the use of artificial fertilizers through the cultivation of what is now considered to be marginal land

4  Increase in extinction rates of biota
5  Reduction in extent of natural ecosystems
6  Impairment of ecosystem services
7  Increase in extent of irrigation

1  Increased soil erosion
2  Increased risks of desertification
3  Reduction in biodiversity

**INCREASED FOSSIL FUEL INPUTS AND THEREFORE A CONTRIBUTION TO GLOBAL WARMING**

**ENVIRONMENTAL DEGRADATION AND LOSS OF GENETIC RESOURCES**

**FIGURE 10.4** The advantages and disadvantages of agricultural biotechnology in relation to the environment. (Mannion, 1997)

improving food content, e.g. the vitamin E content of rice. There are many potential advantages but, as with most technologies, there are also disadvantages. These are summarized in Figure 10.4, which shows that genetic modification of crops and animals could, among other things, accelerate the loss of natural ecosystems.

Nevertheless, genetically modified crops have already replaced conventional crops in several countries. Although only a few such crops are commercially available, notably maize, cotton, soybean and rape (canola), they are already grown over large areas. Table 10.5 shows how rapidly genetically modified crops have spread. This trend is likely to continue at least in the next few years, unless the crops prove to be harmful to the environment or human health.

Genetic modification is also being applied to plant species used in plantation agriculture and forestry (see Section 10.3.4). In time, this may also influence land cover and land use.

### 10.3.4 Forests and forestry

Future land-cover and land-use changes likely to involve forests and forestry relate to the factors discussed above (Sections 10.3.2 to 10.3.4), and to national and international policies and agreements. First, global climatic change, through changing temperature and precipitation regimes, will influence the distribution of plant species and their organization into communities. Thus ecosystems of a different character to those of today will emerge. All remaining natural ecosystems will be affected, as discussed by IPCC (2001a), which also suggests that many already endangered species are likely to become extinct.

**Animal agriculture**
*Increases in productivity*
Somatotropins
Growth hormone-releasing factors
Cloning

*Animal health/therapeutics*
Vaccines
Diagnostics
Immune response modifiers
Parasiticides
Insecticides

*Other*
Bioreactors to produce chemicals of value in human healthcare
Organ production for transplants in humans

**Crop agriculture**
*Plants*
Insect resistance
Herbicide resistance
Disease resistance (to bacteria, viruses)
Altered secondary metabolites, e.g. production of sweeteners
Tolerance to environmental stress, e.g. drought, frost, salinity
Increased productivity through improved use of resources such as water, nutrients and light
Improved nutritional value, e.g. enhanced vitamin content
New crops
Biomass fuels
Production of pharmaceutical compounds

*Biopesticides*
Bioherbicides
Bioinsecticides
Biofungicides

*Diagnostics*
Detection of pathogen incidence

**TABLE 10.4** Some applications of biotechnology in agriculture (Mannion, 2001)

Moreover, disease, pests and incidence of fire will be influenced by climatic change and they, in turn, will influence ecosystem composition and function. Forests may extend to higher latitudes and altitudes than at present and may thus become bigger carbon sinks in the biosphere, though deforestation may offset this overall. In drier regions, increasing water shortages may prohibit tree growth and so productivity, and hence carbon storage, is likely to decrease in the dry middle and low latitudes. As observed by the World Commission on

Forests and Sustainable Development (1999), predictions of the effects of climatic change on the world's forest are equivocal and the larger the magnitude of change the less likely it is that predictions will be accurate, with surprise reactions and outcomes being the norm.

In terms of the human use of forests, rapidly increasing populations in many tropical countries (see Section 10.3.2) will put pressure on existing agricultural systems to expand (see Section 10.3.3). This will contribute to pressures for deforestation. Rates of deforestation are

| | 1997 ha × 10⁶ | 1998 ha × 10⁶ | 1999 ha × 10⁶ |
|---|---|---|---|
| USA | 8.1 | 20.5 | 28.7 |
| Argentina | 1.4 | 4.3 | 6.7 |
| Canada | 1.3 | 2.8 | 4.0 |
| Australia | 0.1 | 0.1 | 0.1 |
| Mexico | <0.1 | <0.1 | <0.1 |
| Spain | 0.0 | <0.1 | <0.1 |
| France | 0.0 | <0.1 | <0.1 |
| South Africa | 0.0 | <0.1 | 0.1 |
| Developed world | 9.5 | 23.4 | 32.72 |
| Developed world | 1.5 | 4.4 | 7.18 |
| Total | 11.0 | 27.8 | 39.9 |

Note: The total area for 2000 is likely to be $44.2 \times 10^6$ ha

**TABLE 10.5** The areas of genetically modified crops 1997–99 (James, 2000)

impossible to predict but pressures will be greatest in those middle- and low-latitude countries where population growth is high and which continue to rely heavily on biomass resources and agricultural goods for export earnings. These will be the least developed nations, mainly in Africa, where little industrialization is occurring. Logging and clear felling for wood and wood products will also contribute to deforestation in these regions. They will continue to be significant in Siberia, Canada, the USA and Scandinavia especially, where industries based on wood are very important. Afforestation programmes will offset loss of forest in these regions to some extent.

Afforestation is also important in other countries; for example, it is widespread in Europe and New Zealand where a great deal of natural forest cover has been lost historically. Indeed, afforestation is likely to increase worldwide for a variety of reasons: to provide wood and wood products locally; to reduce soil erosion; to contribute to water conservation; and to provide a sink for carbon. The latter, the so-called Kyoto forests, are likely to be an important stimulus if agreement is reached on an international convention to allocate greenhouse gas quotas per nation, as is the objective of the Kyoto Protocol established in 1997 (see the United Nations Framework Convention on Climate Change (UNFCCC), 2001, for details), as part of an international initiative on global climatic change (see Section 10.3.8). To date, there is much dissension as to the details of this agreement though it, or something like it, is essential to combat global climate change. Similarly, the designation of nature reserves, protected areas and national parks with defined rules for their use, and adequate enforcement of these rules, will all contribute to forest preservation.

The distribution of the world's forests is thus set to alter considerably in the next few decades with stimuli deriving from a wide range of sources.

## 10.3.5 Industrialization and development

Many nations are going through the process of industrialization, which is a major generator of wealth and a major component of the development process. The Industrial Revolution that characterized the UK and Europe in the eighteenth and nineteenth centuries is now taking place in many Asian and Latin American countries. This is significant in the context of land-cover and land-use change for several reasons. First, fossil-fuel consumption and thus greenhouse gas emissions will increase markedly, as is already evident for nations such as China, India and South Korea. These will influence land cover and land use through fossil-fuel extraction and the climatic repercussions of additional greenhouse gas emissions. Second, increasing standards of living often stimulate changes in dietary preferences; meat consumption and requirements for non-staple crops such as fruit generally increase, with implications for agricultural land use. Third, industrialization encourages urbanization (see Section 10.3.6) as employment is created. Fourth, it encourages further exploitation of mineral, agricultural and forest resources to provide the raw materials for industrial processing.

There are no absolute measures of industrialization or development, or unequivocal projections for the next two to five decades. However, two measures that provide some indication of industrialization are energy consumption and Gross Domestic Product (GDP, i.e. the total value of all goods produced and all services provided by a nation in one year: expressed herein on a per-capita basis). Global energy consumption data for 1973 and 1999 are given in Table 10.6. This shows that the greatest change in energy consumption in the past 25 years has been in the developing nations, especially in China and other parts of Asia. Whilst this in no way means that the developing world

is catching up with the developed world, the increase reflects the rapidity with which energy consumption, especially for industrialization, is occurring. The energy consumption values are paralleled by carbon dioxide emissions (see Table 10.7). Globally, these have increased significantly as energy consumption has increased. In relative terms, emissions from China and other industrializing parts of Asia have increased substantially whilst energy consumption and emissions have declined in the OECD countries, the world's largest energy users.

Energy consumption, and therefore carbon dioxide emissions, are set to increase substantially in the next few decades, as has been discussed by the Energy Information Administration (2001), data from which are given in Table 10.8. This shows that in the next 20 years, world energy demand is predicted to increase by a massive 59 per cent, with several developing regions at the forefront. The EIA points out that such projections are difficult to determine because so many factors contribute to energy-use patterns. However, the EIA's comparison of five different projections by various organizations shows that most estimates anticipate a 2.0–2.3 per cent increase annually. Only particularly slow rates of

| | % 1973 | % 1999 | % change |
|---|---|---|---|
| OECD countries | 62.6 | 52.6 | −16 |
| Non-OECD countries | 2.8 | 0.9 | −68 |
| Former USSR | 14.2 | 9.0 | −37 |
| Middle East | 2.0 | 3.7 | 85 |
| China | 5.7 | 11.4 | 100 |
| Asia* | 4.9 | 11.9 | 143 |
| Latin America | 5.0 | 5.1 | 2 |
| Africa | 2.8 | 5.4 | 93 |
| Total energy consumption | 4539 | 6753 | 49 |

Notes:
M tonnes of oil equivalent
*excludes China
OECD countries are members of the Organization for Economic Co-operation and Development: Australia, Austria, Belgium, Canada, Czech Republic, Denmark, Finland, France, Germany, Greece, Hungary, Iceland, Ireland, Italy, Japan, Korea, Luxembourg, Mexico, Netherlands, New Zealand, Norway, Poland, Portugal, Slovak Republic, Spain, Sweden, Switzerland, Turkey, UK and USA.

**TABLE 10.6** World energy consumption in 1973 and 1999, by region. (Adapted from International Energy Agency, 2001)

| | % 1973 | % 1999 | % change |
|---|---|---|---|
| OECD | 64.3 | 52.8 | −18 |
| Non-OECD Europe | 1.7 | 1.0 | −41 |
| Former USSR | 16.0 | 9.8 | −39 |
| Middle East | 1.1 | 3.8 | 245 |
| China | 5.9 | 13.9 | 197 |
| Asia* | 3.1 | 9.2 | 197 |
| Latin America | 2.6 | 3.7 | 42.3 |
| Africa | 1.7 | 3.2 | 88 |
| Total Emissions: | 16,191 | 23,172 | |
| Mega tonnes of carbon dioxide | | | |

*Notes:*
OECD countries (see notes on Table 10.6)

**TABLE 10.7** Carbon dioxide emissions in 1973 and 1999, by region. (Adapted from International Energy Agency, 2001)

economic growth would result in lower rates of increase. Inevitably, greenhouse gas emissions will also increase; in the case of carbon dioxide there is likely to be an increase from *c.* $6 \times 10^9$ to *c.* $9.8 \times 10^9$ tonnes carbon equivalent. However, these patterns and totals of energy consumption may well be altered if the Kyoto Protocol is adopted. Carbon dioxide quotas per country should encourage energy conservation as well as the provision of carbon sinks such as forests (see Section 10.3.4). Similarly, new energy sources, energy conservation schemes and the varying cost of fossil-fuel energy, notably oil, will all contribute to the exact nature of consumption and greenhouse gas emissions in the future.

| | % change 1999–2020 |
|---|---|
| Developing Asia | 129 |
| Latin America | 123 |
| Middle East | 93 |
| Africa | 77 |
| Eastern Europe/FSU | 43 |
| North America | 35 |
| Western Europe | 22 |
| Industrialized Asia | 22 |
| World | 59 |

**TABLE 10.8** Projected changes in energy demand by region 1999–2000 (Energy Information Administration, 2001)

Potential trends in GDP also provide a crude measure of industrialization, though these are as difficult to predict accurately as future energy-use patterns, to which they are closely linked. Table 10.9 gives data on likely per-capita GDP changes; these data parallel energy usage, with the largest increases being in Asia, notably in China and India, and Latin America. Growth in GDP will be influenced by all the same factors referred to above, as well as boom periods, recession and war. Nevertheless, the data show that the patterns of GDP will mirror those of today; the industrialized countries will remain the strongest performers and the richest, while the Sub-Saharan nations will remain the poorest by this measure.

Overall, the prognosis for the next 20–30 years involves substantial increases in energy requirements, with resulting increases in greenhouse gases, as economic growth continues. This pattern is not, however, globally uniform; rapidly developing regions such as Asia and Latin America will lead the way whilst others, notably many African countries, will show little growth in relative terms.

### 10.3.6 Urbanization

The world is already urbanizing rapidly, especially in developing countries. This trend will continue, partly because of the internal dynamism of population growth and partly

| **A   Gross Domestic Product** US$1000 per capita | | | | | |
|---|---|---|---|---|---|
| | **1990** | **2000** | **2010** | **2020** | **2030** |
| Industrialized nations | 19.5 | 22.87 | 27.22 | 32.08 | 37.61 |
| Reforming economy nations | 2.62 | 1.94 | 2.41 | 3.56 | 5.94 |
| Developing nations | 0.86 | 1.17 | 1.57 | 2.09 | 2.86 |
| World | 3.97 | 4.23 | 4.62 | 5.11 | 5.72 |

| **B   Gross Domestic Product** US$1000 per capita | | | | | **% change 1990–2030** |
|---|---|---|---|---|---|
| Latin America and Caribbean | 2.50 | 2.87 | 3.36 | 4.03 | 4.95 | 98 |
| Middle East and North Africa | 2.12 | 2.25 | 2.44 | 2.67 | 2.98 | 41 |
| Sub-Saharan Africa | 0.54 | 0.55 | 0.58 | 0.64 | 0.82 | 52 |
| Centrally planned Asia and China | 0.38 | 0.62 | 0.88 | 1.19 | 1.65 | 334 |
| Other Pacific-Asia | 1.53 | 2.38 | 3.77 | 5.50 | 7.49 | 390 |
| South Asia | 0.33 | 0.42 | 0.51 | 0.66 | 0.89 | 17 |

**TABLE 10.9** Actual GDP and projections until 2030. (Adapted from International Institute for Applied Systems Analysis, 2001)

because of continued migration from rural areas. In the next 20 years, many of the world's largest cities will be in the developing tropics and subtropics while the population of cities of developed nations will remain more or less static. These trends are, obviously, linked to patterns of global population change, as discussed in Section 10.3.2; they are also linked to trends in industrialization (see Section 10.3.5.).

Table 10.10 gives the Food and Agriculture Organisation's (2001) data on rural–urban population trends between 1975 and 2000, and projections for 2025. Between 1975 and 2000 there was a *c.* 10 per cent increase in urban population globally; a further *c.* 12 per cent increase is forecast for 2025, by which date urban populations will for the first time exceed rural populations. The data also show that urban populations in developed nations will continue to increase, adding a further $128 \times 10^6$ people. The increase in developing nations is, however, projected to be much higher: five times more people will live in urban areas by 2025 than in 1975, and overall there will be *c.* $600 \times 10^6$ more people in urban areas than in rural areas.

Consequently, urban spread will occur, with a resulting loss in agricultural land and remaining natural ecosystems. The production

| $\times 10^6$ | **1975** | | **2000** | | **2025** | |
|---|---|---|---|---|---|---|
| | **Urban** | **Rural** | **Urban** | **Rural** | **Urban** | **Rural** |
| Developed countries | 772.64 | 355.46 | 967.22 | 347.32 | 1095.43 | 264.42 |
| Developing countries | 768.88 | 2168.54 | 1879.32 | 2862.85 | 3519.86 | 2943.99 |
| Total world | 1541.52 | 2524.00 | 2846.54 | 3210.17 | 4615.29 | 3208.41 |
| % world | 37.9 | 62.1 | 46.9 | 53.1 | 58.9 | 41.1 |

**TABLE 10.10** The urban–rural distribution of world population in 1975, 2000 and 2025. (Adapted from Food and Agriculture Organisation, 2001b)

of carbon dioxide and other greenhouse gases will increase markedly just when the capacity of the biosphere to store carbon is reduced. The Ecological Footprint of those nations that experience this rapid urbanization will undoubtedly broaden and deepen. This will happen not only because of increased greenhouse emissions from industry and transport, but also because of the other pressures on the environment that are created by large agglomerations of people. Of these, waste disposal is particularly important; sewage, industrial and domestic waste all require treatment and/or disposal so these too will contribute to land-cover and land-use change. The increased provision of facilities for urban recreation and sport will also have an influence.

## 10.3.7 Communications and information technology

The past one hundred years have witnessed tremendous advances in communications generally. The advent of the telephone, radio, television and video has provided information on a scale hitherto never known. These developments have 'made the world smaller' in so far as knowledge about distant events, people and places is now disseminated rapidly; television facilitated not only the spread of information orally but also visually. Geography and history entered individual homes in Europe and North America. Globalization had taken a new turn and now only the poorest and remotest nations remain outside the influence of this technology. These media promote an understanding of environmental issues and the attractions, or otherwise, of distant places. Indeed, they have transformed the very concept of place. In their roles as broadcasters they promote travel and tourism, as well as highlighting the heterogeneity of people and place. This increased awareness, coupled with the travel experiences it encourages, is important in politicizing the environment and promoting conservation. These are important components of land-cover and land-use change (see Section 10.3.8).

Even more recent developments in communications and information technology are heightening environmental awareness as well as facilitating increased efficiency in energy use, fertilizer use etc. The advent and common use of computers in developed nations has opened up new avenues for obtaining environmental information. For example, international agencies such as the World Bank, the Food and Agriculture Organisation and the United Nations all have websites that promote their work and give free access to a vast range of data relating to the environment and society. Indeed, many of the tables compiled for this book have been derived from these sources. Similarly, governments and local councils as well as environmental pressure groups provide data that can now be annexed in seconds as compared with ten years ago when a visit to a reference library or government archive was essential. Nevertheless, it must be acknowledged that not all the world's people are equally well served. As in the case of most technologies, the developing nations are least well provisioned, though in the industrializing countries in particular, access to telephones and computer ownership are growing rapidly, as Table 10.11 shows. These data also indicate that the world is linking up rapidly through the internet. This growth in communications is likely to continue in the next few decades, with major growth occurring in Latin America, especially in Mexico, Argentina and Brazil, and in industrializing Asia, notably China, India, Thailand and South Korea.

Information technology has many other specific uses that can influence land-cover and land-use change. There are many applications in agriculture. For example, patch by patch or furrow by furrow, artificial fertilizer applications can be determined using probes that relay soil nitrate or phosphate availability to a computer on-board a tractor. Micronutrients can be delivered using the same approach. In addition, irrigation systems can control precisely the amount of water allowed to enter irrigation channels or is deployed by sprinkler systems. Such an approach conserves water as well as reducing waterlogging and salinization. In a broader context, technology such as satellite remote sensing and global positioning

| A Telephones/PCs | Africa | Latin America | North America | Asia | Europe | Oceania | World |
|---|---|---|---|---|---|---|---|
| (i) Mainline telephones per 100 people in 2000 | 2.49 | 24.21 | 68.81 | 9.54 | 39.51 | 40.58 | 16.18 |
| (ii) Cellular telephones per 100 people in 2000 | 1.98 | 11.59 | 34.12 | 6.59 | 36.58 | 35.83 | 12.15 |
| (iii) Estimated PCs per 100 people in 2000 | 1.06 | 7.33 | 48.76 | 2.88 | 16.33 | 42.14 | 7.61 |

**B Numbers of people online in 2000 ($\times 10^6$)**

| | Africa | Latin America | North America | Asia | Europe | Oceania | World |
|---|---|---|---|---|---|---|---|
| (i) According to ITU (2001) per 10,000 people in 2000 | 56.96 | 384.08 | 3797.93 | 329.39 | 1315.86 | 2539.69 | 600.50 |

| | Africa | Asia/ Pacific | Europe | Middle East | Canada and USA | Latin America | World |
|---|---|---|---|---|---|---|---|
| (ii) According to NUA (2001) **Total users** | 4.15 | 153.99 | 154.63 | 4.65 | 180.68 | 25.33 | 513.41 |

**TABLE 10.11** Data on telephone and personal computer ownership, and internet users. (Adapted from International Telecommunications Union, 2001; NUA, 2001)

provides information on rates and degrees of environmental change, e.g. the process of desertification or deforestation, and soil erosion. Such data facilitate the determination of specific areas at risk of degradation or deforestation etc., and thus are invaluable for planning purposes. In this context, projections for the future rely heavily on satellite-collected data and computerized analysis. These include the Intergovernmental Panel on Climate Change (IPCC) projections of the magnitude and impact of global climatic change discussed in Section 10.3.1.

### 10.3.8 Political and legal considerations

The environment has always been a political issue because resource ownership and use is at the heart of political power. In the past 50 years, this relationship has become increasingly overt at local, national and international levels. The same can be said about the relationship between the environment, and laws to use it and protect it. Individual nations contribute to the establishment of land-cover and land-use

patterns by designating National Parks, nature reserves etc., as referred to in Section 10.3.4, and by formulating conservation programmes generally. The establishment of such units and programmes is only effective, however, if there is strong legislation and adequate policing with appropriate deterrents for misuse.

Politically, land cover and land use have been profoundly altered by regional policies such as the European Union's Common Agricultural Policy (CAP), which first came into existence in the 1950s, and government resettlement programmes such as those of Brazil and Indonesia, which commenced in the 1980s and were intended to encourage development in remote regions. There are many other examples referred to in this book. However, the recognition of the environment as an issue of global significance is noteworthy; this began in the 1970s, which thus requires definition as a special decade. Again, this international tone reflects a trend towards increasing globalization but in a more positive sense than is generally understood by the term. The recognition here is not of homogenization but of global interlinkage; different physical environments give rise to different socio-economic systems, but notwithstanding these differences there remain strong connections.

This recognition is important; market forces, political policies, trade agreements and colonial interests in Europe, for example, all shape land cover and land use spatially and alter them on a temporal basis, not only within a given country but also wherever that nation makes an Ecological Footprint. This interlinkage underpins human appropriation of the global carbon cycle and reflects the complexity of the Ecological Footprint as one nation uses the ecological capacity of another within a mutual bond. (There are parallels with the interlinkages espoused in the Gaia hypothesis, see Section 1.3.) Not only does this link nations and regions, it also links individuals within nations. This relationship is epitomized by a conundrum: developed nations in Europe enjoy exotic fruit, houseplants of tropical origin, hardwoods such as mahogany and rosewood, cut flowers (and recreational drugs such as cocaine); they thus encourage the production of all these commodities, all of which are biomass- or carbon-based, whilst the producing countries are encouraged by their governments to continue production in order to earn foreign currency; this occurs even when there is insufficient production of staple crops to support indigenous populations who may become dependent on food aid.

Another aspect of politics/legislation in relation to land-cover and land-use change concerns international agreements and conferences. Such agreements are relatively recent in origin, beginning in the 1970s. The most important of these are listed in Table 10.12. Of these, the United Nations Conference on Environment and Development held in Rio de Janeiro in 1992 is particularly significant because it involved all nations and addressed a wide variety of topics, including climatic change and biodiversity loss, and initiated the formation of government agencies with environmental agendas (see O'Riordan, 2000, for details). The United Nations Framework Convention on Climate Change (UNFCCC) thus came into being and since 1992 there have

been several further meetings (see Table 10.12) focused on the establishment and implementation of greenhouse gas regulation. In 2001, widespread agreement was reached in relation to emission targets per nation but the new Bush administration in the USA refused to ratify the 1997 protocol established in Kyoto and to date there appears to be an impasse (for details of the Kyoto Protocol, see UNFCCC, 2001). In the light of strong evidence for climatic warming being under way (see Section 10.3.1), the sooner the world unites in a concerted effort to mitigate the problem the better!

## 10.4 Envoi

Land-cover change will continue to occur as a component of the Earth's dynamic geological and ecological history. Conversion of natural land cover to land use and subsequent land-use change will continue to occur as components of socio-economic behaviour. This will profoundly affect geological and biological processes at scales from the global to the local; these, in turn, will influence global climatic characteristics. For some types of change, the role of society will be obvious; the deforestation of an area for agriculture is a case in point. Some changes may be fuelled by factors such as an increasing need for export crops or a response to population increase. Distal and proximal causes of land-cover and land-use change are thus many and varied, and are often interlinked in complex ways. Not only are existential or financial, i.e. need or greed, elements involved in driving change, but political policies are often significant as well. Such factors are generally relatively transparent whilst others, e.g. cultural elements such as gender and religion, may be less obvious but no less important. The complex relationship between the driving forces means that predicting the spatial and temporal patterns of land cover and land use, even in the next few decades, is more or less impossible. What can be said, however, is that change rather than stability or continuity is the norm. The human dimension creates a dynamism that operates at much shorter timescales than those

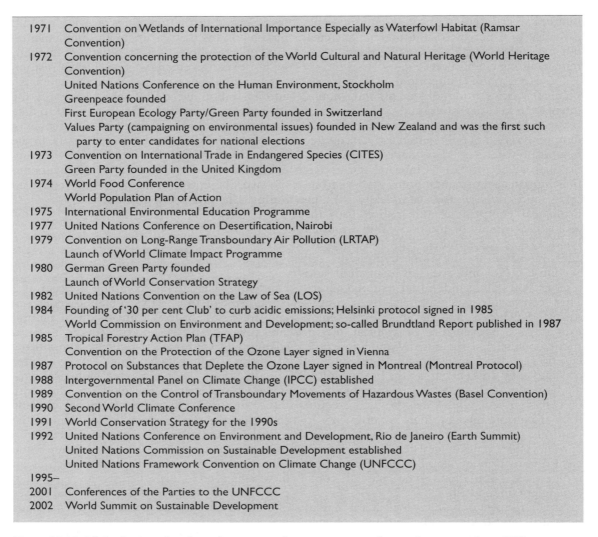

| 1971 | Convention on Wetlands of International Importance Especially as Waterfowl Habitat (Ramsar Convention) |
| 1972 | Convention concerning the protection of the World Cultural and Natural Heritage (World Heritage Convention) |
| | United Nations Conference on the Human Environment, Stockholm |
| | Greenpeace founded |
| | First European Ecology Party/Green Party founded in Switzerland |
| | Values Party (campaigning on environmental issues) founded in New Zealand and was the first such party to enter candidates for national elections |
| 1973 | Convention on International Trade in Endangered Species (CITES) |
| | Green Party founded in the United Kingdom |
| 1974 | World Food Conference |
| | World Population Plan of Action |
| 1975 | International Environmental Education Programme |
| 1977 | United Nations Conference on Desertification, Nairobi |
| 1979 | Convention on Long-Range Transboundary Air Pollution (LRTAP) |
| | Launch of World Climate Impact Programme |
| 1980 | German Green Party founded |
| | Launch of World Conservation Strategy |
| 1982 | United Nations Convention on the Law of Sea (LOS) |
| 1984 | Founding of '30 per cent Club' to curb acidic emissions; Helsinki protocol signed in 1985 |
| | World Commission on Environment and Development; so-called Brundtland Report published in 1987 |
| 1985 | Tropical Forestry Action Plan (TFAP) |
| | Convention on the Protection of the Ozone Layer signed in Vienna |
| 1987 | Protocol on Substances that Deplete the Ozone Layer signed in Montreal (Montreal Protocol) |
| 1988 | Intergovernmental Panel on Climate Change (IPCC) established |
| 1989 | Convention on the Control of Transboundary Movements of Hazardous Wastes (Basel Convention) |
| 1990 | Second World Climate Conference |
| 1991 | World Conservation Strategy for the 1990s |
| 1992 | United Nations Conference on Environment and Development, Rio de Janeiro (Earth Summit) |
| | United Nations Commission on Sustainable Development established |
| | United Nations Framework Convention on Climate Change (UNFCCC) |
| 1995– | |
| 2001 | Conferences of the Parties to the UNFCCC |
| 2002 | World Summit on Sustainable Development |

**TABLE 10.12** Major international conferences and agreements on the environment since 1970. (Adapted from Mannion, 1997, with additions)

of the natural dimension. Superimposition of the two produces a kaleidoscopic world with ever-changing but not always beautiful characteristics. The common denominator linking the past, present and future earth-surface characteristics, the organic with the inorganic, and people and environment is carbon (see Figures 10.1 and 10.2). The more humans appropriate carbon, the more land-cover/land-use change will occur. A corollary of this is that the more humans there are, the more carbon they require.

Globalization is just as much concerned with interlinkages worldwide as with standardization. Recognition of these interlinkages between nations and individuals worldwide, and between people and environment everywhere is a vital component of earth-surface management. The establishment of the environment as a political issue at the international level, with resulting international agreements, represents a major step in acknowledging this collectively. Moreover, full ratification of the Kyoto Protocol will reflect this global approach in so far as its repercussions will be widespread; no nation will remain unaffected by its mitigating effect on climate and the uncertainty of climatic change. Once again, a key component of Kyoto is carbon management.

It is important to recognize that in the past a great deal of technology has been designed to harness carbon, usually as either food or fuel. For these reasons, the initiation of agriculture *c.* 10,000 years ago, and the Industrial Revolution of the eighteenth and nineteenth centuries represent major thresholds in both carbon appropriation and humankind's development. Future technologies must address carbon management to minimize output to the atmosphere and to assist in the conservation of the world's remaining natural ecosystems. These are vital for the services of biogeochemical cycling, including that of carbon; and the preservation of biodiversity is essential to provide opportunities, at species and genetic scales, for the development of new biotic resources. Biotechnology, information technology, alternative fuel technology and many, many other technologies can make a major contribution to carbon management. They themselves will, however, generate land-cover and land-use change.

Finally, heterogeneity characterizes the Earth's surface and is the product of both nature and people. The wonder of nature is its capacity to give rise to so many life forms that are interlinked in a delicate balance; no atom of carbon is out of place. The wonder of humanity is the capacity to use its ingenuity to create order and harness carbon by carving landscapes out of wildscapes. The impact of human ingenuity on the biosphere can be considered either as a tremendous achievement, in so far as *c.* $6 \times 10^9$ people are supported within a wide range of latitudes and altitudes, or as a calamity because so much 'nature' has been lost and the biosphere irreparably degraded, whilst the repercussions of misplaced carbon may result in the uncertainty of climatic change. Land-cover and land-use patterns are the manifestation of a dynamic and irrevocable relationship between people and environment.

# Further reading

**Burroughs, W.J.** 2001: *Climate Change. A Multidisciplinary Approach.* Cambridge University Press, Cambridge.

**Carter, N.** 2001: *The Politics of the Environment.* Cambridge University Press, Cambridge.

**Evenson, R.E., Santaniello, V. and Zuberman, D.** (eds) 2002: *Economic and Social Issues in Agricultural Biotechnology.* CABI, Wallingford.

**Matthews, J.A.** (ed.) 2001: *The Encyclopaedic Dictionary of Environmental Change.* Arnold, London.

**Munn, T.** (ed.) 2001: *Encyclopaedia of Global Environmental Change.* Five vols. John Wiley and Sons, Chichester.

**Reddy, K.R. and Hodges, H.F.** (eds) 2000: *Climate Change and Global Crop Productivity.* CABI, Wallingford.

**Singh, R.B., Fox, J. and Himiyama, Y.** (eds) 2000: *Land Use and Cover Change.* Science Publishers Inc., Enfield, New Hampshire, USA.

# REFERENCES

**1vol.** 1995–97: History of Las Vegas, www.1vol.com (accessed 2 October 2001).

**Aagesen, D.** 2000: Crisis and conservation at the end of the world: sheep ranching in Argentine Patagonia. *Environmental Conservation* **27**, 208–15.

**Adesina, F.A., Siyanbola, W.O., Oketala, F.A., Pelemo, D.A., Momodu, S.A., Adegbulugbe, A.O. and Ojo, L.O.** 1999: Potential of agroforestry techniques in mitigating $CO_2$ emissions in Nigeria: some preliminary estimates. *Global Ecology and Biogeography* **8**, 163–73.

**Adesina, F.A., Mbila, D., Nkamleu, G.B. and Endamana, D.** 2000: Econometric analysis of the determinants of adoption of alley cropping by farmers in the forest zone of southwest Cameroon. *Agriculture, Ecosystems and Environment* **80**, 255–65.

**Akala. V.A. and Lal, R.** 2000: Potential of mine land reclamation for soil organic carbon sequestration in Ohio. *Land Degradation and Development* **11**, 229–97.

**allaboutcancun** 2001: All about Cancun, www.allabout-cancun.com (accessed 2 October 2001).

**Allen, A.** 2001: Containment landfills: the myth of sustainability. *Engineering Geology* **60**, 3–19.

**Andersen, B.G. and Borns, H.W.** 1994: *The Ice Age World.* Scandinavian University Press, Oslo, Copenhagen and Stockholm.

**Andersen, S.Th.** 1966: Interglacial vegetation succession and lake development in Denmark. *Palaeobotanist* **15**, 117–27.

**Anderson, J.R., Hardy, E.E., Roach, J.T. and Witmer, R.E.** 1976: A land use and land cover classification scheme for use with remote sensor data. *Geological Survey Professional Paper* 964. US Government Printing Office, Washington DC.

**Andrieu-Ponel, V., Ponel, P., Bruneton, H., Leveau, P. and de Beaulieu, J-L.** 2000: Palaeoenvironments and cultural landscapes of the last 2000 years reconstructed from pollen and coleopteran records in the Lower Rhône Valley, southern France. *The Holocene* **10**, 341–55.

**Anon.** 2000: *Aluminium and the Australian Economy.* A report to the Australian Aluminium Council, www.aluminium.org.au (accessed 15 May 2001).

**Archibold, O.W.** 1995: *Ecology of World Vegetation.* Chapman and Hall, London.

**Areal, A.C.B.** 1996: General information about Brasilia, www.geocities.com (accessed 31 August 2001).

**Ashley, P.M. and Lottermoser, B.G.** 1999: Arsenic contamination at the Mole River Mine, northern New South Wales. *Australian Journal of Earth Sciences* **46**, 861–74.

**Asselin, H., Fortin, M.-J. and Bergeron, Y.** 2001: Spatial distribution of late-successional coniferous species regeneration following disturbance in southwestern Quebec boreal forest. *Forest Ecology and Management* **140**, 29–37.

**Atherden, M.A. and Hall, J.A.** 1999: Human impact on vegetation in the White Mountains of Crete since AD 500. *The Holocene* **9**, 183–93.

**Atkinson, T.C., Briffa, K.R. and Coope, G.R.** 1987: Seasonal temperatures in Britain during the past 22,000 years, reconstructed using beetle remains. *Nature* **325**, 587–92.

**Atlas of Pakistan** 1986: Survey of Pakistan, Rawalpindi.

**Attorre, F., Bruno, M., Francesconi, F., Valenti, R. and Bruno, F.** 2000: Landscape changes of Rome through tree-lined roads. *Landscape and Urban Planning* **49**, 115–28.

**Avery, D.T.** 1997: Saving nature's legacy through better farming. *Issues in Science and Technology* **14**, 59–64.

**Aweto, A.O.** 2001: Trees in shifting and continuous cultivation farms in Ibadan area, southwestern Nigeria. *Landscape and Urban Planning* **53**, 163–71.

**Baldwin, J.** 2000: Tourism development, wetland degradation and beach erosion in Antigua, West Indies. *Tourism Geographies* **2**, 193–218.

**Bar-Yosef, O. and Belfer-Cohen, A.** 2001: From Africa to Eurasia – early dispersals. *Quaternary International* **75**, 19–28.

**Barnola, J.M., Raynaud, D., Korotkevich, Y.S. and Lorius, C.** 1987: Vostok ice core provides 160,000-year record of atmospheric $CO_2$. *Nature* **329**, 408–14.

**Barrett, J.** 2001: Component ecological footprint: developing sustainable scenarios. *Impact Assessment and Project Appraisal* **19**, 107–18.

**Bassett, T.J. and Koli Bi, Z.** 1999: Environmental discourses and the Ivorian savanna. *Annals of the Association of American Geographers* **90**, 67–95.

**Beach, H.** 1988: *The Saami of Lapland.* Minority Rights Group, London.

**Beerling, D.J.** 1999: New estimates of carbon transfer to terrestrial ecosystems between the last glacial maximum and the Holocene. *Terra Nova* **11**, 162–7.

**Bekenov, A.B., Gracher, A. and Milner-Gulland, E.J.** 1998: The ecology and management of the Saiga antelope in Kazakhstan. *Mammal Review* **28**, 1–52.

**Bell, F.G., Genske, D.D. and Bell, A.W.** 2000:

Rehabilitation of industrial areas: case histories from England and Germany. *Environmental Geology* 40, 121–34.

**Bell, F.G., Bullock, S.E.T., Hälbich, T.F.J. and Lindsay, P.** 2001: Environmental impacts associated with an abandoned mine in the Witbank Coalfield, South Africa. *International Journal of Coal Geology* 45, 195–216.

**Bell, S.J., Barton, A.F.M. and Stocker, L.J.** 2001: Agriculture for health and profit in Western Australia: the western oil mallee project. *Ecosystem Health* 7, 116–21.

**Beresford, Q.** 2001: Developmentalism and its environmental legacy: the western Australian wheatbelt. 47, 403–15.

**Binkley, D. and Resh, S.S.** 1999: Rapid changes in soils following eucalyptus afforestation in Hawaii. *Soil Science Society of America* 63, 222–5.

**Birks, H.J.B.** 1986: Late-Quaternary biotic changes in terrestrial and lacustrine environments, with particular reference to north-west Europe. In Berglund, B.E. (ed.), *Handbook of Holocene Palaeoecology and Palaeohydrology*. John Wiley and Sons, Chichester, 3–65.

**Bottema, S.** 1980: Palynological investigations on Crete. *Review of Palaeobotany and Palynology* 31, 193–217.

**Bouwman, L. and van Vuuren, D.** 1999: *Global Assessment of Acidification and Eutrophication of Natural Ecosystems*. United Nations Environmental Programme/Ruksinstituut voor Volksgezondheid en Milieu, Nairobi and Bilthoven (see also www.rivm.nl/env/int/geo).

**Bowman, D.M.J.S.** 2000: *Australian Rainforests. Islands of Green in a Land of Fire*. Cambridge University Press, Cambridge.

**BP Amoco** 2001: *Statistical Review of World Energy 2001*, www.bp.com (accessed 3 July 2001).

**Bradley, R.S.** 1999: *Paleoclimatology: Reconstructing Climates of the Quaternary*, 2nd edition. Academic Press, San Diego, USA.

**Bradshaw, A.** 2000: The use of natural processes in reclamation – advantages and difficulties. *Landscape and Urban Planning* 51, 89–100.

**Bregman, M.** 2000: Can Alaska heal? (Aftermath of *Exxon Valdez* accident) *Science World*, 10 April.

**Briggs, J. and Mwamfupe, D.** 2000: Peri-urban development in an era of structural adjustment in Africa: the city of Dar es Salaam. *Urban Studies* 37, 797–809.

**British Geological Survey** 2000: *Collation of the Results of the 1997 Aggregate Minerals Survey for England and Wales*. British Geological Survey, Nottingham.

**British Geological Survey** 2001: *World Mineral Statistics 1995–1999*. British Geological Survey, Nottingham.

**Brook, E.J., Sowers, T. and Orchardo, J.** 1996: Rapid variations in atmospheric methane concentration during the past 110,000 years. *Science* 273, 1089–91.

**Bucher, E.H. and Huszar, P.C.** 1999: Sustainable management of the Gran Chaco of South America: ecological promise and economic constraints. *Journal of Environmental Management* 57,99–108.

**Bulte, E.H., Bouman, B.A.M., Plant, R.A.J., Nieuwenhuyse, A. and Jansen, H.G.P.** 2000: The economics of soil nutrient stocks and cattle ranching in the tropics: optimal pasture degradation in humid Costa Rica. *European Review of Agricultural Economics* 27, 207–26.

**Bunting, S.W.** 2001: Appropriation of environmental goods and services by aquaculture: a reassessment employing the ecological footprint methodology and implications for horizontal integration. *Aquaculture Research* 32, 605–9.

**Burger, A.** 2001: Agricultural development and land concentration in a Central European country: a case study of Hungary. *Land Use Policy* 18, 259–68.

**Bush, M.B., Piperno, D.R. and Colinvaux, P.A.** 1989: A 6000-year history of Amazonian maize cultivation. *Nature* 340, 303–5.

**Bush, M.B., Miller, M.C., De Oliveira, P.E. and Colinvaux, P.A.** 2000: Two histories of environmental change and human disturbance in eastern lowland Amazonia. *The Holocene* 10, 543–53.

**Butler, R.W.** 2000: Tourism and the environment: a geographical perspective. *Tourism Geographies* 2, 337–58.

**Caballero, R.** 1999: Castile – La Mancha: a once traditional and integrated cereal-sheep farming system under change. *American Journal of Alternative Agriculture* 14, 188–92.

**Calow, P.** (ed.) 1998: *The Encyclopedia of Ecology and Environmental Management*. Blackwell, Oxford.

**Campbell, D.J., Gichohi, H. Mwangi, A. and Chege, L.** 2000: Land use conflict in Kajiado District, Kenya. *Land Use Policy* 17, 337–48.

**Campbell-Culver, M.** 2001: *The Origin of Plants*. Headline, London.

**Cavalli-Sforza, L.L., Menozzi, P. and Piazza, A.** 1994: *The History and Geography of Human Genes*. Princeton University Press, Princeton, New Jersey.

**Cavalli-Sforza, L.L. and Cavalli-Sforza, F.** 1995: *The Great Human Diasporas*. Addison-Wesley, Reading, USA.

**Chambers, N, Simmons, C. and Wackernagel, M.** 2000: *Sharing Nature's Interest. Ecological Footprints as an Indicator of Sustainability*. Earthscan, London.

**Chappellaz, J., Barnola, J.M., Raynaud, D., Korotkevich, Y.S. and Lorius, C.** 1990: Ice-core record of atmospheric methane over the past 160,000 years. *Nature* 345, 127–31.

**Cherry, D.S., Currie, R.J., Soucek, D.J., Latimer, H.A. and Trent, G.C.** 2001: An integrative assessment of a watershed impacted by abandoned mined land discharges. *Environmental Pollution* 11, 377–88.

**Clement, R.M. and Horn, S.P.** 2001: Pre-Columbian land-use history in Costa Rica: a 3000-year record of forest clearance, agriculture and fires from laguna Zoncho. *The Holocene* 11, 419–26.

**Cocklin, C. and Keen, M.** 2001: Urbanization in the Pacific: environmental change, vulnerability and human security. *Environmental conservation* 27, 392–403.

**Colhoun, E.A.** 2000: Vegetation and climate during the last interglacial–glacial cycle in western Tasmania, Australia. *Palaeogeography Palaeoclimatology Palaeoecology* 155, 195–209.

**Colinvaux, P.A. and De Oliveira, P.E.** 2001: Amazon plant diversity and climate through the Cainozoic. *Palaeogeography, Palaeoclimatology, Palaeoecology* 166, 51–63.

**Commonwealth War Graves Commission** 2001: Commonwealth war graves, www.cwgc.org (accessed 6 September 2001).

**Craig, J.R., Vaughan, D.J. and Skinner, B.J.** 1996: *Resources of the Earth. Origin, Use and Environmental Impact*. Prentice Hall, New Jersey, USA.

**Critchley, C.N.R. and Fowbert, J.A.** 2000: Development of vegetation on set-aside land for up to nine years from a natural perspective. *Agriculture, Ecosystems and Environment* 79, 159–74.

**Crowley, G.M. and Garnett, S.T.** 2000: Changing fire

management in the pastoral lands of Cape York Peninsula of northeast Australia, 1623 to 1996. *Australian Geographical Studies* **38**, 10–26.

**Curragh** 2001: The Curragh racecourse, www.curragh.ie (accessed 28 September 2001).

**De Laet, S.J.** 1994: From the beginnings of food production to the first state. In De Laet, S.J., Dani, A.H., Lorenzo, J.L. and Nunoo, R.B. (eds), *History of Humanity, Volume I: Prehistory and the Beginnings of Civilization*. Routledge, London and UNESCO, Paris, 490–500.

**De Laet, S.J., Dani, A.H., Lorenzo, J.L. and Nunoo, R.B.** (eds) 1994: *History of Humanity, Volume I: Prehistory and the Beginnings of Civilization*. Routledge, London and UNESCO, Paris.

**Degn, H.J.** 2001: Succession to farmland to heathland: a case for conservation of nature and historic farming methods. *Biological Conservation* **97**, 319–30.

**Diamond Registry** 2001: World natural diamond production in 1998 (the Diamond Registry), www.diamondregistry.com (accessed 11 May 2001).

**Donnelly, L.J., De La Cruz, H., Asmar, I., Zapata, O. and Perez, J.D.** 2001: The monitoring and prediction of mining subsidence in the Amaga, Angelopolis, Venecia and Bolombolo Regions, Antioquia, Colombia. *Engineering Geology* **59**, 103–14.

**Dougill, A.J., Thomas, D.S.G. and Heathwaite, A.L.** 1999: Environmental change in the Kalahari: integrated land degradation studies for nonequilibrium dryland environments. *Annals of the Association of American Geographers* **89**, 420–42.

**Downey, P.O. and Smith, J.M.B.** 2000: Demography of the invasive shrub Scotch broom (*cytisus scoparius*) at Barrington Tops, New South Wales: insights for management. *Austral Ecology* **25**, 477–85.

**Dull, R.A.** 1999: Palyological evidence for 19th century grazing-induced vegetation change in the southern Sierra Nevada, California, USA. *Journal of Biogeography* **26**, 899–912.

**Dupont, L.M., Jahns, S., Marret, F. and Ning, S.** 2000: Vegetation change in equatorial West Africa: time slices for the last 150 ka. *Palaeogeography Palaeoclimatology Palaeoecology* **155**, 95–122.

**Eden Project** 2001: Eden Project, www.edenproject.com (accessed 28 June 2001).

**Efendiyeva, I.M.** 2000: Ecological problems of oil exploitation in the Caspian Sea area. *Journal of Petroleum Science and Engineering* **28**, 227–31.

**Elliott, T.I.** 2000: Reclamation of contaminated land affected by shallow mining in the Black Country, England. *Transactions of the Institute of Mining and Metallurgy. Section A: Mining Technology* **109**, 195–201.

*Encarta* 2001: *Toronto*. Microsoft® *Encarta*® online encyclopedia, http://encarta.msn.com (accessed 3 September 2001).

**Energy Information Administration (EIA)** 2001: World energy consumption, www.eia.doe.gov (accessed 7 November 2001).

**Envibase Project** 2001a: The municipality of Athens, www.stadentwicklung.berlin.de (accessed 19 July 2001).

**Envibase Project** 2001b: Rome land use, www.stadentwicklung.berlin.de (accessed 19 July 2001).

**Environmental Protection Agency** 2001: Municipal solid waste, www.epa.gov (accessed 20 September 2001).

**Euromines** 2001: Iron ore production: the world and Europe (Euromines), www.euromines.org (accessed 24 May 2001).

**Faramarzi, M.T.** 1997: *A Travel Guide to Iran*. Yassaman Publications, Tehran.

**Fattorini, M.** 2001: Establishment of transplants on machine-graded ski runs above the timberline in the Swiss Alps. *Restoration Ecology* **9**, 119–26.

**Fearnside, P.M.** 1999: Forests and global warming mitigation in Brazil: opportunities in the Brazilian forest sector for responses to global warming under the 'clean development mechanism'. *Biomass and Bioenergy* **16**, 171–89.

**Fearnside, P.M.** 2001: Soybean cultivation as a threat to the environment in Brazil. *Environmental Conservation* **28**, 23–38.

**Fensham, R.J. and Fairfax, R.J.** 1997: The use of the land survey record to reconstruct pre-European vegetation patterns in the Darling Down, Queensland, Australia. *Journal of Biogeography* **14**, 827–36.

**Fifeguide** 2001: Golf in Fife, www.fifeguide.co.uk (accessed 2 October 2001).

**Firman, T.** 2000: Rural to urban land conversion in Indonesia during boom and bust periods. *Land Use Policy* **17**, 13–20.

**Fisher, A.M. and Harris, S.J.** 1999: The dynamics of tree cover change in a rural Australian landscape. *Landscape and Urban Planning* **45**, 193–207.

**Fjellstad, W.J. and Dramstad, W.E.** 1999: Patterns of change in two contrasting Norwegian agricultural landscapes. *Landscape and Urban Planning* **45**, 177–91.

**Food and Agriculture Organisation** 2001a: *State of the World's Forests*. FAO statistics, www.fao.org (accessed 21 January 2001).

**Food and Agriculture Organisation** 2001b: FAO statistics, www.fao.org (accessed 15 November 2001).

**Fowler, S.V., Syrett and Hill, R.L.** 2000: Success and safety in the biological control of environmental weeds in New Zealand. *Austral Ecology* **25**, 553–62,.

**Francois, L.M., Godderis, Y. Warnant, P., Ramstein, G., de Noblet, N. and Lorenz, S.** 1999: Carbon stocks and isotopic budgets of the terrestrial biosphere at mid-Holocene and last glacial maximum times. *Chemical Geology* **159**, 163–89.

**Frenot, Y., Gloaguen, J.C., Massé, L. and Lebouvier, M.** 2001: Human activities, ecosystem disturbance and plant invasions in subantarctic Crozet, Kerguelen and Amsterdam Islands. *Biological Conservation* **101**, 33–50.

**Fritz, G.J.** 1995: New dates and data on early agriculture. The legacy of complex hunter-gatherers. *Annals of the Missouri Botanical Garden* **82**, 3–15.

**Frolova, A.** 1998: Ecological reasoning: ethical alternatives. *Ecological Economics* **24**, 169–82.

**Fukamachi, K., Oku, H., Kumagai, Y. and Shimomura, A.** 2000: Changes in landscape planning and land management in Arashiyama National Forest in Kyoto. *Landscape and Urban Planning* **52**, 73–87.

**Gad, G. and Holdsworth, D.W.** 1990: The emergence of corporate Toronto. In Kerr, D. and Holdsworth, D.W. (eds), *Historical Atlas of Canada, Volume III: Addressing the Twentieth Century 1991–1961*. University of Toronto Press, Toronto.

**Gamble, C.** 1999: *The Palaeolithic Societies of Europe*. Cambridge University Press, Cambridge.

**Garbini, S. and Schweinfurth, S.P.** 1986: Coal areas of the conterminous United States. *US Geological Survey Circular* 979.

**Geoapikonisis** 2001: www.geoapikonisis.gr (accessed 20 November 2001).

**Geocities** 2001: Islamabad, www.geocities.com/prpakistan_islamabad (accessed 3 September 2001).

**Global Forest Watch: Canada** 2001: Canada's forests, www.globalforestwatch.org/english/canada (accessed 19 March 2001).

**Godwin, H.** 1975: *The History of the British Flora: A Factual Basis for Phytogeography.* Cambridge University Press, Cambridge.

**Guariguata, M.R., Rosales Adame, J.J. and Finegan, B.** 2000: Seed removal and fate in two selectively logged lowland forests with contrasting protection levels. *Conservation Biology* **14**, 1046–54.

**Guevara, J.C., Shasi, C.R., Wuilloud, C.F. and Estevez, O.R.** 1999: Effects of fire on rangeland vegetation in southwestern Mendoza plains (Argentina): composition, frequency, biomass, productivity and carrying capacity. *Journal of Arid Environments* **41**, 27–35.

**Hadley Centre** 2001: Climate change, www.met-office.gov.uk (accessed 7 November 2001).

**Hall, P.** 1999: *Cities in Civilization.* Phoenix Giant, London.

**Hall, T.E. and Farrell, T.A.** 2001: Fuelwood depletion at wilderness campsites: extent and potential ecological significance. *Environmental Conservation* **28**, 241–47.

**Hannah, L., Lohse, D., Hutchinson, C., Carr, J.L. and Lankerani, A.** 1994: A preliminary inventory of human disturbance of world ecosystems. *Ambio* **23**, 246–50.

**Hannah, L., Carr, J.L. and Landerani, A.** 1995: Human disturbance and natural habitat: a biome level analysis. *Biodiversity and Conservation* **4**, 128–55.

**Haraldsson, H.V., Ranhagen, U and Sverdup, H.** 2001: Is eco-living more sustainable than conventional living? Comparing sustainability performances between two townships in southern Sweden. *Journal of Environmental Planning and Management* **44**, 663–79.

**Harlan, J.R.** 1992: *Crops and Man.* American Society of Agronomy and Crop Science Society of America, Madison, USA.

**Harvey, L.D.D.** 2000: *Global Warming. The Hard Science.* Prentice Hall, Harlow.

**Heathcote, R.L.** (ed.) 1988: *The Australian Experience. Essays in Australian Land and Resource Management.* Longman Cheshire, Melbourne.

**Heathcote, R.L.** 1994: *Australia,* 2nd edition. Longman, London.

**Hettler, J., Irion, G. and Lehmann, B.** 1997: Environmental impact of mining waste disposal on a tropical lowland river system: a case study on the Ok Tedi Mine, Papua New Guinea. *Mineralium Deposita* **32**, 280–91.

**Hietala-Koivu, R.** 1999: Agricultural landscape change: a case study in Yläne, southwest Finland. *Landscape and Urban Planning* **46**, 103–8.

**Higham, N.** 1992: *Rome, Britain and the Anglo Saxons.* Sealy, London.

**Hill, R., Griggs, P. and Bubu Ngadimunku Incorporated** 2000: Rainforests, agriculture and Aboriginal fire regimes in wet tropical Queensland, Australia. *Australian Geographical Studies* **38**, 138–57.

**Hillman, G, Hedges, R., Moore, A., Colledge, S. and Pettit, P.** 2001: New evidence of lateglacial cereal cultivation at Abu Hureyra on the Euphrates. *The Holocene* **11**, 383–93.

**Himiyama, Y.** (ed.) 1992: *Land Use Change in Modern Japan.* GIS for Environmental Change Research Project, Ministry of Education and Culture, Tokyo.

**Hinds, P.W.** 2001: Restoration following bauxite mining in Western Australia, www.hort.agri.umn.edu/h5015/99papers/hinds.html (accessed 6 April 2001).

**Ho, P.** 2000: China's rangelands under stress: a comparative study of pasture commons in the Ningxia Hui Autonomous Region. *Development and Change* **31**, 385–412.

**Holmes, P.M., Richardson, D.M., van Wilgen, B.W. and Gelderblom, C.** 2000: Recovery of African fynbos vegetation following alien woody plant clearing and fire implications for restoration. *Austral Ecology* **25**, 631–9.

**Hong Kong Government** 2001a: Overview of sewage infrastructure, www.info.gov.hk (accessed 20 September 2001).

**Hong Kong Government** 2001b: Recreation and sports facilities, www.ksd.gov.hk (accessed 28 September 2001).

**Hooghiemstra, H.** 1989: Quaternary and upper Pliocene glaciations and forest development in the tropical Andes: evidence from a long high-resolution pollen record from the sedimentary basin of Bogotá, Colombia. *Palaeogeography Palaeoclimatology Palaeoecology* **72**, 11–26.

**Houghton, R.A. and Hackler, J.L.** 1999: Emissions of carbon from forestry and land-use change in tropical Asia. *Global Change Biology* **5**, 481–92.

**Houghton, R.A. and Hackler, J.L.** 2000, Changes in terrestrial carbon storage in the United States I: The roles of agriculture and forestry. *Global Ecology and Biogeography* **9**, 125–44.

**Houghton, R.A., Hackler, J.L. and Lawrence, K.T.** 2000: Changes in terrestrial carbon storage in the United States 2: The role of fire and fire management. *Global Ecology and Biogeography* **9**, 145–70.

**Houghton, R.A., Skole, D.L., Nobre, C.A., Hackler, J.L., Lawrence, K.T. and Chomentowster, W.H.** 2000: Annual fluxes of carbon from deforestation and regrowth in the Brazilian Amazon. *Nature* **403**, 301–4.

**Imbernon, J.** 1999: Changes in agricultural practice and landscape over a 60-year period in North Lampung, Sumatra. *Agriculture, Ecosystems and Environment* **76**, 61–6.

**Inogwabini, B.-I., Hall, J.S., Vedder, A., Curran, B., Yamagiwa, J. and Basabose, K.** 2000: Status of large mammals in the mountain sector of Kahuzi-Biega National Park, Democratic Republic of Congo. *East African Wildlife Society* **38**, 269–76.

**Intergovernmental Panel on Climate Change** 2001a: *Climate Change 2001: The Scientific Basis.* Cambridge University Press, Cambridge (see also www.ipcc.ch).

**Intergovernmental Panel on Climate Change** 2001b: *Climate Change 2001: Impacts, Adaptation and Vulnerability.* Cambridge University Press, Cambridge (see also www.ipcc.ch).

**Intergovernmental Panel on Climate Change** 2001c: *Climate Change 2001: Mitigation.* Cambridge University Press, Cambridge (see also www.ipcc.ch).

**International Energy Agency (IEA)** 2001: Key world energy statistics from the IEA, www.iea.org (accessed 7 November 2001).

**International Institute for Applied Systems Analysis** 2001: Indicators: GDP/PPP per capita, www.iiasa.ac.at (accessed 22 November 2001).

**International Tanker Owners Pollution Federation Ltd** 2000: Tanker oil spill statistics, www.ITOPF.com (accessed 3 July 2001).

**International Telecommunications Union** 2001: Basic indicators, cellular subscribers and information technology, www.itu.int (accessed 16 November 2001).

**Inwood, S.A.** 1998: *History of London*. Macmillan, London.

**Irish Racing** 2001: Irish racecourses, www.irish-racing.com (accessed 28 September 2001).

**Islam, K.R., Kamaluddin, M., Bhuiyan, M.K. and Badruddin, A.** 1999: Comparative performance of exotic and indigenous forest species for tropical semievergreen degraded forest land reforestation in Chittagong, Bangladesh. *Land Degradation and Development* **10**, 241–9.

**Iversen, J.** 1958: The bearing of glacial and interglacial epochs on the formation and extinction of plant taxa. *Uppsala Universiteit Arsk* **6**, 210–15.

**Jain, A., Rai, S.C. and Sharma, E.** 2000: Hydro-ecological analysis of a sacred lake watershed system in relation to land-use/cover change from Sikkim Himalaya. *Catena* **40**, 263–78.

**James, C.** 2000: *Global Status of Commercialized Transgenic Crops: 1999*. International Service for the Acquisition of Agri-biotech Applications, Brief No. 17.

**Jeans, D.N.** (ed.) 1987: *Australia: A Geography, Vol. 2: Space and Society*. Sydney University Press, Sydney.

**Jessen, K. and Milthers, V.** 1928: Stratigraphical and palaeontological studies of interglacial fresh water deposits in Jutland and northwest Germany. *Danmarks Geologiske Undersøgelse, Series III*, **48**, 1–379.

**Ji, C.Y., Liu, Q., Sun, D., Wang, S., Lin, P. and Li, X.** 2001: Monitoring urban expansion with remote sensing in China. *International Journal of Remote Sensing* **22**, 1441–55.

**Johnson, G.E. and Woon, Y.-F.** 1997: Rural development patterns in post-reform China: the Pearl River delta region in the 1990s. *Development and Change* **28**, 731–51.

**Jolly, D., Prentice, I.C., Bonnefille, R., Ballouche, A., Bengo, M., Brenac, P., Buchet, G., Burney, D., Cazet, J.-P., Cheddadi, R., Edorh, T., Elenga, H., Elmoutaki, S., Guiot, J., Laarif, F., Lamb, H., Lezine, A.-M., Maley, J., Mbenza, M., Peyron, O., Reille, M., Reynaud-Farrera, I., Riollet, G., Ritchie, J.C., Roche, E., Scott, L., Ssemmanda, I., Straka, H., Umer. M., Van Campo, E., Vilimumbalo, S., Vincens, A. and Waller, M.** 1998: Biome reconstruction from pollen and plant macrofossil data for Africa and the Arabian peninsula at 0 and 6000 years. *Journal of Biogeography* **25**, 1007–27.

**Jones, M and Brown, T.** 2000: Agricultural origins: the evidence of modern and ancient DNA. *The Holocene* **10**, 769–76.

**Jouzel, J., Lorius, C., Petit, J.-R., Genthon, C., Barkov, N.I., Kotlyakov, V.M. and Petrov, V.M.** 1987: Vostok ice core: a continuous isotope temperature record over the last climatic cycle (160,000 years), *Nature* **329**, 403–8.

**Kalipeni, E. and Feder, D.** 1999: A political ecology perspective on environmental change in Malawi with the Blantyre Fuelwood Project Area as a case study. *Politics and the Life Sciences* **18**, 37–54.

**Kalyuzhnova, Y., Jaffe, A.M., Lynch, D. and Sickles, R.** (eds) 2001. *Energy in the Caspian Region: Present and Future*. Palgrave, Basingstoke.

**Kammerbauer, J. and Ardon, C.** 1999: Land use dynamics and landscape change pattern in a typical watershed in the hillside region of central Honduras. *Agriculture, Ecosystems and Environment*, **75**, 93–100.

**Khan, N.A.** 1998: Land tenurial dynamics and partici-pating forestry management in Bangladesh. *Public Administration and Development* **18**, 335–47.

**Kibreab, G.** 1997: Environmental causes and impact of refugee movements: a critique of the current debate. *Disasters* **21**, 20–38.

**Klooster, D.** 1999: Community-based forestry in Mexico: can it reverse processes of degradation? *Land Degradation and Development* **10**, 365–81.

**Knight, J.** 2000: From timber to tourism: recommoditizing the Japanese forest. *Development and Change* **31**, 341–59.

**Knoche, D., Embacher, A. and Katzur, J.** 2000: Element dynamics of oak ecosystems on acid-sulphurous mine soils in the Lusatian lignite mining district (Eastern Germany). *Landscape and Urban Planning* **51**, 113–22.

**Koch, J.M., Ward, S.C., Grant, C.D. and Ainsworth, G.L.** 1996: Effects of bauxite mine restoration operations on topsoil seed reserves in the jarrah forest of Western Australia. *Restoration Ecology* **4**, 368–76.

**Kumpula, J., Colpaert, A. and Nieminen, M.** 2000: Condition, potential recovery rate, and productivity of Lichen (*Cladonia* spp.) ranges in the Finnish reindeer management area. *Arctic* **53**, 152–60.

**Lal, R.** (ed.) 1999: *Soil Quality and Soil Erosion*. Soil and Water Conservation Society, Ankeny and Iowa, and CRC Press, Boca Raton, USA.

**Laurance, W.F.** 2000: Tropical logging and human invasions. *Conservation Biology* **15**, 4–5.

**Laurance, W.F., Vasconcelos, H.L. and Lovejoy, T.E.** 2000: Forest loss and fragmentation in the Amazon: implications for wildlife conservation. *Oryx* **34**, 39–45.

**Leopardstown** 2001: Track/history, www.leopardstown.com (accessed 28 September 2001).

**Lewin, R.** 1999: *Human Evolution*, 4th edition. Blackwell, Oxford.

**Lilieholm, R.J. and Romney, L.R.** 2000: Tourism, National Parks and wildlife. In Butler, R.W. and Boyd, S.W. (eds), *Tourism and National Parks. Issues and Implications*. John Wiley and Sons, Chichester, 137–51.

**Lillesand, T.M. and Kiefer, R.W.** 2000: *Remote Sensing and Image Interpretation*, 4th edition. John Wiley and Sons, New York, USA.

**Lin, G.C.S.** 2001: Metropolitan development in a transitional socialist economy: spatial restructuring in the Pearl River Delta, China. *Urban Studies* **38**, 383–406.

**Lindenmayer, D.B. and McCarthy, M.A.** 2001: The spatial distribution of non-native plant invaders in a pine-eucalypt landscape mosaic in south-eastern Australia. *Biological Conservation* **102**, 77–87.

**Lonely Planet** 2001: *Buenos Aires*. City Map Series, Lonely Planet.

**Lopez, E., Bocco, G., Mendoza, M. and Duhau, E.** 2001: Predicting land-cover and land-use change in the urban fringe. A case in Morelia City, Mexico. *Landscape and Urban Planning* **55**, 271–85.

**Lorius, C., Jouzel, J., Ritz, C., Merlivat, L., Barkov, N.I., Korotkevich, Y.S. and Kotlyakov, V.M.** 1985: A 150,000-year climatic record from Antarctic ice. *Nature* **316**, 591–6.

**Lovelock, J.E.** 1995: *The Ages of Gaia*. Oxford University Press, Oxford.

**Lowe, J.J. and Walker, M.J.C.** 1997: *Reconstructing Quaternary Environments*, 2nd edition. Longman, Harlow.

**Lowe, J.J., Hoek, W.Z. and INTIMATE group** 2001: Inter-regional correlation of palaeoclimatic records for the last glacial–interglacial transition: a protocol for improved precision recommended by the INTIMATE project group. *Quaternary Science Reviews* **20**, 1175–87.

**Lütz, M. and Bastian, O.** 2002 (in press): Implementation of landscape planning and nature conservation in the agricultural landscape – a case study from Saxony. *Agriculture, Ecosystems and Environment.*

**McClain, H.M.** 2001: Oil fires and spills leave hazardous legacy, http://europe.cnn.com (accessed 21 September 2001).

**McGee, T.T.** 1989: Urbanisasi or kotadesasi? Evolving patterns of urbanization in Asia. In Costa, F.J., Dutt, A.K., Ma, L.T.C. and Noble, A.G. (eds), *Urbanization in Asia: Spatial Dimensions and Policy Issues.* University of Hawaii Press, Honolulu, 93–108.

**McNally, G.H.** 2000: Geology and Mining practice in relation to shallow subsidence in the Northern Coalfield, New South Wales. *Australian Journal of Earth Sciences* **47**, 21–34.

**Madsen, L.M.** 2002 (in press): Afforestation plans: new challenges. The Danish afforestation programme and spatial planning. *Landscape and Urban Planning.*

**Maisels, C.K.** 1999: *Early Civilizations of the Old World.* Routledge, London.

**Maki, S., Kalliola, R. and Vuorinen, K.** 2001: Road construction in the Peruvian Amazon: process, causes and consequences. *Environmental Conservation* **28**, 199–214.

**Mannion, A.M.** 1995: *Agriculture and Environmental Change. Temporal and Spatial Dimensions.* John Wiley and Sons, Chichester.

**Mannion, A.M.** 1997: *Global Environmental Change. A Natural and Cultural Environmental History*, 2nd edition. Longman, Harlow.

**Mannion, A.M.** 1999a: *Natural Environmental Change.* Routledge, London.

**Mannion, A.M.** 1999b: Domestication and the origins of agriculture: an appraisal. *Progress in Physical Geography* **23**, 37–56.

**Mannion, A.M.** 1999c: Acid precipitation. In Pacione, M. (ed.), *Applied Geography: Principles and Practice.* Routledge, London, 36–50.

**Mannion, A.M.** 2000: Carbon: the link between environment, culture and technology. Department of Geography, University of Reading, *Geographical Paper* **154**, 25.

**Mannion, A.M.** 2001: Biotechnology and sustainable agriculture. *AgBiotechNet* **3**, 1–4.

**Marsh, G.P.** 1864: *Man and Nature; Or, the Earth as Modified by Human Action.* Scribner, New York.

**Mather, A.S.** 1992: The forest transition. *Area* **24**, 367–79.

**Mather, A.S. and Chapman, K.** 1995: *Environmental Resources.* Longman, Harlow.

**Mather, A.S., Neddle, C.L. and Fairbairn, J.** 1998: The human drivers of global land cover change: the case of forests. *Hydrological Processes* **12**, 1983–94.

**Mathijs, E. and Swinnen, J.F.M.** 2001: Production organization and efficiency during transition: an empirical analysis of East German agriculture. *Review of Economics and Statistics* **83**, 100–7.

**Mayle, F.E., Bell, M., Birks, H.H., Brooks, S.J., Coope, G.R., Lowe, J.J., Sheldrick, C., Shijie, L., Turney, C.S.M. and Walker, M.J.C.** 1999: Climate variations in Britain during the last glacial–Holocene transition (15.0–11.5 cal ka BP): comparison with the GRIP ice-core record. *Journal of the Geological Society, London* **156**, 411–23.

**Mellaart, J.** 1994: Western Asia during the Neolithic and Chalcolithic (about 12,000–5000 years ago). In DeLaet, S.J., Dani, A.H., Lorenzo, J.L. and Nunoo, R.B. (eds), *History of Humanity, Volume I: Prehistory and the Beginnings of Civilization.* Routledge, London and UNESCO, Paris, 425–40.

**Meyer-Arendt, K.J.** 2001: Recreational development and shoreline modification along the north coast of Yucatan, Mexico. *Tourism Geographies* **3**, 87–104.

**Miao, Z. and Marrs, R.** 2000: Ecological restoration and land reclamation in open-cast mines in Shanxi Province, China. *Journal of Environmental Management* **59**, 205–15.

**Mikhailova, E.A., Bryant, R.B., Cherney, D.J.R., Post, C.J. and Vassener, I.I.** 2000: Botanical composition, soil and forage quality under different management regimes in Russian grasslands. *Agriculture, Ecosystems and Environment* **80**, 213–26.

**Mining Technology** 2001a: Thunder Basin, Wyoming, www.mining-technology.com (accessed 30 May 2001).

**Mining Technology** 2001b: Bingham Canyon, Utah, www.mining-technology.com (accessed 31 May 2001).

**Ministry of Agriculture, Forestry and Food, Japan** 2001: Forestry, MAFF, Tokyo, www.maff.go.jp. (accessed 4 April 2001).

**Ministry of Agriculture and Forestry, New Zealand** 2001: The New Zealand forestry industry, MAF, New Zealand, www.maf.govt.nz (accessed 4 April 2001).

**Ministry of Agriculture, Fisheries and Food, UK** 2000: *Hill Farm Allowance Scheme.* Explanatory booklet for 2001. MAFF, London.

**Misgav, A., Perl, N. and Avnimelech, Y.** 2001: Selecting a compatible open space use for a closed landfill site. *Landscape and Urban Planning* **55**, 95–111.

**Miyoshi, N., Fujiki, T. and Morita, Y.** 1999: Palynology of a 250 m core from Lake Biwa: a 430,000-year record of glacial–interglacial vegetation change in Japan. *Review of Palaeobotany and Palynology* **104**, 267–83.

**Moody, J., Rackham, O. and Rapp, G.** 1996: Environmental archaeology of prehistoric NW Crete. *Journal of Field Archaeology* **23**, 273–97.

**Morello, J., Buzai, G.D., Baxendale, C.A., Rodriguez, A.F., Matteucci, S.D., Godagnone, R.E. and Casas, R.R.** 2000: Urbanization and the consumption of fertile land and other ecological changes: the case of Buenos Aires. *Environment and Urbanization* **12**, 119–31.

**Morris, A.E.J.** 1994: *History of Urban Form Before the Industrial Revolution*, 3rd edition. Longman, Harlow.

**Mugica, F.F., Anton, M.G., Ruiz, J.M. Juaristi, C.M. and Ollero, H.S.** 2001: The Holocene history of *Pinus* forests in the Spanish Northern Meseta. *The Holocene* **11**, 343–58.

**Murray, W.E.** 2000: Neoliberal globalisation, 'exotic' agro-exports, and local change in the Pacific Islands: a study of the Fijian kava sector. *Singapore Journal of Tropical Geography* **21**, 355–73.

**Nakamura, T. and Short, K.** 2001: Land-use planning and distribution of threatened wildlife in a city of Japan. *Landscape and Urban Planning* **53**, 1–15.

**National Capital Authority** 2001: National Capital Plan, www.nationalcapital.gov.au (accessed 2001).

**Nepstad, D.C., Verissimo, A., Alencar. A., Nobre, C., Lima, E., Lefebvre, P., Schlesinger, P., Potter, C., Moutinho, P., Mendoza, E., Cochrane, M. and Brooks, V.** 1999: Large-scale impoverishment of Amazonian forests by logging and fire. *Nature* **398**, 505–8.

**Neupane, R.P. and Thapa, G.B.** 2001: Impact of agroforestry intervention on soil fertility and farm income under the subsistence farming system of the middle hills, Nepal. *Agriculture, Ecosystems and Environment* **84**, 157–67.

**New Zealand Forest Service** 1977: *Forests of New Zealand.* Forest Mapping Series 9, Auckland.

**New Zealand Government** 2001: Solid waste in landfills, 1998, www.environment.govt.nz (accessed 12 September 2001).

**Newsome, T.H. and Comer, J.C.** 2000: Changing inter-urban location patterns of major-league sports facilities. *Professional Geographer* **52**, 105–20.

**Nielsen, T.L. and Zöbisch, M.A.** 2001: Multi-factorial causes of land-use change: land-use dynamics in the agropastoral village of Im Mial, northwestern Syria. *Land Degradation and Development* **12**, 143–61.

**Noël, H., Garboolino, E., Brauer, A., Lallier-Vergès, E., de Beaulieu, J.-L. and Disnar, J.-R.** 2001: Human impact and soil erosion during the last 5000 years as recorded in lacustrine sedimentary organic matter at Lac d'Annecy, the French Alps. *Journal of Paleolimnology* **25**, 229–44.

**Normile, D.** 1997: Yangtze seen as earliest rice site. *Science* **275**, 309.

**Nowak, D.L., Noble, M.H., Sisinni, S.M. and Dwyer, J.F.** 2001: Assessing the US urban forest resource. *Journal of Forestry* **99**, 37–42.

**NUA** 2001: How many online?, www.nua.ie (accessed 16 November 2001).

**O'Riordan, T.** 2000: The sustainability debate. In O'Riordan, T. (ed.), *Environmental Science for Environmental Management*. Prentice Hall, Harlow, 29–62.

**Oguz, D.** 2000: User surveys of Ankara's urban parks. *Landscape and Urban Planning* **52**, 165–71.

**Ohtsuka, T.** 1999: Early stages of secondary succession on abandoned cropland in north-east Borneo Island. *Biological Research* **14**, 281–90.

**Okuda, M., Yasuda, Y. and Setoguchi, T.** 2001: Middle to late Pleistocene vegetation history and climatic changes at Lake Kopais, Southwest Greece. *Boreas* **30**, 73–82.

**Olsson, E.G.A., Austrheim, G. and Grenne, S.N.** 2000: Landscape change patterns in mountains, land use and environmental diversity, mid-Norway 1960–1993. *Landscape Ecology* **15**, 155–70.

*Oxford Hammond Atlas of the World* 1993: Oxford University Press, Oxford.

**Paine, R.** 1994: *Herds of the Tundra: A Portrait of Reindeer Pastoralism*. Smithsonian Institution Press, Washington DC.

**Paniagua, A., Kammerbauer, J., Aredillo, M. and Andrews, M.** 1999: Relationship of soil characteristics to vegetation successions on a sequence of degraded rehabilitated soils in Honduras. *Agriculture, Ecosystems and Environment* **72**, 215–25.

**Parker, A.G., Goudie, A.S., Anderson, D.E., Robinson, M.A. and Bonsall, C.** 2002: A review of the mid-Halocene elm decline in the British Isles. *Progress in Physical Geography* 26, 1–45.

**Parry, R.B. and Perkins, C.R.** 2000: *World Mapping Today*, 2nd edition. Bowker Saur, London.

**Pauleit, S. and Duhme, F.** 2000: Assessing the environmental performance of land cover types for urban planning. *Landscape and Urban Planning* **52**, 1–20.

**Pearce, D.** 1995: *Tourism Today. A Geographical Analysis*, 2nd edition. Longman, Harlow.

**Pearson, O.M.** 2001: Postcranial remains and the origin of modern humans. *Evolutionary Anthropology: Issues, News and Views* **9**, 229–47.

**Peberdy, K.J.** 1998: Wetland creation for nature conservation – example from the post-industrial landscapes – example from the UK. In McComb, A.J. and Davis, J.A. (eds), *Wetlands for the Future*. Gleneagles Publishing, Adelaide, Australia, 719–37.

**Peglar, S.M.** 1993: The development of the cultural landscape around Diss Mere, Norfolk, UK, during the past 7000 years. *Review of Paleobotany and Palynology* **76**, 1–47.

**Pelissier, R., Pascal, J.-P., Houllier, D. and Laborde, H.** 1998: Impact of selective logging on the dynamics of a low elevation dense moist evergreen forest in the Western Ghats (South India). *Forest Ecology and Management* **105**, 107–19.

**Peralta, P. and Mather, P.** 2000: An analysis of deforestation patterns in the extractive reserves of Acre, Amazonia from satellite imagery: a landscape ecological approach. *International Journal of Remote Sensing* **21**, 2555–70.

**Perkins, C.R. and Parry, R.B.** 1996: *Mapping the UK.* Bowker Saur, London.

**Peterson, G.D. and Heemskirk, M.** 2001: Deforestation and forest regeneration following small-scale gold mining in the Amazon: the case of Suriname. *Environmental Conservation* **28**, 117–26.

**Petit, J.R., Jouzel, J., Raynaud, D., Barkov, N.I., Basile, I., Bender, M., Chappellaz, J., Davis, J. Delaygue, G., Delmotte, M., Kotlyakov, V.M., Legrand, M., Lipenkov, V., Lorius, C., Pepin, L., Ritz, C., Saltzman, E. and Stievenard, M.** 1999: 420,000 years of climate and atmospheric history revealed by the Vostok deep Antarctic ice core. *Nature* **399**, 429–36.

**Pettit, N.E. and Froend, R.H.** 2001: Long-term changes in the vegetation after the cessation of livestock grazing in *Eucalyptus marginata* (jarrah) woodland remnants. *Austral Ecology* **26**, 22–31.

**Pilbeam, C.J., Tripathi, B.P., Sherchan, D.P. Gregory, P.J. and Gaunt, J.** 2000: Nitrogen balances for households in the mid-hills of Nepal. *Agriculture, Ecosystems and Environment* **79**, 61–72.

**Pingali, P.L.** 2001: Environmental consequences of agricultural commercialisation in Asia. *Environment and Development Economics* **6**, 483–502.

**Pinto-Correia, T.** 2000: Future development in Portuguese rural areas: how to manage agricultural support for landscape conservation. *Landscape and Urban Planning* **50**, 95–106.

**Pogue, D.W. and Schnell, G.D.** 2001: Effects of agriculture on habitat complexity in a prairie-forest ecotone in the Southern Great Plains of North America. *Agriculture, Ecosystems and Environment* **87**, 287–98.

**Population Reference Bureau** 2001: Data sheet 1, www.prb.org (accessed 10 November 2001).

**Power, J.P. and Campbell, B.M.S.** 1992: Cluster analysis and the classification of medieval demesne-farming systems. *Transactions of the Institute of British Geographers New Series* **17**, 227–45.

**Preece, R.C. and Bridgland, D.R.** 2000: Holywell Coombe, Folkestone: A 13,000-year history of an English Chalkland Valley. *Quaternary Science Reviews* **18**, 1075–125.

**Prentice, I.C. and Webb, III, T.** 1998: BIOME 6000: reconstructing global and mid-Holocene vegetation patterns from palaeoecological records. *Journal of Biogeography* **25**, 997–1005.

**Prentice, I.C., Guiot, J., Huntley, B., Jolly, D. and**

**Cheddadi, R.** 1996: Reconstructing biomes from palaeoecological data: a general method and its application to European pollen data at 0 and 6 ka. *Climate Dynamics* **12**, 185–94.

**Price, A.R.G.** 1998: Impact of the 1991 Gulf War on the coastal environment and ecosystems: current status and future prospects. *Environmental International* **24**, 91–6.

**Prieur-Richard, A.-H. and Lavorel, S.** 2000: Invasions: the perspective of diverse plant communities. *Austral Ecology* **25**, 1–7.

**Punchestown** 2001: Punchestown racecourse, www.punchestown.com (accessed 28 September 2001).

**Putz, F.E., Dykstra, D.P. and Heinrich, R.** 2000: Why poor logging practices persist in the tropics. *Conservation Biology* **14**, 951–6.

**Putz, F.E., Blate, G.M., Redford, K.H., Fimbel, R. and Robinson, J.** 2001: Tropical forest management and conservation of biodiversity: an overview. *Conservation Biology* **15**, 7–20.

**Pyne, S.J.** 1998: Forged in fire: history, land and anthropogenic fire. In Balée, W. (ed.), *Advances in Historical Ecology*. Columbia University Press, New York, USA, 64–103.

**Qadeer, M.A.** 2000: Ruralopolises: the spatial organisation and residential land economy of high-density rural regions in South Asia. *Urban Studies* **37**, 1583–1603.

**Radford, I.J., Nicholas, D.M., Brown, J.R. and Kriticos, D.J.** 2001: Paddock-scale patterns of seed production and dispersal in the native shrub *Acacia nilotica* (Mimosaceae) in northern Australian rangeland. *Austral Ecology* **26**, 338–48.

**Rainforest Action Network** 2001: Beyond Oil campaign, oil in the Amazon, www.ran.org (accessed 14 June 2001).

**Ramankutty, N. and Foley, J.A.** 1999: Estimating historical changes in land cover: croplands from 1700 to 1992. *Global Biogeochemical Cycles* **13**, 997–1027.

**Ramrath, A., Sadori, L. and Negendank, J.F.W.** 2000: Sediments from Lago di Mezzano, central Italy: a record of lateglacial/Holocene climatic variations and anthropogenic impact. *The Holocene* **10**, 87–95.

**Rao, K.S. and Pant, R.** 2001: Land use dynamics and landscape change pattern in a typical micro watershed in the mid elevation zone of central Himalaya, India. *Agriculture, Ecosystems and Environment* **86**, 113–23.

**Raynaud, D., Barnola, J.M., Chappellaz, J., Blunier, T., Indermühle, A. and Stauffer, B.** 2000: The ice record of greenhouse gases: a view in the context of future changes. *Quaternary Science Reviews* **19**, 9–17.

**Reenberg, A.** 2001: Agricultural land use pattern dynamics in the Sudan – Sahel – towards an event-driven framework. *Land Use Planning* **18**, 309–19.

**Rees, W.** 1992: Ecological footprints and appropriated carrying capacity: what urban economics leaves out. *Environment and Urbanization* **4**, 120–30.

**Ren, G.** 2000: Decline of the mid- to late Holocene forests in China: climatic change or human impact? *Journal of Quaternary Science* **15**, 273–81.

**Ren, G. and Beug, H.-J.** 2002 (in press): Mapping Holocene pollen data and vegetation of China. *Quaternary Science Reviews*.

**Republic of South Africa Cricket Organization** 2001: South Africa, www.rsa.cricket.org (accessed 28 September 2001).

**Richards, J.F.** 1990: Land transformation. In Turner II, B.L., Clark, W.C., Kates, R.W., Richards, J.F, Mathews, J.T. and Meyer, W.B. (eds), *The Earth as Transformed by Human Action*. Cambridge University Press, Cambridge, with Clark University, 163–78.

**Richardson, D.M.** 1998: Forestry trees as invasive aliens. *Conservation Biology* **12**, 18–26.

**Richardson, D.M., Pysek, P., Rejmarick, M., Barbour, M.G., Panetta, F.D. and West, C.J.** 2000: Naturalisation and invasion of alien plants: concepts and definitions. *Diversity and Distributions* **6**, 93–107.

**Ries, J.B.** 1996: Landscape damage by skiing at the Schauinsland in the Black Forest, Germany. *Mountain Research and Development* **16**, 27–40.

**Rigg, J.** 1998: Rural–urban interactions, agriculture and wealth: a southeast Asian perspective. *Progress in Human Geography* **22**, 497–522.

**Rooney, D.** 1994: *A Guide to Angkor*. Asia Books, Bangkok.

**Rössing** 2000: Overview: uranium as a source of energy, www.rossing.com (accessed 29 May 2001).

**Routledge, L.M.** 1999: The impact of EU agricultural policy on the conservation of the English Pennines. *Environmental Conservation* **26**, 2–6.

**Rowe, J.S.** 1977: *Forest regions of Canada*. Canadian Forestry Service, Department of Environment, Natural Resources, Canada, Ottawa.

**Royal Parks** 2001: The Royal Parks, www.royalparks. gov.uk (accessed 7 September 2001).

**Rydberg, D. and Falck, J.** 2000: Urban forestry in Sweden from a silvicultural perspective: a review. *Landscape and Urban Planning* **47**, 1–18.

**Saiko, T.** 2001 *Environmental Crises*. Prentice Hall, Harlow.

**Santos, L., Romani, J.R.V. and Jalut, G.** 2000: History of vegetation during the Holocene in the Courel and Queixa Sierras, Galicia, northwest Iberian Peninsula. *Journal of Quaternary Science* **15**, 621–32.

**Sato, Y.J.** 1997: Cultivated rice was born in the Lower and Middle Yangtze River. *Nikkei Science* **1**, 32–42.

**Sax, D.F.** 2001: Latitudinal gradients and geographic ranges of exotic species: implications for biogeography. *Journal of Biogeography* **28**, 139–50.

**Sax, D.F. and Brown, J.H.** 2000: The paradox of invasion. *Global Ecology and Biogeography* **9**, 363–72.

**Schimel, D.S.** 1995: Terrestrial ecosystems and the carbon cycle. *Global Change Biology* **1**, 77–91.

**Schimel, D.S., House, J.I., Hibbard, K.A., Bousquet, P., Ciais, P., Peylin, P., Braswell, B.H., Apps, M.J., Baker, D., Bondeau, A., Canadell, J., Churkina, G., Cramer, W., Denning, A.S., Field, C.B., Friedlingstein, P., Goodale, C., Heimann, M., Houghton, R.A., Melillo, J.M., Moore III, B., Murdiyarso, D., Noble, I., Pacala, S.W., Prentice, I.C., Raupach, M.R., Rayner, P.J., Scholes, R.J., Steffen, W.L. and Wirth C.** 2001: Recent patterns and mechanisms of carbon exchange by terrestrial ecosystems. *Nature* **414**, 169–72.

**Schlesinger, W.H.** 1997: *Biogeochemistry: An Analysis of Global Change*. Academic Press, San Diego, USA.

**Schulz, F. and Wiegleb, G.** 2000: Development options of natural habitats in a post-mining landscape. *Land Degradation and Development* **11**, 99–110.

**Schulze, E.-D., Lloyd, J., Kelliher, F.M., Wirth, C., Rebmann, C., Lühker, B., Munn, M., Knohl, A., Milyukova, I.M., Schulze, W., Ziegler, W., Varlagin, A.B., Sogachev, A.F., Valentini, R., Dove, S., Grigoriev, S., Kolle,O., Panfyorov, M.I., Tchebakova, N. and**

**Vygodskaya, N.N.** 1999: Productivity of forests in the Eurosiberian boreal region and their potential to act as a carbon sink – a synthesis. *Global Change Biology* **5**, 703–22.

**Scotland's Golf Courses** 2001: St Andrews, the Old Course, www.scotlands-golf-courses.com (accessed 26 October 2001).

**Seddon, N., Tobias, J., Yount, J.W., Ramanampanmonjy, J.R., Butchart, S. and Randrianizahana, H.** 2000: Conservation issues and priorities in the Mikea Forest of south-west Madagascar. *Oryx* **34**, 287–304.

**Serebryanny, L.** 2002 (in press). Mixed and deciduous forests. In Shahgedanova, M., Goudie, A.S. and Orme, A.R (eds), *The Physical Geography of Northern Euarasia: Russia and Neighbouring States*. Oxford University Press, Oxford.

**Serneels, S. and Lambin, E.F.** 2001: Proximate causes of land-use change in Narok District, Kenya: a spatial statistical model. *Agriculture, Ecosystems and Environment* **85**.

**Sharma, K.D., Kumar, S. and Gough L.P.** 2000: Rehabilitation of lands mined for limestone in the Indian desert. *Land Degradation and Development* **11**, 563–74.

**Shaw, R.P.** 1989: Rapid population growth and environmental degradation: ultimate versus proximate factors. *Environmental Conservation* **16**, 1999–2008.

**Shell** 2001: 2000 highlights, www.shellnigeria.com.

**Siegert, F., Ruecker, G., Hinrichs, A. and Hoffman, A.A.** 2001: Increased damage from fires in logged forests during droughts caused by El Niño. *Nature* **414**, 437–40.

**Silver, W.L., Ostertag, R. and Lugo, A.E.** 2000: The potential for carbon sequestration through reforestation of abandoned tropical agricultural and pasture lands. *Restoration Ecology* **8**, 394–407.

**Smethurst, D.** 1999: Land degradation and the decline in ranching in the Sierra Nevada foothills, California. *Land Degradation and Development* **10**, 161–75.

**Smith, B.D.** 1995: *The Emergence of Agriculture*. W.H. Freeman, New York, USA.

**Smith, J., van de Kop, P., Reategui, K., Lombardi, I., Sabogal, C. and Diaz, A.** 1999: Dynamics of secondary forests in slash-and-burn farming: interactions among land use types in the Peruvian Amazon. *Agriculture, Ecosystems and Environment* **76**, 85–98.

**Smith, M.A., Loneragan, W.A., Grant, C.D. and Koch, J.M.** 2000: Effect of fire on the topsoil seed banks of rehabilitated bauxite mine sites in the jarrah forest of Western Australia. *Ecological Management and Restoration* **1**, 50–60.

**Smithers, J. and Blay-Palmer, A.** 2001: Technology innovation as a strategy for climate adaptation in agriculture. *Applied Geography* **21**, 175–97.

**Sobel, D.** 1995: *Longitude: The True Story of a Genius who solved the Greatest Scientific Problem of his Time*. Walker, New York.

**Solis, R.S., Haas, J. and Creamer, W.** 2001: Dating Caral, a preceramic site in the Supe Valley on the central coast of Peru. *Science* **292**, 723–6.

**Speranza, A., Hanke, J., van Geel, B. and Fanta, J.** 2000: Late-Holocene human impact and peat development in the Cerna Hora bog, Krkonose Mountains, Czech Republic. *The Holocene* **10**, 575–85.

**St Andrews** 2001: The official website of the home of golf, www.standrews.org.uk (accessed 1 October 2001).

**Steininger, M.K., Tucker, C.J., Townshend, J.R.G., Killeed, T.J., Desch, A., Bell, V. and Ersto, P.** 2001: Tropical deforestation in the Bolivian Amazon. *Environmental Conservation* **28**, 127–34.

**Stéphenne, N. and Lambin, E.F.** 2001: A dynamic simulation model of land-use changes in Sudano-Sahelian countries of Africa (SALU). *Agriculture, Ecosystems and Environment*. **85**, 145–61.

**Stringer, C. and McKie, R.** 1996: *African Exodus: The Origins of Modern Humanity*. Pimlico, London.

**Sui, D.Z. and Zeng, H.** 2001: Modeling the dynamics of landscape structure in Asia's emerging desakota regions: a case study of Shenzhen. *Landscape and Urban Planning* **53**, 37–52.

**Sunderlin, W.D., Angelsen, A., Resosudarmo, D.P. and Dermawan, A.** 2001: Economic crisis, small farmer well being, and forest cover change in Indonesia. *World Development* **29**, 767–782.

**Sykes, B.** 1999: The molecular genetics of European ancestry. *Philosophical Transactions of the Royal Society of London* **354**, 131–9.

**Sykes, B.** 2001: *The Seven Daughters of Eve*. Bantam Press, London.

**Tarasov, P.E., Webb III, T., Andreev, A.A., Afanas'eva, N.B., Berezina, N.A., Bezusko, L.G., Blyakharchuk, T.A., Bolikhovskaya, N.S., Cheddadi, R., Chernavskaya, M.M., Chernove, G.M., Dorofeyuk, N.I., Dirksen, V.G., Elina, G.A., Filimonova, L.V., Glebov, F.Z., Guiot, J., Gunova, V.S., Harrison, S.P., Jolly, D., Khomutova, V.I., Kvavadze, E.V., Osipova, I.M., Panova, N.K., Prentice, I.C., Saarse, L., Sevastyanov, D.V., Volkova, V.S. and Zernitskaya, V.P.** 1998: Present-day and mid-Holocene biomes constructed from pollen and plant macrofossil data from the former Soviet Union. *Journal of Biogeography* **25**, 1029–53.

**Taylor, D., Robertshaw, P. and Marchant, R.A.** 2000: Environmental change and political-economic upheaval in precolonial western Uganda. *The Holocene* **10**, 527–36.

**Taylor, J.E. and Fox, B.J.** 2001: Assessing the disturbance impact on vegetation and lizard communities of fluoride pollution interacting with fire and mining in eastern Australia. *Austral Ecology* **26**, 321–37.

**Thomas, D.S.G., Sporton, D. and Perkins, J.** 2000: The environmental impact of livestock ranches in the Kalahari, Botswana: natural resource use, ecological change and human response in a dynamic dryland system. *Land Degradation and Development* **11**, 327–41.

**Thomas, W.L. Jr** (ed.) 1956: *Man's Role in Changing the Face of the Earth*. University of Chicago Press, Chicago, USA.

*The Times Concise Atlas of the World* 1995: Times Books, London.

*The Times History of London* 1999: Times Books, London.

*The Times History of the World* 1999: Times Books, London.

**Tiwari, P.C.** 2000: Land-use changes in Himalaya and their impact on the plains ecosystem: need for sustainable land use. *Land Use Policy* **17**, 101–11.

**Toft, R.J., Harris, R.J. and Williams, P.A.** 2001: Impacts of the weeds *Tradescantia fluminensis* on insect communities in fragmented forests in New Zealand. *Biological Conservation* **102**, 31–46.

**Toivonen, K.** 2000: Integrating forestry and conservation in boreal forests: ecological, legal and socio-economic aspects. *Forestry* **73**, 129–36.

**Torbert, J.L., Burger, J.A., Schoenholtz, S.H. and Kreh, R.E.** 2000: Growth of three pine species after eleven years on reclaimed mine-soils in Virginia. *Northern Journal of Applied Forestry* **17**, 95–9.

**Troels-Smith, J.** 1960: Ivy, mistletoe and elm: climatic indicators – fodder plants: a contribution to the interpretation of the pollen zone border VII–VIII. *Danmarks Geologiske Undersøgelse* Series 4, **4**, 1–32.

**Turner II, B.L., Clark, W.C., Kates, R.W., Richards, J.F., Mathews, J.T. and Meyer, W.B.** (eds) 1990: *The Earth as Transformed by Human Action.* Cambridge University Press, Cambridge, with Clark University.

**Turner, M.D.** 1999: Merging local and regional analyses of land-use change: the case of livestock in the Sahel. *Annals of the Association of American Geographers* **89**, 191–219.

**Udvardy, M.D.F.** 1975: *Classification of the Biogeographical Provinces of the World*, Occasional Paper No. 18. International Union for Conservation of Nature and Natural Resources, Switzerland.

**UNEP World Conservation Monitoring Centre** 2001: www.unep-wcmc.org (accessed 10 November 2001).

**United Nations (UN)** 2001: World population prospects. The 2000 revision, www.un.org (accessed 11 October 2001).

**United Nations Framework Convention on Climate Change (UNFCCC)** 2001: Understanding climate change, www.unfccc.de (accessed 9 November 2001).

**United States Geological Survey** 2001: Mineral commodity summaries, http://minerals.usgs.gov/minerals/pubs/mcs (accessed 15 June 2001).

**United States Office of Surface Mining** 2001: US coal production statistics by state, www.OSMRE.GOV/coal-prodindex.htm (accessed 15 May 2001).

**Uranium Information Centre** 2001a: World uranium mining, nuclear issues, Briefing Paper No. 41, www.uic.com.au (accessed 31 May 2001).

**Uranium Information Centre** 2001b: Australia's uranium mines, www.uic.com.au (accessed 31 May 2001).

**Uranium Institute** 2001: Environmental aspects of uranium mining, Information Paper No. 25, www.uilondon.org (accessed 4 June 2001).

**Van den Bergh, J.C.J.M. and Verbruggen, H.** 1999: Spatial sustainability, trade and indicators; an evaluation of the 'ecological footprint'. *Ecological Economics* **29**, 61–72.

**van der Knaap, W.O., van Leeuwen, J.F.N. and Ammann, B.** 2000: Palynostratigraphy of the last centuries in Switzerland based on 23 lake and mire deposits: chronostratigraphic pollen markers, regional patterns and local histories. *Review of Palaeobotany and Palynology* **108**, 85–98.

**van der Merwe, J.P.A. and Kellner, K.** 1999: Soil disturbance and increase in species diversity during rehabilitation of degraded arid rangelands. *Journal of Arid Environments* **41**, 323–33.

**Van Krunkelsven, E., Inogwabini, B.-I. and Draulans, D.** 2000: A survey of bonobos and other large mammals in the Salonga National Park, Democratic Republic of Congo. *Oryx* **34**, 180–7.

**Velichko, A.A., Zelikson, E.M. and Borisova, O.K.** 1999: Vegetation, phytomass and carbon storage in Northern Eurasia during the last glacial–interglacial cycle and the Holocene. *Chemical Geology* **159**, 191–204.

**Vera, R.R., Hoyos, P. and Moya, M.C.** 1998: Pasture renovation practices of farmers in the neotropical savannahs. *Land Degradation and Development* **9**, 47–56.

**Verburg, P.H., Chen Y. and Veldkamp, T.** 2000: Spatial explorations of land use change and grain production in China. *Agriculture, Ecosystems and Environment* **82**, 333–54.

**Verburg, P.M. and van der Gon, H.A.C.D.** 2001: Spatial and temporal dynamics of methane emissions from agricultural sources in China. *Global Change Biology* **7**, 31–47.

**Vernadsky, V.I.** 1945: The biosphere and the noosphere. *American Scientist* **33**, 1–12.

**Viglizzo, E.F., Lertora, F., Pordomingo, A.J., Bernardos, J.N., Roberto, Z.E. and Del Valle, H.** 2001: Ecological lessons and applications from one century of low external-input farming in the pampas of Argentina. *Agriculture, Ecosystems and Environment* **83**, 65–81.

**Vilchek, G.** 2002 (in press): Environmental impacts of oil and gas development. In Shahgedanova, M., Goudie, A.S. and Orme, A.R. (eds), *The Physical Geography of Northern Eurasia: Russia and Neighbouring States.* Oxford University Press, Oxford.

**Vincentian Missionaries** 1998: The Payatas Environmental Development Programme: microenterprise promotion and involvement in solid waste management in Quezon City. *Environment and Urbanization* **10**, 55–68.

**Vitousek, P.M., Mooney, H.A., Lubchenco, J. and Melillo, J.** 1997: Human domination of Earth's ecosystems. *Science* **277**, 494–9.

**Von Grafenstein, U., Erlenkauser, H., Brauer, A., Jouzel, J. and Johnsen, S.J.** 1999: A mid-European decadal isotope-climate record from 15,500 to 5000 years BP. *Science* **284**, 1654–7.

**Wackernagel, M. and Rees, W.E.** 1996: *Our Ecological Footprint. Reducing Human Impact on the Earth.* New Society Publishers, Philadelphia.

**Wackernagel, M., Onisto, L. Linares, A.C., Falfan, I.S.L., Garcia, J.M., Guerrero, A.I.S. and Guerrero, M.G.S.** 1997: Ecological footprints of nations: how much nature do they use and how much nature do they have? Commissioned by the Earth Council of Cost Rica for the Rio+ 5 Forum. International Council for Local Environmental Initiatives, Toronto.

**Wackernagel, M., Onisto, L., Bello, P., Linares, A.C., Falfan, I.S.L., Mendez, J., Guerrero, A.I.S. and Guerrero, M.G.S.** 1999: National natural capital accounting with the ecological footprint concept. *Ecological Economics* **29**, 375–90.

**Walker, M.J.C., Björck, S. and Lowe, J.J.** 2001: Integration of ice core, marine and terrestrial records (INTIMATE) from around the North Atlantic region: an introduction. *Quaternary Science Reviews* **20**, 1169–74.

**Walker, R., Moran, E. and Anselin, L.** 2000, L. 2000: Deforestation and cattle ranching in the Brazilian Amazon: external capital and household processes. *World Development* **28**, 683–99.

**Walks, R.A.** 2001: The social ecology of the post-Fordist/global city? Economic restructuring and socio-spatial polarization in the Toronto urban region. *Urban Studies* **38**, 407–47.

**Waller, M.P. and Hamilton, S.** 2000: Vegetation history of the English chalklands: a mid-Holocene pollen sequence from Caburn, East Sussex. *Journal of Quaternary Science* **15**, 253–72.

**Wang, Y., Dawson, D., Han, J., Peng, Z., Liv, Z. and Ding, Y.** 2001: Landscape ecological planning and design of degraded mining land. *Land Degradation and Development* **12**, 449–59.

**Ward, C.** 2000: Soil development on rehabilitated bauxite mines in south-west Australia. *Australian Journal of Soil Research* **38**, 453–64.

**Weber, M.G. and Stokes, B.J.** 1998: Forest fires and sustainability in the boreal forests of Canada. *Ambio* 545–50.

**Whitmore, T.C.** 1997: Tropical forest disturbance, disappearance and species loss. In Laurance, W.F. and Bierregaard, R.O. (eds), *Tropical Forest Remnants: Ecology, Management and Conservation of Fragmented Communities*. University of Chicago Press, 3–12.

**Wilkie, D., Shaw, E., Rotberg, F., Morelli, G. and Auze, P.** 2000: Roads, development, and conservation in the Congo Basin. *Conservation Biology* **14**, 1614–22.

**Wilkinson, A.G.** 1999: Poplars and willows for soil erosion control in New Zealand. *Biomass and Bioenergy* **16**, 263–74.

**Willcox, G.** 1996: Evidence for plant exploitation and vegetation history from three Early Neolithic pre-pottery sites on the Euphrates (Syria). *Vegetation History and Archaeobotany* **5**, 143–52.

**Williams, J.A. and West, C.J.** 2000: Environmental weeds in Australia and New Zealand: issues and approaches to management. *Austral Ecology* **25**, 425–44.

**Williams, M.** 1989: *Americans and their Forests: A Historical Geography*. Cambridge University Press, New York.

**Williams, M.** 1990: Forests. In Turner II, B.L., Clark, W.C., Kates, R.W., Richards, J.F, Mathews, J.T. and Meyer, W.B. (eds), *The Earth as Transformed by Human Action*. Cambridge University Press with Clark University, 179–201.

**Wilson, E.J., McDougall, F.R. and Willmore, J.** 2001: Euro-trash: searching Europe for a more sustainable approach to waste management. *Resources, Conservation and Recycling* **31**, 327–34.

**Wimble, G., Wells, C.E. and Hodgkinson, D.** 2000: Human impact on mid- and late Holocene vegetation in South Cumbria, UK. *Vegetation History and Archaeobotany* **9**, 17–30.

**Witt, G.B., Berghammer, L.J., Becton, R.J.S. and Moll, F.J.** 2000: Retrospective monitoring of rangeland vegetation change: ecohistory from deposits of sheep dung associated with shearing sheds. *Austral Ecology* **25**, 260–7.

**Wood, B.** 1992: Origin and evolution of the genus Homo. *Nature* **355**, 783–90.

**World Commission on Forests and Sustainable Development** 1999: *Our Forests. Our Future*. Cambridge University Press, Cambridge.

**World Resources Institute** 1996: *World Resources 1996–7*. Oxford University Press, Oxford.

**World Resources Institute** 1997: *The Last Frontier Forests*. World Resources Institute, Washington DC, www.wri.org/ffi/lff-eng/russeuro (accessed 20 March 2001).

**World Resources Institute** 1998: *World Resources 1998–9*. Oxford University Press, Oxford.

**World Tourism Organization** 2001: International tourism, www.world-tourism.org (accessed 2 October 2001).

**Worldwide Fund for Nature** 1998: *Living Planet Report 1998*. Worldwide Fund for Nature, Gland, Switzerland, New Economics Foundation, London and World Conservation Monitoring Centre, Cambridge.

**Worldwide Fund for Nature** 1999: Living Planet Report 1999, www.panda.org (accessed 23 January 2000).

**Wright, R.T. and Nebel, B.J.** 2002: *Environmental Science. Toward a Sustainable Future*, 8th edition. Pearson Education, New Jersey.

**Wyatt, B.K., Greatorex Davies, N., Bunce, R.G.H., Fuller, R.M. and Hill, M.O.** 1993: *Comparison of Land Cover Definitions. Final Report. Dictionary of Surveys and Classifications of Land Cover and Land Use*. Institute of Terrestrial Ecology Report to the Department of Environment.

**Xu, H., Wang, X. and Xiao, G.** 2000: A remote sensing and GIS integrated study on urbanization with its impact on arable lands: Fuqing City, Fujan Province, China. *Land Degradation and Development* **11**, 301–14.

**Yang, H. and Li, X.** 2000: Cultivated land and food supply in China. *Land Use Policy* **17**, 73–88.

**Ye, Z.H., Wong, J.W.C., Wong, M.H., Baker, A.J.M., Shu, W.S. and Lan, C.Y.** 2000a: Revegetation of Pb/Zn mine tailings, Guangdong Province, China. *Restoration Ecology* **8**, 87–92.

**Ye, Z.H., Wong, J.W.C. and Wong, M.H.** 2000b: Vegetation response to lime and manure compost amendments on acid lead/zinc mine tailings: a greenhouse study. *Restoration Ecology* **8**, 289–95.

**Young, A.R.M.** 1996: *Environmental Change in Australia since 1788*. Oxford University Press, Melbourne.

**Yu, G., Prentice, C., Harrison, S.P. and Sun, X.** 1998: Pollen-based biome reconstructions for China at 0 and 6000 years. *Journal of Biogeography* **25**, 1055–69.

**Zhao, Z.J.** 1998: The middle Yangtze region in China is one place where rice was domesticated: phytolith evidence from the Diaotonghuan cave, northern Jiangxi. *Antiquity* **278**, 885–97.

**Zhao, Z.J. and Piperno, D.R.** 2000: Late Pleistocene/Holocene environments in the Middle Yangtze River Valley, China, and rice (*Oryza sativa* L.) domestication: the phytolith evidence. *Geoarchaeology* **15**, 203–22.

# INDEX